# Genomics, Proteomics, and Clinical Bacteriology

# METHODS IN MOLECULAR BIOLOGY™

## John M. Walker, SERIES EDITOR

METHODS IN MOLECULAR BIOLOGY™

# Genomics, Proteomics, and Clinical Bacteriology

*Methods and Reviews*

Edited by

## Neil Woodford

and

## Alan P. Johnson

*Specialist and Reference Microbiology Division*
*Health Protection Agency–Colindale*
*London, UK*

HUMANA PRESS ✳ TOTOWA, NEW JERSEY

© 2010 Humana Press Inc.
999 Riverview Drive, Suite 208
Totowa, New Jersey 07512

**www.humanapress.com**

This publication is printed on acid-free paper. ∞
ANSI Z39.48-1984 (American Standards Institute)

Permanence of Paper for Printed Library Materials.

Production Editor: Jessica Jannicelli.
Cover design by Patricia F. Cleary.

For additional copies, pricing for bulk purchases, and/or information about other Humana titles, contact Humana at the above address or at any of the following numbers: Tel.: 973-256-1699; Fax: 973-256-8341; E-mail: humana@humanapr.com; or visit our Website: www.humanapress.com

Printed in the United States of America. 10 9 8 7 6 5 4 3 2 1
E-ISBN: 978-1-59259-763-5

Library of Congress Cataloging in Publication Data

Genomics, proteomics, and clinical bacteriology : methods and reviews / edited by Neil Woodford and Alan Johnson.
    p. ; cm. -- (Methods in molecular biology ; 266)
  Includes bibliographical references and index.
  ISBN 978-1-61737-428-9
  1. Molecular microbiology. 2. Bacterial genetics. 3. Genomics.
4. Proteomics. 5. Diagnostic bacteriology.
  [DNLM: 1. Genomics--methods. 2. Bacteriological
Techniques--methods. 3. Proteomics--methods. QU 58.5 G3355 2004]
I. Woodford, Neil. II. Johnson, Alan P. (Alan Patrick), 1951- III.
Series: Methods in molecular biology (Clifton, N.J.) ; v. 266.
  QR74.G46 2004
  616.9'201--dc22
                                2004000332

# Preface

Gazing into crystal balls is beyond the expertise of most scientists. Yet, as we look further into the 21st century, one does not have to be Nostradamus to predict that the current genomics and proteomics "revolution" will have an immense impact on medical bacteriology. This impact is already being realized in many academic departments, and although encroachment on routine diagnostic bacteriology, particularly in the hospital setting, is likely to occur at a slower pace, it remains nonetheless inevitable. Therefore, it is important that no one working in bacteriology should find themselves distanced from these fundamental developments. The involvement of all clinical bacteriologists is essential if the significant achievements of genome sequencing and analysis are to be turned into tangible advances, with resulting benefits for patient care and management. It is our hope that *Genomics, Proteomics, and Clinical Bacteriology: Methods and Reviews* will play a part in bringing such a development to fruition.

The advances in genomics and proteomics have already given us frequent opportunities to reassess our knowledge and understanding of established bacterial adversaries, and have provided us with the means to identify new foes. The new knowledge gained is enabling us to reconsider, for example, our concepts of bacterial pathogenicity, phylogeny and novel targets for antibacterial chemotherapy. These topics, and others, are considered in *Genomics, Proteomics, and Clinical Bacteriology: Methods and Reviews*.

We have convened a group of internationally renowned authors, all of whom have prepared state-of-the-art reviews. Although we admit the choice of topics was influenced by personal interests, we nonetheless believe it to be fairly comprehensive. The 16 chapters have been divided loosely into three sections: Principles of Bacterial Genomics (Chapters 1–6); Application of Genomics to Diagnostic Bacteriology (Chapters 7–10); and Interrogating Bacterial Genomes (Chapters 11–16). We have attempted to make the chapters easy to read, and hope that they will not be considered esoteric. Throughout this project, our intention was to create a volume that would be beneficial to everyone working in medical bacteriology, whatever their background. We also hope it would serve as a natural successor to our previous volume, *Molecular Bacteriology: Protocols and Clinical Applications*, published by Humana Press in 1998.

It has been a fascinating experience editing the volume, and we hope that our readers will learn as much as we did.

*Neil Woodford*
*Alan Johnson*

# Contents

# Contributors

MATTHEW B. AVISON • *Department of Biochemistry, University of Bristol, School of Medical Sciences, Bristol, UK*

PETER M. BENNETT • *Department of Pathology and Microbiology, University of Bristol, School of Medical Sciences, Bristol, UK*

DONALD L. N. CARDY • *British BioCell International, Golden Gate, Cardiff, UK*

STUART J. CORDWELL • *Australian Proteome Analysis Facility, Macquarie University, Sydney, Australia*

DAWN FIELD • *Oxford Centre for Ecology and Hydrology, Oxford, UK*

MICHAEL N. GWYNN • *Department of Microbiology, Microbial, Musculoskeletal, and Proliferative Diseases Center of Excellence and Drug Discovery, GlaxoSmithKline, Collegeville, PA*

KEIICHI HIRAMATSU • *Department of Bacteriology, Juntendo University, Tokyo, Japan*

DAVID J. HOLMES • *Department of Microbiology, Microbial, Musculoskeletal, and Proliferative Diseases Center of Excellence and Drug Discovery, GlaxoSmithKline, Collegeville, PA*

JENNIFER HUGHES • *Department of Ecology and Evolutionary Biology, Brown University, Providence, RI*

ALAN P. JOHNSON • *Antibiotic Resistance Monitoring and Reference Laboratory, Specialist and Reference Microbiology Division, Health Protection Agency–Colindale, London, UK*

MAKOTO KURODA • *Department of Bacteriology, Juntendo University, Tokyo, Japan*

IRENA KUKAVICA-IBRULJ • *Centre de Recherche sur la Fonction Structure et Ingénierie des Protéines, Université Laval, Québec, Canada*

ANDREW J. LAWSON • *Laboratory of Enteric Pathogens, Specialist and Reference Microbiology Division, Health Protection Agency–Colindale, London, UK*

DARIO E. LEHOUX • *Centre de Recherche sur la Fonction Structure et Ingénierie des Protéines, Université Laval, Québec, Canada*

ROGER C. LEVESQUE • *Centre de Recherche sur la Fonction Structure et Ingénierie des Protéines, Université Laval, Québec, Canada*

PETER MARSH • *Health Protection Agency South East, Southampton Laboratory, Southampton General Hospital, Southampton, UK*

B. CHERIE MILLAR • *Northern Ireland Public Health Laboratory, Department of Bacteriology, Belfast City Hospital, Belfast, Northern Ireland*

JOHN E. MOORE • *Northern Ireland Public Health Laboratory, Department of Bacteriology, Belfast City Hospital, Belfast, Northern Ireland*

E. RICHARD MOXON • *Molecular Infectious Diseases Group, Weatherall Institute of Molecular Medicine and University of Oxford Department of Paediatrics, John Radcliffe Hospital, Oxford, UK*

ROBERT J. OWEN • *Laboratory of Enteric Pathogens, Specialist and Reference Microbiology Division, Health Protection Agency–Colindale, London, UK*

MARK J. PALLEN • *Bacterial Pathogenesis and Genomics Unit, Division of Immunity and Infection, Medical School, University of Birmingham, Birmingham, UK*

DAVID J. PAYNE • *Department of Microbiology, Microbial, Musculoskeletal, and Proliferative Diseases Center of Excellence and Drug Discovery, GlaxoSmithKline, Collegeville, PA*

ERIC POTVIN • *Centre de Recherche sur la Fonction Structure et Ingénierie des Protéines, Université Laval, Québec, Canada*

MARTY ROSENBERG • *Department of Microbiology, Microbial, Musculoskeletal, and Proliferative Diseases, GlaxoSmithKline, Collegeville, PA*

FRANÇOIS SANSCHAGRIN • *Centre de Recherche sur la Fonction Structure et Ingénierie des Protéines, Université Laval, Québec, Canada*

NICHOLAS A. SAUNDERS • *Genomics, Proteomics, and Bioinformatics Unit, Specialist and Reference Microbiology Division, Health Protection Agency–Colindale, London, UK*

BRIAN G. SPRATT • *Department of Infectious Disease Epidemiology, Faculty of Medicine, Imperial College, St. Mary's Hospital, London, UK*

CAROLA VAN IJPEREN • *Sexually Transmitted and Bloodborne Virus Laboratory, Specialist and Reference Microbiology Division, Health Protection Agency–Colindale, London, UK*

NEIL WOODFORD • *Antibiotic Resistance Monitoring and Reference Laboratory, Specialist and Reference Microbiology Division, Health Protection Agency–Colindale, London, UK*

# I

## PRINCIPLES OF BACTERIAL GENOMICS

# 1

## Bacterial Genomes for the Masses

### Relevance to the Clinical Laboratory

**Mark J. Pallen**

### Summary

Bacterial genome sequencing has revolutionized the research landscape and promises to deliver important changes to the clinical microbiology laboratory, through the identification of novel diagnostic targets and through the birth of a new discipline or "genomic epidemiology." Current progress and future prospects for exploitation of genome sequences in clinical bacteriology are reviewed.

**Key Words:** Genome sequencing; bacteriology; genomics; molecular epidemiology; genomic epidemiology; diagnosis.

## 1. Introduction

> "All changed, changed utterly:
> A terrible beauty is born."

> W. B. Yeats—Easter 1916

The world of bacterial genome sequencing "changed utterly" when the first two complete bacterial genome sequences were published by scientists at The Institute for Genomic Research (TIGR) in Maryland in 1995 *(1–3)*. Sequencing of the *Haemophilus influenzae* genome was followed quickly by that of *Mycoplasma genitalium*—both using a whole-genome shotgun sequencing strategy that shattered the dogma that genomes should be sequenced using a top-down approach, where detailed mapping and creation of an ordered library had to precede genome sequencing. Consequently, aside from a handful of

From: *Methods in Molecular Biology, vol. 266: Genomics, Proteomics, and Clinical Bacteriology:*
*Methods and Reviews*
Edited by: N. Woodford and A. Johnson © Humana Press Inc., Totowa, NJ

projects already underway in 1995, nearly all other bacterial genome sequencing projects have relied on the quick, easy, and efficient approach of making a random shotgun library, automatically sequencing tens of thousands of clones, and then using computer power to assemble this molecular jigsaw puzzle into a handful of fragments ("contigs"), followed by a "finishing phase" to close every gap and remove every ambiguity (*see* Chapter 3).

The success of whole-genome shotgun sequencing has been stupendous—at the time of writing over 60 bacterial genome sequences have been completed and over 200 more are underway (*see* http://wit.integratedgenomics.com/ GOLD/). The genome of every significant bacterial pathogen of humans, plants, and animals has already been or soon will be sequenced (**Table 1**), including, in many cases, the genomes of multiple strains, pathovars, or serovars from the same species (**Table 2**).

## 2. Why Sequence Genomes?

The clinical bacteriologist may well ask, "What's the use of a genome sequence?", to which the scientist might reasonably reply with a quote variously attributed to Faraday or Franklin: "What's the use of a baby?" Although a completed genome may consist simply of millions of As, Cs, Ts, and Gs, genome sequencing has already amply justified its existence as an enabling technology in the research laboratory, energizing novel, hypothesis-driven research on bacterial pathogenesis and physiology, and spawning a whole range of high-throughput, genomic-based technologies that fall under the umbrella term "functional genomics" (*4,5*). For example, efforts are already underway to make knockout mutants, express proteins, survey protein–protein interactions, solve structures, and measure gene expression on a whole-genome scale in several different microorganisms (*6–11*). And thanks to the synergies between genome sequencing, the Internet, and the polymerase chain reaction (PCR), any scientist who wants any gene from any organism can have it within a few days for the effort of a few Web searches (*see* Chapter 2) and for the price of a pair of oligonucleotides. Genome sequencing has transformed the research landscape!

One overarching theme to emerge from the genome-sequencing revolution is how little we still know about bacteria. In every new genome sequence, we usually find that at least 30–50% of the genes have no known function—even in the laboratory workhorse *Escherichia coli* K12, at least 20% of the genes are without a functional assignment (*12*). For organisms that we cannot grow in the laboratory (or cannot grow easily), such as *Treponema pallidum* or *Mycobacterium leprae*, genome sequencing is the only way to gain insights into their biochemistry, physiology, and virulence mechanisms (*13–19*). In addition, functional genomics approaches are allowing us to probe aspects of physiol-

**Table 1**
**Landmarks in Bacterial Pathogen Genome Sequencing**

| Year | Landmarks |
|---|---|
| 1995 | *Haemophilus influenzae* |
| | *Mycoplasma genitalium* |
| 1996 | *Mycoplasma pneumoniae* |
| 1997 | *Helicobacter pylori* |
| | *Borrelia burgdorferi* |
| 1998 | *Mycobacterium tuberculosis* |
| | *Treponema pallidum* |
| | *Chlamydia trachomatis* |
| | *Rickettsia prowazekii* |
| 1999 | A second *H. pylori* genome |
| | *Chlamydia pneumoniae* |
| 2000 | *Campylobacter jejuni* |
| | *Neisseria meningitidis* serogroup A |
| | *Neisseria meningitidis* serogroup B |
| | Two more *C. pneumoniae* genomes |
| | A second *C. trachomatis* genome |
| | *Vibrio cholerae* |
| | *Pseudomonas aeruginosa* |
| | *Ureaplasma urealyticum* |
| 2001 | *Escherichia coli* O157:H7 (two strains) |
| | *Mycobacterium leprae* |
| | *Pasteurella multocida* |
| | *Streptococcus pyogenes* |
| | *Staphylococcus aureus* (two strains) |
| | *Streptococcus pneumoniae* (two strains) |
| | *Rickettsia conori* |
| | *Yersinia pestis* |
| | *Salmonella typhi* |
| | *Salmonella typhimurium* |
| | *Listeria monocytogenes* |
| 2002 | *Brucella melitensis* |
| (to August) | *Clostridium perfringens* |
| | *Streptococcus pyogenes* (two strains) |
| | *Fusobacterium nucleatum* |
| | A third *S. aureus* strain |
| | *Bacillus anthracis* (two strains) |

For references and other information on these projects, *see* the Genomes Online Database: http://wit.integratedgenomics.com/GOLD/.

**Table 2**
**Genome Sequencing Projects on Multiple Strains in the Same Species**

| Species | Completed projects | Underway or planned |
|---|---|---|
| *Bacillus anthracis* | Two strains | Several more |
| *Bacillus cereus* | | Two strains |
| *Campylobacter jejuni* | One strain | One strain |
| *Chlamydia pneumoniae* | Three strains | |
| *Chlamydia trachomatis* | Two strains | |
| *Clostridium perfringens* | One strain | One strain |
| *Escherichia coli* | Two strains of O157:H7 One strain of K12 | At least five pathovars |
| *Haemophilus influenzae* | One strain | Two strains |
| *Helicobacter pylori* | Two strains | |
| *Listeria mononcytogenes* | One strain | One strain |
| *Mycobacterium avium* | | Two strains (includes *paratuberculosis*) |
| *Mycobacterium tuberculosis* | Three strains (includes *M. bovis*) | One strain (BCG) |
| *Neisseria meningitidis* | Serogroup A strain Serogroup B strain | Serogroup C strain |
| *Salmonella enterica* | Serovar typhi Typhimurium | >10 more serovars |
| *Staphylococcus aureus* | Three strains | At least six strains |
| *Staphylococcus epidermidis* | | At least two strains |
| *Streptococcus pneumoniae* | Two strains | At least two strains |
| *Streptococcus pyogenes* | Two strains | |
| *Tropheryma whippelii* | | Two strains |

For references and other information on these projects, *see* the Genomes Online Database: http://wit.integratedgenomics.com/GOLD/.

ogy and assign functions to genes that may not be obvious when bacteria are grown under laboratory conditions on agar plates *(20–23)*. Also, genome sequencing allows a complete "naming of parts"—once we know every gene in the organism, finding all the genes associated with a particular pathway, structure, or phenotype becomes a finite, tractable task. One early example of this was the identification and characterization of all lipopolysaccharide biosynthesis genes in *H. influenzae* from scrutiny of the genome *(24)*.

On a more applied note, a major driving force in the genomics revolution has been the desire of the pharmaceutical industry to find novel targets for antimicrobial drugs *(25)* (*see* Chapter 11). Comparing microbial and animal genomes allows the identification of genes or pathways that are present in pathogens but absent from humans, i.e., plausible targets for compounds showing selective toxicity *(26,27)*. Similarly, genome sequences have been used as the starting point for finding new vaccine targets *(4,19,28,29)*. The best example of this was a stepwise sieving of genes from the meningococcal genome, starting with several hundred surface-associated proteins identified through bioinformatics, through to a handful of highly credible vaccine candidates *(30)*.

## 3. Relevance to the Clinical Laboratory

Aside from the indirect effects that might flow from vaccine and drug discovery, genome sequencing has already had some impact in the clinical microbiology laboratory, most noticeably in the field of epidemiological typing, and, to a lesser degree, in diagnostic bacteriology and the study of antimicrobial resistance.

### *3.1. Molecular Epidemiology*

The least adventurous use of genome sequences in molecular epidemiology is an adjunct to established methods. For example, having a genome sequence can assist in setting up a multilocus sequence-typing method for a new species *(31,32)* (*see* Chapter 15). In a similar vein, Desai and colleagues reported using the genome sequence of *Campylobacter jejuni* strain NCTC 11168 to develop a fluorescent amplified fragment length polymorphism (fAFLP) typing method for this organism, with the aim of establishing a molecular epidemiological database *(33)*.

More exciting is when genome sequencing challenges an accepted dogma of epidemiological typing. Studies in Wren's group arising from the genome sequence of *C. jejuni* NCTC 11168 showed that the established Penner serotyping method for this bacterium depended not on lipo-oligosaccharide determinants, but instead on variation in polysaccharides from a previously unsuspected capsule, overturning a decades-old preconception *(34–39)*.

Yet, more innovative is the use of complete genome sequences to prime novel methods that would have been impossible in the pregenomic era. The most commonly used method for this "genomotyping" is hybridization to a microarray (*see* Chapter 10). Here, oligonucleotides or PCR products representing all genes in the genome of a given bacterium are arrayed as microscopic spots on a solid surface (typically a glass slide). Fluorescently labeled genomic DNA from a test strain is then hybridized to the array, and a confocal microscope is used to determine the pattern of hybridization. Spots that do not show hybridization represent genes that are present in the genome-sequenced strain used to create the array, but absent from the test strain. Several groups have used this approach to look at strain-to-strain variation:

- Dorrell et al. performed whole-genome microarray comparisons of 11 *C. jejuni* strains of diverse origin to show that at least 21% of genes in the sequenced strain were absent or highly divergent in one or more of the isolates tested. This study also provided further evidence that the capsule accounts for Penner serotype specificity and could potentially lay the foundations for novel epidemiological typing methods for *C. jejuni (35)*.

- Workers in Mekalanos's laboratory exploited a microarray built from the genome sequence of the *Vibrio cholerae* El Tor O1 strain N16961 to compare the gene content of N16961 with that of other *V. cholerae* isolates *(40)*. Although they found a high degree of conservation among the strains tested, they were able to identify genes unique to all pandemic strains as well as genes specific to the seventh-pandemic El Tor and related O139 serogroup strains.

- Hakenbeck and colleagues *(41)* used a microarray to investigate genomic variation in 20 *Streptococcus pneumoniae* isolates representing major antibiotic-resistant clones and 10 different capsular serotypes. They found that up to 10% of genes varied between individual isolates and the reference strain. Among the variable loci were mosaic genes encoding antibiotic resistance determinants and gene clusters related to bacteriocin production. They also performed genomic comparisons between *S. pneumoniae* and commensal *Streptococcus mitis* and *Streptococcus oralis* strains.

- Fitzgerald and colleagues *(42)* used a microarray to study genomic diversity in *Staphylococcus aureus.* They concluded that the most prevalent clone of *S. aureus* producing toxic-shock syndrome toxin evolved long ago and that the epidemic of toxic-shock syndrome that arose in the 1970s was caused by a change in the host environment rather than by rapid intercontinental dissemination of a new hyper-virulent clone.

Unfortunately, all of the previous studies have been performed on archived strain collections and none have used microarrays to type strains prospectively, in real time, "in anger." Although it is only a matter of time before microarray-based typing methods reach the clinical bacteriology laboratory (or at least the reference laboratory), one study, yet to be definitively published, stands out in its timely use of microarrays in an outbreak investigation. Shafi and coworkers *(43)* at the University of Leicester used a whole-genome *Mycobacterium tuberculosis* microarray to detect differences between the outbreak strain from the recent Leicester Crown Hills Community College outbreak *(44)*, the genome-sequenced strain H37Rv, and other *M. tuberculosis* isolates. Having quickly defined which genes were missing in the outbreak strain using the microarray, they then went on to design a rapid PCR-based method to distinguish the outbreak strain from other strains prevalent in Leicester in time to influence management of the outbreak *(44)*.

One problem with microarray-based approaches to the study of genomic diversity is that they can tell you only what is absent from the test strain compared with the genome-sequenced strain, rather than what new genes might be present. In my own laboratory, we have been exploring the use of long PCR to determine differences between strains of *E. coli*. So far we have successfully amplified fragments of up to 32 kb in length, and have detected large-scale genomic differences in a handful of chromosomal regions *(45)*. In the near future, we plan to scale up to achieve genome-wide coverage.

The ultimate in molecular epidemiological typing is to sequence the entire genomes of the strains one wants to type. Already multiple strains have been sequenced for several bacterial species **(Table 2)**. With the price of genome sequencing falling dramatically—around $0.10 per base pair for a fully finished and annotated sequence *(46)* (i.e., approx $200,000 for an average genome)—and now clearly within the resources of a research project grant, it is easy to envisage this approach becoming ever more popular over the coming years, and it becoming routine to sequence a particularly tricky outbreak strain of a hospital or community pathogen. As is often the case with technical innovation, the most futuristic molecular epidemiological use of genome sequencing per se—rather than derivative technologies such as microarrays—has come from defense applications. Read et al. sequenced the entire genome of a *Bacillus anthracis* strain isolated from a victim of a recent bioterrorist anthrax attack and compared it with that of a recently sequenced reference strain *(46)*. Their study revealed 60 new epidemiological markers that included single nucleotide polymorphisms, inserted or deleted sequences, and tandem repeats *(46)*. Selected markers were tested on a collection of anthrax isolates and were used to show that the "bioterrorism" strain was indistinguishable from laboratory isolates of the Ames strain, but distinct from the only other known natural occurrence of the Ames genotype (a goat isolate) *(46)*.

### 3.2. Diagnostic Bacteriology

Just as comparative genome sequencing can reveal differences useful for typing strains within a species, it can also be used to provide molecular diagnostic targets that can distinguish closely related species, subspecies, or pathovars, or to find novel targets for immunodiagnosis. The close similarity between *Mycobacterium avium* subsp. *paratuberculosis* and other mycobacterial species has hindered the development of specific diagnostic reagents. A comparison of an incomplete *M. avium* subsp. *paratuberculosis* genome sequence with that of *M. avium* subsp. *avium* revealed 21 genes that were present in all *M. avium* subsp. *paratuberculosis* isolates and absent from all other mycobacterial species tested, thus providing novel targets, e.g., for diagnostic PCR *(47)*.

Raczniak and colleagues used genome sequences to classify organisms according to the class of lysyl-tRNA synthetase that they employed *(26)*. This led to the realization that the Lyme disease agent, *Borrelia burgdorferi,* possessed a class of enzyme distinct from the human host and from almost all other bacteria, thus providing a potential new target for either DNA-based or immunologically based diagnosis.

### 3.3. Antimicrobial Resistance

Genome sequencing has the potential to reveal novel resistance determinants or new arrangements of known resistance genes. Researchers in Hiramatsu's laboratory *(48)* compared the genome sequence of a methicillin-resistant *S. aureus* (MRSA) strain isolated in 1982 with that of an MRSA strain exhibiting low-level vancomycin resistance (minimum inhibitory concentration, MIC, 8 µg/mL) isolated in 1997. They reported that most of the antibiotic resistance genes were carried either by plasmids or by mobile genetic elements including a unique resistance island. They subsequently sequenced the genome of an apparently highly virulent, community-acquired MRSA and reported that the *mecA* gene, which encodes the novel penicillin-binding protein (PBP2') responsible for methicillin resistance, was carried in this strain by a novel form of Staphylococcal chromosomal cassette *mec* (SCC*mec*), and that the strain carried 19 additional virulence genes *(49)*.

Fitzgerald and colleagues *(42)* used a microarray to study genomic diversity in *S. aureus,* including MRSA. They discovered that MRSA strains fell into five or more highly divergent groups that differed by as much as several hundred genes. They therefore concluded that MRSA strains have arisen independently by lateral transfer of the SCC*mec* element into methicillin-susceptible precursors.

A highly novel use of whole-genome microarrays for determining which genes might contribute to resistance has recently been reported by Gill et al. *(50)*. Their strategy was to create a genomic library of *E. coli* sequences expressed at high copy number in a wild-type *E. coli* host, then to select for clones that showed increased resistance, and finally to use DNA microarrays simultaneously to isolate and identify the enriched gene inserts *(50)*. They recovered 19 genes conferring tolerance to the cleaning agent Pine-Sol, and 27 genes conferring sensitivity, and suggested that the method could have wide applications in identifying resistance determinants *(50)*.

### 4. Conclusions

The exploitation of genome sequences in the clinical laboratory remains in its infancy. Two impediments limit progress here:

1. The very success of traditional agar-based methods. However, the ability of molecular, genome-based methods to compete successfully with established approaches will improve as automation and molecular methods make inroads into the laboratory, so that the marginal costs of setting up a new molecular assay steadily decrease.
2. The lack of individuals dually trained in clinical bacteriology and molecular biology/genomics. Would anyone interested in doing an MD or PhD or collaborating in this area please contact the author!

Perhaps the ideal outcome of genome sequencing in the clinical laboratory would be the development of simple biochemical tests that could assist in diagnosis and typing. For example, something as simple as the use of sorbitol MacConkey agar and beta-glucuronidase fermentation to distinguish *E. coli* O157:H7 from harmless commensals. Although in this case the frame-shift mutation that inactivates the beta-glucuronidase gene in *E. coli* O157:H7 was identified without using the genome sequence *(51)*, detailed comparisons of genome sequences from closely related species or pathovars might reveal similar kinds of differences that could then be detected in the clinical laboratory by simple biochemistry.

In closing, one can conclude that, although the genomics revolution has already had a decisive impact on microbiology research, it has yet to have a major effect in the clinical laboratory. However, it is only 7 yr since the first bacterial genome was sequenced. One is reminded of the dialogue between US President Nixon and the Chinese Prime Minister Chou En-Lai during Nixon's historic visit to China. Nixon asked Chou, a keen student of history, what impact he thought the French Revolution had had on Western Civilization. To which Chou replied "It's too early to tell!"

## Acknowledgments

I thank the Biotechnology and Biological Sciences Research Council (BBSRC) for funding bioinformatics and long-PCR studies in my research group.

## References

1. Fleischmann, R. D., Adams, M. D., White, O., Clayton, R. A., Kirkness, E. F., Kerlavage, A. R., et al. (1995) Whole-genome random sequencing and assembly of *Haemophilus influenzae* Rd. *Science* **269,** 496–512.
2. Fraser, C. M., Gocayne, J. D., White, O., Adams, M. D., Clayton, R. A., Fleischmann, et al. (1995) The minimal gene complement of *Mycoplasma genitalium*. *Science* **270,** 397–403.
3. Moxon, E. R. (1995) Whole genome sequencing of pathogens: a new era in microbiology. *Trends Microbiol.* **3,** 335–337.

4. Moxon, E. R., Hood, D. W., Saunders, N. J., Schweda, E. K., and Richards, J. C. (2002) Functional genomics of pathogenic bacteria. *Philos. Trans. R Soc. Lond. B Biol. Sci.* **357,** 109–116.

5. Schoolnik, G. K. (2002) Functional and comparative genomics of pathogenic bacteria. *Curr. Opin. Microbiol.* **5,** 20–26.

6. Giaever, G., Chu, A. M., Ni, L., Connelly, C., Riles, L., Veronneau, S., et al. (2002) Functional profiling of the *Saccharomyces cerevisiae* genome. *Nature* **418,** 387–391.

7. Winzeler, E. A., Liang, H., Shoemaker, D. D., and Davis, R. W. (2000) Functional analysis of the yeast genome by precise deletion and parallel phenotypic characterization. *Novartis Found. Symp.* **229,** 105–111.

8. Kim, S. H. (2000) Structural genomics of microbes: an objective. *Curr. Opin. Struct. Biol.* **10,** 380–383.

9. Rain, J. C., Selig, L., De Reuse, H., Battaglia, V., Reverdy, C., Simon, S., et al. (2001) The protein-protein interaction map of *Helicobacter pylori. Nature* **409,** 211–215.

10. Dieckman, L., Gu, M., Stols, L., Donnelly, M. I., and Collart, F. R. (2002) High throughput methods for gene cloning and expression. *Protein Expr. Purif.* **25,** 1–7.

11. Schoolnik, G. K. (2002) Microarray analysis of bacterial pathogenicity. *Adv. Microb. Physiol.* **46,** 1–45.

12. Hinton, J. C. (1997) The *Escherichia coli* genome sequence: the end of an era or the start of the FUN? *Mol. Microbiol.* **26,** 417–422.

13. Eiglmeier, K., Parkhill, J., Honore, N., Garnier, T., Tekaia, F., Telenti, A., et al. (2001) The decaying genome of *Mycobacterium leprae. Lepr. Rev.* **72,** 387–398.

14. Cole, S. T., Eiglmeier, K., Parkhill, J., James, K. D., Thomson, N. R., Wheeler, P. R., et al. (2001) Massive gene decay in the leprosy bacillus. *Nature* **409,** 1007–1011.

15. Fraser, C. M., Norris, S. J., Weinstock, G. M., White, O., Sutton, G. G., Dodson, R., et al. (1998) Complete genome sequence of *Treponema pallidum,* the syphilis spirochete. *Science* **281,** 375–388.

16. Norris, S. J., Fraser, C. M., and Weinstock, G. M. (1998) Illuminating the agent of syphilis: the *Treponema pallidum* genome project. *Electrophoresis* **19,** 551–553.

17. Norris, S. J., Cox, D. L., and Weinstock, G. M. (2001) Biology of *Treponema pallidum:* correlation of functional activities with genome sequence data. *J. Mol. Microbiol. Biotechnol.* **3,** 37–62.

18. Weinstock, G. M., Hardham, J. M., McLeod, M. P., Sodergren, E. J., and Norris, S. J. (1998) The genome of *Treponema pallidum:* new light on the agent of syphilis. *FEMS Microbiol. Rev.* **22,** 323–332.

19. Weinstock, G. M., Smajs, D., Hardham, J., and Norris, S. J. (2000) From microbial genome sequence to applications. *Res. Microbiol.* **151,** 151–158.

20. Ye, R. W., Tao, W., Bedzyk, L., Young, T., Chen, M., and Li, L. (2000) Global gene expression profiles of *Bacillus subtilis* grown under anaerobic conditions. *J. Bacteriol.* **182,** 4458–4465.

21. Helmann, J. D., Wu, M. F., Kobel, P. A., Gamo, F. J., Wilson, M., Morshedi, M. M., et al. (2001) Global transcriptional response of *Bacillus subtilis* to heat shock. *J. Bacteriol.* **183,** 7318–7328.
22. Okinaka, Y., Yang, C. H., Perna, N. T., and Keen, N. T. (2002) Microarray profiling of *Erwinia chrysanthemi* 3937 genes that are regulated during plant infection. *Mol. Plant Microbe Interact.* **15,** 619–629.
23. Khil, P. P. and Camerini-Otero, R. D. (2002) Over 1000 genes are involved in the DNA damage response of *Escherichia coli. Mol. Microbiol.* **44,** 89–105.
24. Hood, D. W., Deadman, M. E., Allen, T., Masoud, H., Martin, A., Brisson, J. R., et al. (1996) Use of the complete genome sequence information of *Haemophilus influenzae* strain Rd to investigate lipopolysaccharide biosynthesis. *Mol. Microbiol.* **22,** 951–965.
25. McDevitt, D. and Rosenberg, M. (2001) Exploiting genomics to discover new antibiotics. *Trends Microbiol.* **9,** 611–617.
26. Raczniak, G., Ibba, M., and Soll, D. (2001) Genomics-based identification of targets in pathogenic bacteria for potential therapeutic and diagnostic use. *Toxicology* **160,** 181–189.
27. Fairlamb, A. H. (2002) Metabolic pathway analysis in trypanosomes and malaria parasites. *Philos. Trans. R. Soc. Lond. B Biol. Sci.* **357,** 101–107.
28. Fraser, C. M., Eisen, J. A., and Salzberg, S. L. (2000) Microbial genome sequencing. *Nature* **406,** 799–803.
29. Saunders, N. J. and Moxon, E. R. (1998) Implications of sequencing bacterial genomes for pathogenesis and vaccine development. *Curr. Opin. Biotechnol.* **9,** 618–623.
30. Pizza, M., Scarlato, V., Masignani, V., Giuliani, M. M., Arico, B., Comanducci, M., et al. (2000) Identification of vaccine candidates against serogroup B meningococcus by whole-genome sequencing. *Science* **287,** 1816–1820.
31. Dingle, K. E., Colles, F. M., Wareing, D. R., Ure, R., Fox, A. J., Bolton, F. E., et al. (2001) Multilocus sequence typing system for *Campylobacter jejuni. J. Clin. Microbiol.* **39,** 14–23.
32. Homan, W. L., Tribe, D., Poznanski, S., Li, M., Hogg, G., Spalburg, E., et al. (2002) Multilocus sequence typing scheme for *Enterococcus faecium. J. Clin. Microbiol.* **40,** 1963–1971.
33. Desai, M., Logan, J. M., Frost, J. A., and Stanley, J. (2001) Genome sequence-based fluorescent amplified fragment length polymorphism of *Campylobacter jejuni,* its relationship to serotyping, and its implications for epidemiological analysis. *J. Clin. Microbiol.* **39,** 3823–3829.
34. Parkhill, J., Wren, B. W., Mungall, K., Ketley, J. M., Churcher, C., Basham, D., et al. (2000) The genome sequence of the food-borne pathogen *Campylobacter jejuni* reveals hypervariable sequences. *Nature* **403,** 665–668.
35. Dorrell, N., Mangan, J. A., Laing, K. G., Hinds, J., Linton, D., Al-Ghusein, H., et al. (2001) Whole genome comparison of *Campylobacter jejuni* human isolates using a low-cost microarray reveals extensive genetic diversity. *Genome Res.* **11,** 1706–1715.

36. Wren, B. W., Linton, D., Dorrell, N., and Karlyshev, A. V. (2001) Post genome analysis of *Campylobacter jejuni. Symp. Ser. Soc. Appl. Microbiol.* **30,** 36S–44S.
37. Linton, D., Karlyshev, A. V., and Wren, B. W. (2001) Deciphering *Campylobacter jejuni* cell surface interactions from the genome sequence. *Curr. Opin. Microbiol.* **4,** 35–40.
38. Karlyshev, A. V. and Wren, B. W. (2001) Detection and initial characterization of novel capsular polysaccharide among diverse *Campylobacter jejuni* strains using alcian blue dye. *J. Clin. Microbiol.* **39,** 279–284.
39. Karlyshev, A. V., McCrossan, M. V., and Wren, B. W. (2001) Demonstration of polysaccharide capsule in *Campylobacter jejuni* using electron microscopy. *Infect. Immun.* **69,** 5921–5924.
40. Dziejman, M., Balon, E., Boyd, D., Fraser, C. M., Heidelberg, J. F., and Mekalanos, J. J. (2002) Comparative genomic analysis of *Vibrio cholerae:* genes that correlate with cholera endemic and pandemic disease. *Proc. Natl. Acad. Sci. USA* **99,** 1556–1561.
41. Hakenbeck, R., Balmelle, N., Weber, B., Gardes, C., Keck, W., and de Saizieu, A. (2001) Mosaic genes and mosaic chromosomes: intra- and interspecies genomic variation of *Streptococcus pneumoniae. Infect. Immun.* **69,** 2477–2486.
42. Fitzgerald, J. R., Sturdevant, D. E., Mackie, S. M., Gill, S. R., and Musser, J. M. (2001) Evolutionary genomics of *Staphylococcus aureus:* insights into the origin of methicillin-resistant strains and the toxic shock syndrome epidemic. *Proc. Natl. Acad. Sci. USA* **98,** 8821–8826.
43. Shafi, J. (2002) Oral Presentation: Rapid Genomic and Biological Characterisation Of The Leicester School Tuberculosis Outbreak Strain, at Acid-FaSt Club Summer Meeting, Cardiff, UK.
44. Watson, J. M. and Moss, F. (2001) TB in Leicester: out of control, or just one of those things? *BMJ* **322,** 1133–1134.
45. Alli, O. A. T., McBride, A., Wilkinson, K., Hinks, L., Day, I. N. M., and Pallen, M. J. (2002) Poster presentation: long PCR as a tool for exploring genomic diversity in *Escherichia coli,* at Genome-based Pathogen Biology: The First 25 Years and Beyond, Hinxton Hall Conference Centre, The Wellcome Trust Genome Campus, Hinxton, Cambridgeshire, England.
46. Read, T. D., Salzberg, S. L., Pop, M., Shumway, M., Umayam, L., Jiang, L., et al. (2002) Comparative genome sequencing for discovery of novel polymorphisms in *Bacillus anthracis. Science* **296,** 2028–2033.
47. Bannantine, J. P., Baechler, E., Zhang, Q., Li, L., and Kapur, V. (2002) Genome scale comparison of *Mycobacterium avium* subsp. *paratuberculosis* with *Mycobacterium avium* subsp. *avium* reveals potential diagnostic sequences. *J. Clin. Microbiol.* **40,** 1303–1310.
48. Kuroda, M., Ohta, T., Uchiyama, I., Baba, T., Yuzawa, H., Kobayashi, I., et al. (2001) Whole genome sequencing of meticillin-resistant *Staphylococcus aureus. Lancet* **357,** 1225–1240.

49. Baba, T., Takeuchi, F., Kuroda, M., Yuzawa, H., Aoki, K., Oguchi, A., et al. (2002) Genome and virulence determinants of high virulence community-acquired MRSA. *Lancet* **359,** 1819–1827.
50. Gill, R. T., Wildt, S., Yang, Y. T., Ziesman, S., and Stephanopoulos, G. (2002) Genome-wide screening for trait conferring genes using DNA microarrays. *Proc. Natl. Acad. Sci. USA* **99,** 7033–7038.
51. Monday, S. R., Whittam, T. S., and Feng, P. C. (2001) Genetic and evolutionary analysis of mutations in the *gusA* gene that cause the absence of beta-glucuronidase activity in *Escherichia coli* O157:H7. *J. Infect. Dis.* **184,** 918–921.

# 2

# Public Databases

*Retrieving and Manipulating Sequences for Beginners*

## Neil Woodford

### Summary

This chapter outlines the basic requirements for finding and exploring sequences of interest in public databases, such as GenBank. As such, it is not aimed at experienced sequencers, for whom this will be "second nature," but at the many clinical bacteriologists who rarely have need of DNA sequences in their usual work, and who would like to develop their interest in what can appear to be a daunting area. The topics discussed include finding and retrieving sequences from GenBank, identifying homologous sequences using BLAST searches, resources for accessing microbial genomes, and the Protein Data Bank. Finally, recommendations are made for useful software (freeware) and online sequence manipulation resources.

**Key Words:** Bioinformatics; GenBank database; BLAST search; homology; Protein Data Bank; TIGR; Sanger Institute; freeware; online resources.

## 1. Introduction

*Bioinformatics* is a current buzzword, but it means different things to different people. Generally, it is the discipline where maximum biological information is derived from a nucleotide or amino acid sequence. Very often, this requires science to be performed *in silico*—a phrase that emphasizes the fact that many molecular biologists spend increasing amounts of their time in front of a computer screen, generating hypotheses that can subsequently be tested and (hopefully) confirmed in the laboratory. In my very first undergraduate microbiology practical (only 20 yr ago), I was told that there was no such thing as theoretical microbiology. However, this is becoming increasingly questionable and, in a few years time, it may no longer be true! At a fundamental level,

From: *Methods in Molecular Biology, vol. 266: Genomics, Proteomics, and Clinical Bacteriology:
Methods and Reviews*
Edited by: N. Woodford and A. Johnson © Humana Press Inc., Totowa, NJ

"bioinformaticians" write computer programs to perform searches, align sequences, and so on. Such tasks obviously require detailed understanding of the underlying mathematical algorithms and statistical approaches (computational biology), and are beyond the scope of this chapter (or this book, for that matter). However, far more people use bioinformatics at a more practical level, by accessing and utilizing the vast number of resources available to analyze and derive useful information from nucleic acid or peptide sequences. These sequences may have been generated in their own research, or may have been downloaded from public databases. This chapter provides a beginner's guide to this second, more pragmatic approach.

## 2. Retrieving DNA Sequences From Public Databases

There are several databases in which DNA sequences may be deposited (e.g., GenBank, EMBL) and there is regular and frequent exchange of new sequences between them. In consequence, researchers deposit sequences with only one database and, similarly, anyone searching for a sequence needs to interrogate only one database. Many bacteriologists use PubMed (a medical literature search tool) regularly and, conveniently, the GenBank search engine (which can be accessed via http://www.ncbi.nlm.nih.gov/Entrez/index.html) shares the same front end (**Fig. 1A**). Hence, the GenBank sequence depository is suitable for many beginners because it provides a familiar screen layout. Ultimately, the choice of which database to search is a matter of personal preference.

The easiest way of finding a specific sequence in a database is by using its unique accession number; this is constant in all sequence databases. Most journals require published sequences to be accessible in a public database, unless they are patented. Each paper should include the unique database accession numbers for all deposited sequences (usually found at the end of the **Methods** or **Discussion** sections). Alternatively, if an accession number is not available, searches may be performed on key words or author names. The GenBank search engine will return either a single sequence record (if an accession number was used) (**Fig. 1B**), or a numbered list of sequence records (if key words were used in the search). In the returned listings, the accession numbers are hyperlinks that can be used to open the actual deposition records for each sequence. Within each record there are further hyperlinks to relevant PubMed entries (i.e., corresponding publications) and, if the sequence contains multiple open-reading frames (ORFs), to individual genes and to their translated products. For example, if a search is performed on accession number M97297, the search engine will return the record for the prototype VanA glycopeptide resistance transposon, Tn*1546* of *Enterococcus faecium* (**Fig. 1B**) *(1)*. This resistance element is 10,851 bp long and contains nine ORFs. The GenBank record

**A**

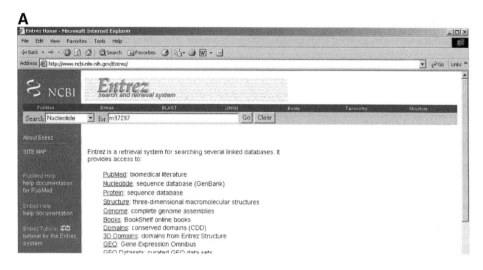

**B**

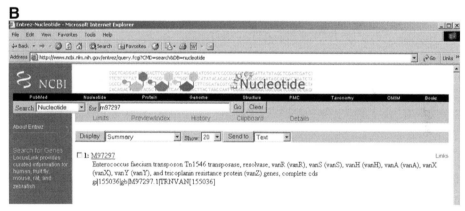

Fig. 1. **(A)** The front end of the GenBank search engine (which also provides access to PubMed and other resources). To obtain sequences, select "nucleotide" in the "Search" dropdown menu, enter the accession number (or author name, key words, and so on) in the "for" field and click on "Go." **(B)** An example of a retrieved entry. The accession number acts as a hyperlink to open the full GenBank record. The check box may be used to store the file on the clipboard or to save it to a file on a local drive. In author or key word searches, multiple sequences are usually retrieved.

gives nucleotide positions of each ORF and the peptide sequence of the predicted product, and hyperlinks to the sequences of individual genes and peptides. The full 10,851-bp sequence appears at the end of the record (**Fig. 2**).

A search may fail to retrieve a required sequence for several reasons. First, it may not yet have been released for public access. Authors have to deposit

Fig. 2. Part of the GenBank record for accession number M97297 (the 10,851-bp transposon Tn*1546* of *E. faecium*). After details of the sequenced organism, PubMed citation etc., there is a full list of ORFs within a record. Nucleotide positions within the sequence are given (e.g., the *vanA* gene spans from nt 6979 to 8010 in the screenshot shown). Adjacent to this information are hyperlinks to the sequence of each specific gene and coding sequence (CDS), which avoids the need to "count and crop" it from the full 10,851-bp sequence at the end of the entry.

sequences with databases prior to submitting a paper so that they can quote accession numbers in manuscripts. However, such unpublished sequences may be withheld from the publicly accessible part of the database until after publication. Once work has been published, the authors have to notify the GenBank administrators so that release may proceed. Authors may forget to do this. Also, the huge explosion in sequencing has caused a corresponding increase in database submissions. Database administrative staff cannot keep pace with requests for sequence modification and release, so there is often a delay (usually only a few days) before instructions are acted upon. Second, there may be a typographical error in the published accession number. To test for this, it is often advisable to try a keyword or author search (when performing such searches in GenBank, remember to keep "Nucleotide" in the search field, so that you query GenBank rather than PubMed). If problems accessing a published sequence persist, inform the GenBank administrator (gb-admin@ncbi.nlm.nih.gov), quoting the accession number and journal reference in which it appears.

## 2.1. Saving a Sequence

When the required sequence has been retrieved from the database it may be viewed online or, for subsequent manipulation, a copy may be saved to a local drive. When using the GenBank database, the procedure for saving sequence files follows exactly the same steps as for saving literature citations from PubMed. Thus, to save a sequence, (1) check the box next to its accession number, (2) choose "GenBank" or "FASTA" format in the dropdown "Display" option box, (3) choose "File" in the dropdown "Send to" option box, and (4) click "Send to." Follow the standard Windows instructions to save to the required place on a local drive.

For more complex analyses, it is necessary to retrieve and save multiple sequences, often as a single file. Usually subsequent manipulation is intended (e.g., multiple alignments), and saving the sequences in FASTA (rather than GenBank) format is most appropriate. If the required sequences are retrieved in a single search (e.g., on a particular author or key word) they can be saved simply by selecting the box next to each required accession number. However, saving multiple sequences from separate searches is not a problem (again, it's the same as saving different author citations from PubMed): (1) Perform the first search, select the boxes for all required sequences, and store the sequence on the search engine's clipboard (this is achieved by choosing "Clipboard" in the dropdown "Send to" option box, and then clicking "Send to"); (2) perform the next search, check the boxes for all required sequences, and store them on the search engine's clipboard; (3) repeat the previous step as many times as necessary (the clipboard will store up to 500 items); (4) when all searches are complete, click "Clipboard" to reveal all of your stored sequences; (5) select the boxes again, choose "FASTA" format in the dropdown "Display" option box; (6) choose "File" in the dropdown "Send to" option box; and (7) click "Send to." Follow standard Windows instructions to save all to a single file in the required place on a local drive.

## 2.2. Viewing a Downloaded Sequence File

Even with no bioinformatics software on a PC, saved GenBank files can be opened using WordPad (a standard Windows program) or a similar text viewer. Opening the saved file as text in Microsoft Word or a similar processing program is not recommended, as such packages will try to reformat the file on closing. If the sequence was saved in GenBank format, WordPad shows the entire record (including genes, peptides, Medline citations and so on), although the hyperlinks present in the original record no longer function. If the file was saved in FASTA format (the usual option for many subsequent manipulations), WordPad reveals a leading descriptive line (in the format ">sequence title") followed by the nucleotide sequence only.

Although using WordPad is arguably one of the best ways of simply viewing sequence files, any manipulation of the sequences necessitates the use of bioinformatics software. It is not essential, but it makes life a whole lot simpler (an analogy would be creating a Web page using a specific Web design package, compared with using just a text editor and knowledge of HTML; not impossible, but unnecessarily time-consuming). Many commercial bioinformatics packages are available, but there are many freeware programs available and they are often just as good. Indeed, the freeware can frequently do everything one is likely to need day to day. The exception to this seems to be the absence of primer design freeware for local installation; there are DOS-based programs, but very little that is Windows based, as far as I know (*see* **Subheading 6.8.** for an online solution to this problem). BioEdit (*see* **Subheading 6.2.**) is an excellent, easy to use program. It is simple to install and goes a long way toward providing all required bioinformatics applications in a single package.

## 3. Basic Local Alignment Search Tool (BLAST)

BLAST *(2)* is an online program that searches for significant homology between query sequences and those in the GenBank database. These searches can be initiated from within some bioinformatics packages, such as BioEdit (*see* **Subheading 6.2.**), or directly from the main BLAST site (http://www.ncbi.nlm.nih.gov/BLAST/). The BLAST tool is actually a group of programs that can be used to compare nucleotide sequences against other nucleotide sequences (blastn), peptide sequences against peptide sequences (blastp), or peptide sequences against translated nucleotide sequences (tblastn). Using the tool is simply a matter of cutting and pasting the required query sequence into the appropriate box, selecting the most appropriate program, and clicking on "Blast."

If searching for homologs in diverse species, where genes may not show high levels of nucleotide similarity, the tblastn option can be particularly useful. This is because proteins with similar functions often retain conserved domains and motifs (i.e., groups of conserved amino acids) (*see* Chapter 3) even though variation in codon usage, for example, means that the corresponding genes are less similar. These characteristic protein family "signatures" can serve to identify potential homologs. Hence, unless alleles of particular genes are sought, it is usually good practice first to perform a tblastn search with a peptide sequence; the results are often more fruitful than those of a direct blastn search with the nucleotide sequence.

The results of BLAST searches are usually available in a short time (seconds to a few minutes), but can be returned by e-mail at particularly busy periods. The results indicate the degree of similarity between your sequence and those in the database. The results include score ($S$) and probability ($E$) values. The score for each sequence is determined by the algorithms used in the program. The $E$ value indicates how likely an equal or greater score is to have arisen by chance (i.e., is the homology likely to be a real or chance occurrence). Results are returned in order of decreasing $E$ value, so the most significant hits appear first.

The list of sequences returned in a BLAST report will consist of true-positives and false-positives. True-positives are sequences that share an evolutionary origin with the query sequence (are true homologs), or are examples of evolutionary convergence. False-positives show sequence similarity owing to chance. There are no mathematical algorithms that can distinguish absolutely between these possibilities. Negatives are homologous sequences that were not flagged by the search *(3)*. All algorithms that assess sequence similarity must reach a compromise between two conflicting goals: (1) to produce a list that includes all significant hits (i.e., avoiding negatives); (2) to avoid generating long lists of meaningless chance homologies (i.e., avoiding false-positives).

Thus, in a BLAST report, only some of the returned sequences are genuine homologs. For many searches, common sense can indicate which results should be pursued and which discarded. For example, a returned sequence showing 100% identity to the query sequence over 15 nt is unlikely to be meaningful if that query sequence was 800-bp long! $E$ values $<10^{-50}$ are commonly generated by BLAST searches and indicate likely close relationships; these result from good homology and a long overlap between the subject and query sequences.

The returned BLAST report includes a graphical representation of homologous sequences (**Fig. 3A**) in which each line is hyperlinked to the alignment between the query and a particular retrieved sequence. The GenBank accession numbers of the retrieved sequences are given next to each alignment (**Fig. 3B**), and these are hyperlinked to the corresponding GenBank entries. There is also an option to select sequence boxes and save retrieved sequence files.

## 4. Sources of Microbial Genome Sequences

Many chapters in this volume describe the extensive and ever-increasing number of complete microbial genome sequences. Many of these are freely available, and there are appropriate resources to facilitate their use. Indeed, much of the drive toward improved bioinformatics software comes from the large genome-sequencing corporations.

The Institute for Genomic Research (TIGR) has compiled and maintains a Comprehensive Microbial Resource (CMR) tool that allows researchers to

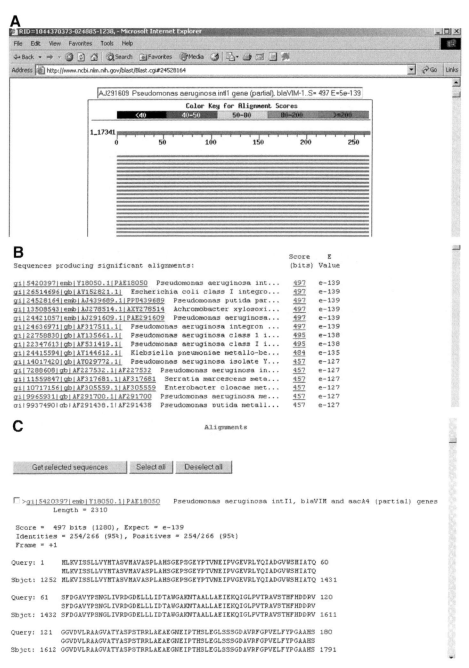

Fig. 3. Three components of a BLAST search report. A tblastn search was performed with the peptide sequence of VIM-1 carbapenemase. (**A**) Graphic representation of returned sequences. Each line in the graphic is hyperlinked to specific query sequence–retrieved sequence alignment further down the report. As the mouse moves over each line, details of the retrieved sequence appear in the title bar. (**B**) List of

access all bacterial genome sequences completed to date (http://www.tigr.org/ tigr-scripts/CMR2/CMRHomePage.spl) *(4)*. The Sanger Institute also has extensive interests in genome sequencing, and its data are also freely available (http://www.sanger.ac.uk/Projects/Microbes/). Data from the Sanger Institute and TIGR can be accessed in various ways. Importantly, there are BLAST facilities incorporated within these websites. Rather than searching the whole of GenBank, BLAST searches from the TIGR or Sanger sites allow specified genomes to be interrogated using a query sequence (remember that tblastn is often the best first search option; *see* **Subheading 3.**). The matching contigs (or even complete genomes) can be retrieved and downloaded to local drives for further manipulation (often via FTP).

## 5. The Protein Data Bank (PDB)

The Protein Data Bank (http://www.rcsb.org/pdb/) is a publicly accessible repository for proteins of known 3D structure. The structures were determined experimentally, by X-ray crystallography or solution nuclear magnetic resonance. As with DNA sequences in GenBank, each deposited structure has a unique accession number, which typically consists of a number followed by three characters (e.g., 1FOF is the accession number of the class D β-lactamase OXA-10 [PSE-2]; *5*). The homepage has search and retrieval facilities using either accession numbers or keywords. As with GenBank, searches performed on accession numbers will return the specific record, while those using key words may return multiple records. Each record has hyperlinks to relevant PubMed citations. The returned PDB files can be downloaded to a local drive and viewed or manipulated with suitable software, such as Deep View (*see* **Subheading 6.5.**) or RasMol (http://www.bernstein-plus-sons.com/software/rasmol/).

For researchers studying protein structure, the value of this database is obvious, but it can be of great use to bacteriologists generally. First, it can be used simply as a resource to discover more about proteins of interest. Second, known protein structures can be used as templates to perform comparative modeling of homologs whose 3D structures have not been determined *(6)*.

## 6. Recommended Downloadable Freeware and Online Resources

The software listed below represents a very small fraction of the available resources. The author has found the selected programs useful, and they will

---

Fig. 3. *(continued)* returned sequences, with hyperlinks to the full GenBank record. **(C)** An example of a query sequence–retrieved sequence alignment. The accession number acts as a hyperlink to open the full GenBank record. Note also the check box for saving retrieved sequence files.

provide a good starting point for any new bioinformatics user. All URLs were accurate at the time of writing, but they are prone to change. If a link gets broken, keyword searches on the particular program name using a search engine such as Google should identify the new homepage rapidly.

### 6.1. Artemis

Artemis *(7)* is a genome viewer and annotation tool that allows visualization of sequence features and the results of analyses within the context of the sequence and its six-frame translation. Available from: http://www.sanger.ac.uk/Software/Artemis/

### 6.2. BioEdit

BioEdit is an excellent general bioinformatics program that can handle most of the manipulations that researchers need regularly. It can be used simply to view downloaded GenBank files, or for a variety of more complex procedures. Available from: http://www.mbio.ncsu.edu/BioEdit/bioedit.html

### 6.3. BLAST

BLAST *(2)* is likely the most commonly used bioinformatics tool. It performs online searches for sequences in GenBank that show significant homology to a query sequence. Cut and paste your sequence (DNA or peptide), check relevant option boxes, and away you go. BLAST is frequently incorporated into genome websites to allow specific genomes to be interrogated. Available at: http://www.ncbi.nlm.nih.gov/BLAST/

### 6.4. ClustalX

ClustalX *(8,9)* is a Windows interface for the ClustalW multiple sequence alignment program *(10,11)*. It provides an integrated environment for performing multiple sequence and profile alignments and analyzing the results. Alignments can be generated in various formats, and can be imported into GeneDoc (*see* **Subheading 6.6.**) or PHYLIP (*see* **Subheading 6.7.**). The tree files may be viewed with TreeView (*see* **Subheading 6.10.**). Available from: ftp://ftp-igbmc.u-strasbg.fr/pub/ClustalX/

### 6.5. Deep View (Swiss-Pdb Viewer)

Swiss-PdbViewer *(12)* is a powerful application with a fairly user-friendly interface that allows analysis of PDB files (3D protein structures), including several proteins at the same time. The proteins can be superimposed in order to deduce structural alignments and compare their active sites, and so on. The program is linked to Swiss-Model (http://swissmodel.expasy.org/), an automated homology modeling server. There is an excellent tutorial (http://

www.usm.maine.edu/~rhodes/SPVTut/index.html) to get you started with
Deep View. Available from: http://ca.expasy.org/spdbv/

### 6.6. GeneDoc

GeneDoc *(13)* is a full-featured multiple-sequence alignment editor, ana-
lyzer, and shading utility for Windows. Excellent for formatting ClustalX-
generated alignments into publication-quality figures. Available from: http://
www.psc.edu/biomed/genedoc/

### 6.7. PHYLIP

PHYLIP is a powerful package of programs for inferring phylogenies (evolutionary trees). PHYLIP is the most widely distributed phylogeny package,
but it does not have a Windows interface and can appear daunting. Trees can
be calculated using parsimony, distance matrix, and likelihood methods.
Bootstrapping and preparation of consensus trees is also available. An excellent overview of the package is recommended *(14)*. Available from: http://
evolution.genetics.washington.edu/phylip.html

### 6.8. Primer 3

An online resource for designing PCR and sequencing primers. Cut and paste
your DNA sequence, check relevant option boxes, and away you go. Available
at: http://www-genome.wi.mit.edu/cgi-bin/primer/primer3_www.cgi

### 6.9. SeqVISTA

SeqVISTA *(15)* enables annotated nucleotide or protein sequences to be
viewed graphically. Once installed, it may be launched from your Web
browser's toolbar to visualize GenBank records. SeqVISTA aims to display
results from diverse sequence analysis tools in an integrated fashion. Available
from: http://zlab.bu.edu/SeqVISTA/

### 6.10. TreeView

TreeView *(16)* is a simple program for displaying phylogenies. It can read
many different tree file formats (including those generated by ClustalX and
PHYLIP) and can produce publication-quality trees simply. Available from:
http://taxonomy.zoology.gla.ac.uk/rod/treeview.html

### 6.11. WebCutter

An online resource for identifying restriction endonuclease cutting sites
within a sequence. Cut and paste your DNA sequence, check relevant option
boxes, and away you go. Available at: http://www.firstmarket.com/cutter/
cut2.html

## References

1. Arthur, M., Molinas, C., Depardieu, F., and Courvalin, P. (1993) Characterization of Tn1546, aTn3-related transposon conferring glycopeptide resistance by synthesis of depsipeptide peptidoglycan precursors in *Enterococcus faecium* BM4147. *J. Bacteriol.* **175,** 117–127.
2. Altschul, S. F., Gish, W., Miller, W., Myers, E. W., and Lipman, D. J. (1990) Basic local alignment search tool. *J. Mol. Biol.* **215,** 403–410.
3. Pagni, M. and Jongeneel, C. V. (2001) Making sense of score statistics for sequence alignments. *Brief. Bioinform.* **2,** 51–67.
4. Peterson, J. D., Umayam, L. A., Dickinson, T., Hickey, E. K., and White, O. (2001) The Comprehensive Microbial Resource. *Nucleic Acids Res.* **29,** 123–125.
5. Paetzel, M., Danel, F., de Castro, L., Mosimann, S. C., Page, M. G., and Strynadka, N. C. (2000) Crystal structure of the class D β-lactamase OXA-10. *Nat. Struct. Biol.* **7,** 918–925.
6. Mullan, L. J. (2002) Protein 3D structural data—where it is, and why we need it. *Brief. Bioinform.* **3,** 410–412.
7. Rutherford, K., Parkhill, J., Crook, J., Horsnell, T., Rice, P., Rajandream, M. A., and Barrell, B. (2000) Artemis: sequence visualization and annotation. *Bioinformatics* **16,** 944–945.
8. Jeanmougin, F., Thompson, J. D., Gouy, M., Higgins, D. G., and Gibson, T. J. (1998) Multiple sequence alignment with Clustal X. *Trends Biochem. Sci.* **23,** 403–405.
9. Thompson, J. D., Gibson, T. J., Plewniak, F., Jeanmougin, F., and Higgins, D. G. (1997) The CLUSTAL_X windows interface: flexible strategies for multiple sequence alignment aided by quality analysis tools. *Nucleic Acids Res.* **25,** 4876–4882.
10. Thompson, J. D., Higgins, D. G., and Gibson, T. J. (1994) CLUSTAL W: improving the sensitivity of progressive multiple sequence alignment through sequence weighting, position-specific gap penalties and weight matrix choice. *Nucleic Acids Res.* **22,** 4673–4680.
11. Higgins, D. G. and Sharp, P. M. (1988) CLUSTAL: a package for performing multiple sequence alignment on a microcomputer. *Gene* **73,** 237–244.
12. Guex, N. and Peitsch, M. C. (1997) SWISS-MODEL and the Swiss-Pdb Viewer: an environment for comparative protein modeling. *Electrophoresis* **18,** 2714–2723.
13. Nicholas, K. B., Nicholas, H. B. Jr., and Deerfield, D. W. II. (1997) GeneDoc: analysis and visualization of genetic variation. *EMBNEW News* **4,** 14.
14. Retief, J. D. (2000) Phylogenetic analysis using PHYLIP. *Methods Mol. Biol.* **132,** 243–258.
15. Hu, Z., Frith, M. C., Niu, T., and Weng, Z. (2003) SeqVISTA: a graphical tool for sequence feature visualization and comparison. BMC. *Bioinformatics* **4,** 1.
16. Page, R. D. M. (1996) TREEVIEW: an application to display phylogenetic trees on personal computers. *Comp. Appl. Biosci.* **12,** 357–358.

# 3

## Genome Sequencing and Annotation

*An Overview*

### Makoto Kuroda and Keiichi Hiramatsu

#### Summary

Many microbial genome sequences have been determined, and more new genome projects are ongoing. Shotgun sequencing of randomly cloned short pieces of genomic DNA can provide a simple way of determining whole genome sequences. This process requires sequencing of many fragments, compilation of the separate sequences into one contiguous sequence, and careful editing of the assembled sequence. The genes present on the microbial genome are then predicted using clues derived from typical gene features, such as codon usage, ribosomal binding sequences, and bacterial initiation codons. Function of genes is predicted by homology searches performed against either public or well-established protein databases. This chapter discusses each of these stages in a genome-sequencing project.

**Key Words:** Shotgun sequencing; contig assembly; gap closing; gene annotation; gene prediction; homology searches; motif analysis.

### 1. Introduction

This chapter summarizes the experimental procedures that must be undertaken to determine whole bacterial genome sequences. Most bacterial genomes range from 0.5 to 7.0 Mb in size, and a random shotgun technique is the most effective sequencing strategy, except for organisms whose genomes contain multiple DNA repeats.

From: *Methods in Molecular Biology, vol. 266: Genomics, Proteomics, and Clinical Bacteriology: Methods and Reviews*
Edited by: N. Woodford and A. Johnson © Humana Press Inc., Totowa, NJ

## 2. Whole-Genome Sequencing

### 2.1. Construction of a Genomic DNA Library

**Figure 1** illustrates the experimental procedures for shotgun cloning. Genome DNA, which may sometimes contain extrachromosomal DNA such as plasmids, is randomly sheared by ultrasonication or by passage under high pressure through a narrow slit, and subjected to agarose gel electrophoresis to yield short (1–2 kb), medium (4–10 kb), and long (approx 100 kb) fragments. Short DNA fragments are useful for determining a draft sequence of the whole genome. The sheared genome DNA fragments may have cohesive (sticky) ends, and it is necessary to "polish" them with a DNA exonuclease (e.g., BAL31 nuclease) and the Klenow fragment of DNA polymerase. The resulting blunted DNA fragments are ready for insertion into a cloning vector (e.g., pUC18). Several host strains of *Escherichia coli* are available commercially for genomic library construction; the strain used should have a high transformation efficiency ($>1 \times 10^8$ transformants/µg sample DNA), and be able to accept methylated DNA. Natural bacteria often have restriction-modification systems involving DNA digestion and methylation that protect their genome integrity from exogenous DNA. Strains of *E. coli*, such as DH5αMCR or DH10B, carry mutations in *mcr*ABC and *mrr,* and are suitable hosts for cloning methylated DNA. Once the fragments have been cloned, they are amplified from transformants in the genomic library by polymerase chain reaction (PCR), using vector-specific primers (e.g., M13 primers) and purified to act as templates for sequencing.

### 2.2. Assembly of Sequence Reads into Contigs

The large number of reads (individual sequences; one per transformant from the library) is assembled with computer programs, such as the base-calling program phred and the assembler program phrap *(1,2),* into larger fragment sequences called *contigs*. This is achieved by making use of overlapping or common sequences within reads, and results in the production of an integrated map of the chromosome. At first the number of contigs increases during the assembly of reads, but overlapping sequences between contigs allows them to be further assembled, resulting in a subsequent gradual decrease in number (**Fig. 2**). During analysis of the *Staphylococcus aureus* genome, this decrease in the number of contigs was attenuated after 40,000 reads, indicating that any

Fig. 1. (*opposite page*) Method for shotgun sequencing. A genome is randomly sheared, trimmed with exonuclease, and cloned into a vector. Fragments are sequenced without knowledge of their chromosomal location, and the sequence reads are assembled into contigs from any overlaps found.

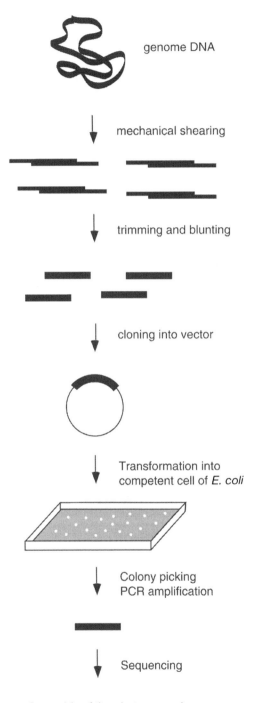

genome DNA

mechanical shearing

trimming and blunting

cloning into vector

Transformation into
competent cell of *E. coli*

Colony picking
PCR amplification

Sequencing

Assembly of the shotgun reads

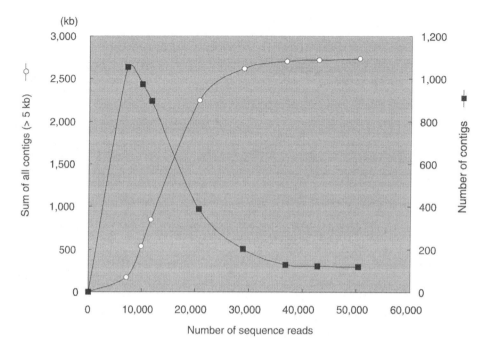

Fig. 2. Assembly of contigs in *Staphylococcus aureus* N315 genome sequencing. Initially, the more sequence reads completed, the larger the number of assembled contigs. However, assembly of contigs will attenuate owing to interference with bacterial specific repeat units, even though more sequence reads are completed.

further sequencing of shotgun clones would not be expected to add new sequences.

## 2.3. Finishing and Gap Closing

Although shotgun sequencing of short genomic fragments is a fast method for determining a crude image of genome structure, some gaps will remain, due to the presence in the genome of multiple-copy repeat units, such as ribosomal RNA operons, transposons, insertion sequences, bacteriophages, and gene duplications. The average length of each individual read obtained with current DNA sequencing methods is approx 500–700 bp. Repeat units longer than this will be misassembled into one locus, leaving gaps (**Fig. 3**). Gaps in the true genome sequence may also occur owing to the lack of a clone containing a particular genomic DNA insert, either because it was not represented in the library or because of gene toxicity for the *E. coli* strain used as the cloning host. Physical mapping and long-PCR amplification are effective methods to

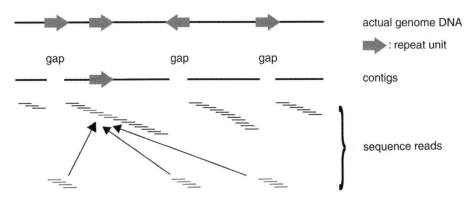

Fig. 3. Multiple repeats—such as rRNA operons, transposons, and insertion sequences—generate gaps during assembly of sequence reads into contigs. Although repeats are located at different sites on chromosome, they have identical sequences, and reads from different copies are assembled into one contig.

fill in remaining gaps between contigs (**Fig. 4**). Medium-length (4–10 kb) shotgun sequencing may also be helpful.

The multicopy repeat unit most commonly found in bacteria is the rRNA operon *(3)*. The number of copies varies among species; for example, *S. aureus* possesses five or six copies, while mycobacteria and some other genera contain only a single copy. The operon is composed of genes encoding 5S, 16S, and 23S rRNA, and tRNAs. The detailed compositions of the rRNAs and tRNAs differ in each operon, and therefore sequencing templates containing individual rRNA operons should be obtained by phage subcloning or long-PCR amplification, to determine their exact structure and localization in the whole-genome sequence. The resulting DNA fragments are subjected to DNA sequencing on both strands by primer walking, and gaps are closed one by one (**Fig. 4**). Gaps caused by other multiple-copy units can also be closed using these procedures. Some bacterial surface proteins may contain internal repeats with long core lengths (encoded by DNA many hundreds of base pairs in size) and each suspected open reading frame (ORF) needs careful evaluation for successful assembly.

### 2.4. Confirmation of Whole-Genome Sequence and Structure

After gap closing, it is necessary to confirm whether the contigs have been assembled correctly. The final genome sequence must coincide with actual restriction maps obtained by pulsed-field gel electrophoresis (PFGE) (**Fig. 5**), using several restriction enzymes. Long-PCR "walking" along the whole genome sequence may compensate for lack of information on physical map-

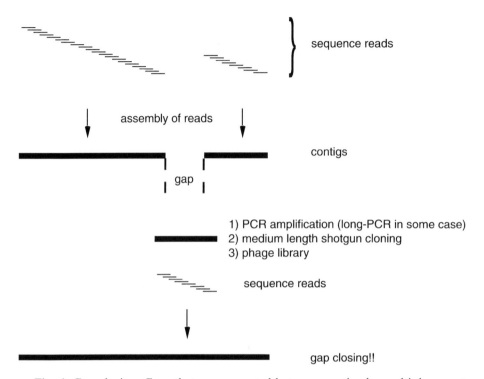

Fig. 4. Gap closing. Gaps that are generated between contigs by multiple repeats are closed with the sequence obtained by the DNA fragments bridging the gaps.

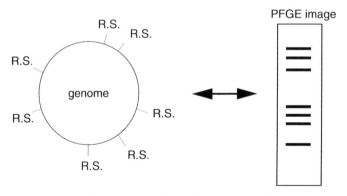

Fig. 5. Predicted restriction enzyme sites (RS) of the final genome sequence must agree with the patterns obtained by PFGE analysis performed with several restriction enzymes to increase reliability of the physical map.

ping, and restriction pattern analysis of long-PCR fragments (approx 16 kb) will further increase the reliability of the final sequence.

A high level of redundancy is required for optimal DNA sequencing, i.e., each area of the genome should be represented several times in the library and hence sequenced several times; up to 10 levels of redundancy may be needed to solve the problem in a small fraction of abnormal clones. In the case of the *S. aureus* genome sequence (2.8 Mb), 56,000 sequence reads were sufficient to obtain an average 10 levels of redundancy. To increase accuracy, it may be necessary to edit manually sequencing errors following visual inspection by using editor programs, such as *Consed (4)*.

## 3. Genome Annotation

An entire genome structure may be determined by physical mapping and DNA sequencing as described above. The next step is to annotate it. Annotation is the determination or prediction of sequences in a genome that correspond to genes, including protein-encoding and RNA-encoding regions. This generally involves the use of computer programs that scan genomic DNA sequences to predict coding sequences and assign functional roles to predicted proteins. There are many programs freely available for use on personal computers or through websites (*see* Chapter 2). The features of these programs are beyond the scope of this chapter, and we have limited the discussion to the essential steps required for annotating a newly sequenced bacterial genome.

### 3.1. Gene Prediction

The simplest method of predicting genes encoding proteins is to search the DNA sequence for ORFs. An ORF is a DNA sequence that contains a contiguous set of codons (i.e., three nucleotides that specify a particular amino acid). There are six potential reading frames in every DNA sequence, with three possible frames on one strand, and another three on the complementary strand. All ORFs read in the 5' to 3' direction on a given sequence. An example of a search of the *S. aureus dnaA* gene is shown in **Fig. 6.** The genetic code used to translate nucleotide sequences into amino acids differs slightly between organisms; four codons of vertebrate mitochondrial DNA differ from the universal genetic code (codes are available at http://www.ncbi.nlm.nih.gov/htbin-post/Taxonomy/wprintgc?mode=t). The code that is most appropriate for the target species must be selected for ORF prediction, and the bacterial genetic code should be used for bacterial genomes. Bacterial translation is initiated by recruitment of a 70S ribosomal complex, initiation factors (IFs), and formylmethionine-tRNA$^{Met}$ (fMet-tRNA$^{Met}$). fMet-tRNA$^{Met}$ has a higher binding affinity for IFs than methionine-tRNA$^{Met}$, and it recognizes not only the universal initiation codon AUG in mRNA, but also GUG and UUG as alternatives. A ribosome-

```
Initiation Codons  : ATG,GTG,TTG,
Termination Codons : TAA,TAG,TGA,
```

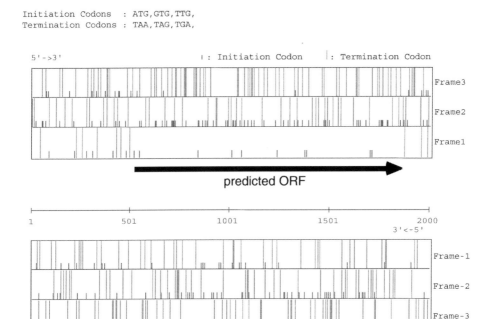

Fig. 6. ORF map of a portion of the *Staphylococcus aureus dnaA* gene using the GenetyxMac program. Bacterial initiation codons (ATG, GTG, and TTG) and termination codons (TAA, TAG, and TGA) are shown in all six potential reading frames as one-third and full vertical bars, respectively. The *dnaA* gene, shown as an arrow, is extracted as an ORF in frame 1 of the plus strand.

binding sequence (also known as a Shine-Dalgarno [SD] sequence) is sought to evaluate the most probable location of predicted genes. SD sequences comprise purine-rich (A and G) sequences with a strong similarity to the complementary sequence near the 3' end of 16S rRNA. An SD sequence located 6–11 bp upstream of an initiation codon is important for translation initiation. **Fig. 7** shows an example of a search of candidate SD sequences and initiation codons; the pairs with the most appropriate length of spacer region would be the most probable SD sequence and initiation site of the ORF.

The order of the codons in a gene determine the order of the amino acids in the protein encoded by that gene. The order of the codons also reflects the evolutionary origin of the gene, and constraints associated with its expression. Each species has a characteristic pattern of use of synonymous codons; the preference of codon usage reflects species-specific % GC content and the relative absence or abundance of some kinds of tRNA species *(5)*. **Table 1** shows the codon usage for alanine and leucine in the low–% GC Gram-positive bacte-

```
SA0006 (7005..9674)

  6781 TTGGAGAAAT GAATGCAGAT CAATTATGGG AAACAACAAT GAACCCTGAG CACCGCGCTC

  6841 TTTTACAAGT AAAACTTGAA GATGCGATTG AAGCGGACCA AACATTTGAA ATGTTAATGG

  6901 GTGACGTTGT AGAAAACCGT AGACAATTTA TAGAAGATAA TGCAGTTTAT GCAAACTTAG
                                           *****
  6961 ACTTCTAAGC GCTGTGAACT GAACTTTTGA AGGAGGAACT CTTGATGGCT GAATTACCTC
                                                           M   A   E   L   P   Q
  7021 AATCAAGAAT AAATGAACGA AATATTACCA GTGAAATGCG TGAATCATTT TTAGATTATG
        S   R   I   N   E   R   N   I   T   S   E   M   R   E   S   F   L   D   Y   A
  7081 CCATGAGTGT TATCGTTGCT CGTGCATTGC CAGATGTTCG TGACGGTTTA AAACCAGTAC
        M   S   I   V   A   R   A   L   P   D   V   R   D   G   L   K   P   V   H
  7141 ATCGTCGTAT ACTATATGGA TTAAATGAAC AAGGTATGAC ACCGGATAAA TCATATAAAA
        R   R   I   L   Y   G   L   N   E   Q   G   M   T   P   D   K   S   Y   K   K
  7201 AATCAGCACG TATCGTTGGT GACGTAATGG GTAAATATCA CCCTCATGGT GACTCATCTA
        S   A   R   I   V   G   D   V   M   G   K   Y   H   P   H   G   D   S   S   I
```

Fig. 7. Identification of the most probable initiation site of translation. The sequence of *gyrA* of *Staphylococcus aureus* N315 is shown. Possible initiation codons (ATG, GTG, and TTG) and SD sequences (purine-rich sequence) are highlighted. To determine the most probable initiation codon, the length of the spacer region between the initiation codon and SD sequence is considered. In the case of *gyrA,* the GGAGG sequence positioned from 6992 to 6996 shown as asterisks above the sequence is most likely to be the SD sequence, and the ATG at 7005 to 7007, downstream of the SD sequence (shown as colons above the sequence), is the probable initiation codon.

rium *S. aureus,* the enteric gram-negative bacterium *E. coli,* and the high–% GC Gram-positive bacterium *Mycobacterium tuberculosis. S. aureus* has a low GC content (33.5%) and has a strong codon bias in favor of A and T/U nucleotides. Leucine is encoded by six possible codons, but *S. aureus* prefers UUA and UUG (74% of all six codons), rather than other GC-rich codons (CUA, CUC, CUG, and CUU). On the other hand, *E. coli* and *M. tuberculosis* prefer GC-rich codons; again codon usage and % GC content of the genome correlate with each other. Codon usage analysis for other organisms is available through the website http://www.kazusa.or.jp/codon/ *(6).*

Computer programs such as GeneMark *(7)* and GLIMMER *(8)* use hidden Markov models (HMMs) and interpolated Markov models (both complex statistical algorithms) to identify potential genes in DNA sequences, based on the characteristics of protein-encoding sequences described above. Such programs have been developed to identify genes without human intervention. However, the predictions often include false-positive or false-negative ORFs. Thus, revision by visual inspection may increase the reliability of ORF identification.

### 3.2. Homology Searches

The most reliable way to identify a gene in a new genome is to find a close homolog in another organism. Homologs are proteins that share significant

**Table 1**
**Codon Preference**

| Codon | Amino acid | Proportion (%) | | |
|-------|-----------|----------------------------------|----------------------------------|----------------------------------------------|
| | | S. aureus N315 coding GC% (33.5) | E. coli K12 coding GC% (51.8) | M. tuberculosis H37Rv coding GC% (65.9) |
| GCA | Alanine | 47 | 21 | 10 |
| GCC | Alanine | 7 | 27 | 45 |
| GCG | Alanine | 15 | 36 | 37 |
| GCU | Alanine | 31 | 16 | 8 |
| CUA | Leucine | 9 | 4 | 5 |
| CUC | Leucine | 2 | 10 | 18 |
| CUG | Leucine | 3 | 50 | 52 |
| CUU | Leucine | 12 | 10 | 6 |
| UUA | Leucine | 59 | 13 | 2 |
| UUG | Leucine | 15 | 13 | 18 |

sequence similarity, which can be attributed to a common evolutionary ancestor. Homology searches can be done effectively using programs such as BLAST *(9)* to search all the entries in the GenBank database (http://www.ncbi.nlm.nih.gov/BLAST/) (*see* Chapter 2). Gene annotation is based on finding significant alignment of predicted ORFs to sequences of known function *(10)*. Gene products are classified on the basis of function, biological role, and cellular location. However, many of the genes in new genomes show no significant homology to GenBank entries. Matches of lesser significance provide only a tentative prediction and should be used as a working hypothesis of gene function *(11)*. Thus, further analyses of motifs, structure, and localization can help to assign characteristics to hypothetical proteins. Websites that may be used to search the sequences of predicted genes for regions encoding possible motifs, transmembrane helices, and protein sorting signals are listed in **Table 2.**

In comparisons between the proteomes of organisms, each protein is used as a query in a database similarity search against another proteome, or combined set of proteomes, in a database such as GenBank, SWISSPROT, or Protein Information Resource (PIR). A pair of proteins in organisms of two species that align along most of their lengths with a highly significant alignment score are likely to be orthologs (**Fig. 8**), i.e., proteins that share a common ancestry and that have kept the same function following an evolutionary speciation process. These proteins perform the core biological functions shared by all organisms, including DNA replication, transcription, translation, and intermediary metabolism. They do not include proteins unique to the biology of a particular

**Table 2**
**Websites for Genome Analysis**

| Theme | Source | Link | Reference |
|---|---|---|---|
| Reference | PubMed | http://www.ncbi.nlm.nih.gov/entrez/query.fcgi | – |
| Nucleotide | GenBank | http://www.ncbi.nlm.nih.gov/entrez/query.fcgi?db=Nucleotide | – |
| | EMBL | http://www.srs.ebi.ac.uk/ | – |
| Genome | GenBank | http://www.ncbi.nlm.nih.gov:80/entrez/query.fcgi?db=Genome | – |
| | TIGR | http://www.tigr/org/tdb/ | – |
| Protein | GenBank | http://www.ncbi.nlm.nih.gov:80/entrez/query.fcgi?db=Protein | – |
| | SwissProt | http://www.expasy.ch/sprot/ | – |
| Codon | Genetic code | http://www.ncbi.nlm.nih.gov/htbin-post/Taxonomy/ wprintgc?mode=t | *See* website |
| | Codon usage | http://www.kazusa.or.jp/codon/ | *6* |
| RNA | rRNA | http://rdp.cme.msu.edu/html/ | *3* |
| | tRNAscan-SE | http://www. genetics.wustl.edu/eddy/tRNAscan-SE/ | *5* |
| | tmRNA | http://www.indiana.edu/~tmrna/ | *16* |
| Homology search | Blast | http://www.ncbi.nlm.nih.gov/BLAST/ | *See* website |
| Motif search | InterPro | http://www.ebi.ac.uk.interpro/index.html | *15* |
| | Prosite | http://www.expasy.ch/prosite/ | *17* |
| Multiple alignment | ClustalW, X | ftp.ebi.ac.uk./pub/software | *13,14* |
| Classification | COGs | http://www.ncbi.nlm.nih.gov/COG/ | *12* |
| Metabolic pathway | KEGG | http://www.genome.ad.jp/kegg/kegg2.html | *18* *See* website |
| | PathDB | http://www.ncgr.org/pathdb/ | |
| Localization | PSORT | http://psort.nibb.ac.jp/ | *See* website |
| Useful link | PEDRO | http://www.public.iastate.edu/~pedro/research_tools.html | – |

organism. The Clusters of Orthologous Groups (COGs) database may aid assignment of gene function (http://www.ncbi.nlm.nih.gov/COG/) *(12)*. This database is used by performing an all-by-all genome comparison across a spectrum of prokaryotic organisms, and a portion of the yeast proteome. Each COG consists of individual orthologous proteins or orthologous sets of paralogs from at least three lineages. All-against-all self-comparison may reveal paralogs with a high degree of sequence similarity that have almost certainly originated from gene duplication events within a species.

Sequence alignments found in database searches indicate the presence of one or more conserved domains in each cluster or group of clusters. The most similar regions in a multiple set of proteins may represent conserved functional or structural domains. Hence, such clusters are next analyzed for the presence of known domains, by searches of a domain database. Functional searches with global alignments can be performed through the Internet (*see* **Table 2**), and alignments can be performed with locally installed software, such as

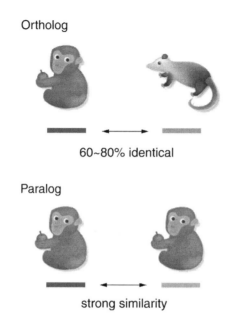

Fig. 8. Homologs comprise orthologs and paralogs. Orthologs: A pair of genes found in two species are orthologous when the encoded proteins are 60–80% identical in an alignment. The proteins almost certainly have the same three-dimensional structure, domain structure, and biological function. The encoding genes have originated from a common ancestor gene, and are also predicted by a very close phylogenetic relationship between sequences or by a cluster analysis. Paralogs: Genes in the same species that are related through gene duplication events. These events may lead to the production of a family of related proteins with similar biological functions within a species.

CLUSTALW *(13)* or CLUSTALX *(14)*, which has a graphic interface (**Fig. 9**). The software is available from ftp.ebi.ac.uk/pub/software. A multiple alignment of a set of sequences may be viewed as an evolutionary history of those sequences. If they align well, they are likely to have been derived recently from a common ancestral sequence. The task of aligning a set of sequences of varying relatedness is identical to that of discovering the evolutionary relationships among the sequences.

### 3.3. Motif Search Analysis: Prediction of Enzyme Catalytic Domains and Localization

InterPro is a useful resource for whole-genome analysis, and has been used to analyze the proteomes of a number of organisms for which the genome sequences are known (http://www.ebi.ac.uk/interpro/index.html) *(15)*. InterPro

**A**

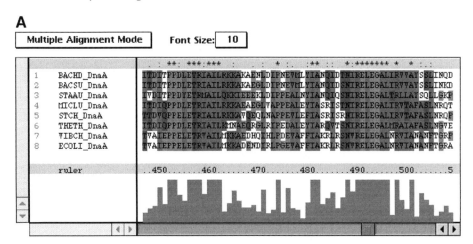

**B**

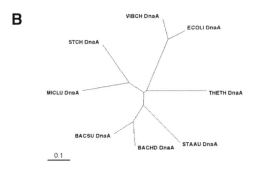

Fig. 9. **(A)** Multiple sequence alignment of DnaA proteins generated with the CLUSTALX program. One portion of the alignment is shown. Asterisks and colons above the aligned sequences indicate conserved and similar amino acids among the sequences, respectively. BACHD, *Bacillus halodurans;* BACSU, *Bacillus subtilis;* STAAU, *Staphylococcus aureus;* MICLU, *Micrococcus luteus;* STCH, *Streptomyces chrysomallus;* THETH, *Thermus thermophilus;* VIBCH, *Vibrio cholerae;* ECOLI, *Escherichia coli.* **(B)** An unrooted evolutionary tree showing relationships among DnaA protein sequences predicted by the multiple sequence alignment with CLUSTALX.

combines a number of databases (Pfam, PRINTS, PROSITE, ProDom, SMART, TIGRFAMs, SWISSPROT, and TrEMBL) that use different methodologies, and a varying degree of biological information, on well-characterized proteins to derive protein signatures. This kind of analysis identifies the number and types of domains that are shared among organisms, or that have been duplicated in proteomes to produce paralogs.

Prediction of protein localization may assist functional assignment of the products of predicted ORFs. The PSORT program (**Table 2**) predicts the pres-

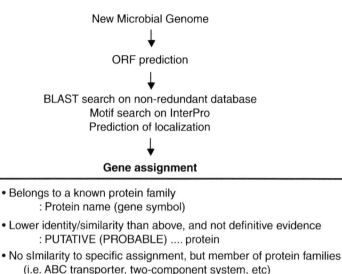

New Microbial Genome
↓
ORF prediction
↓
BLAST search on non-redundant database
Motif search on InterPro
Prediction of localization
↓
**Gene assignment**

- Belongs to a known protein family
  : Protein name (gene symbol)
- Lower identity/similarity than above, and not definitive evidence
  : PUTATIVE (PROBABLE) .... protein
- No sImilarity to specific assignment, but member of protein families
  (i.e. ABC transporter, two-component system, etc)
  : .... protein family
- Similarity to the conceptual translation of gene
  : conserved hypothetical protein
- No database match
  : hypothetical protein

Fig. 10. Flow chart of the stages of genome annotation from ORF prediction to gene assignment.

ence of signal sequences that contain an N-terminal basically charged region, and a central hydrophobic region. It also predicts probable transmembrane segments and the signal sequences of lipoproteins. These predictions can be used to suggest the most probable location of the ORF product (cytoplasm, periplasmic space, extracellular, or membrane-associated). For example, some surface proteins in staphylococci are anchored to the peptidoglycan chains with a LPXTG amino acid motif that is recognized by the sortase SrtA protein. Finding such a well-characterized motif in an unknown predicted ORF will strongly suggest the probable protein localization.

### 3.4. Gene Assignment

The function of predicted ORFs can be determined through homology and motif searches. The products of predicted genes are classified into five types based upon the extent of their similarity to sequence entries in the various databases searched (**Fig. 10**).

- An ORF encoding a product with good similarity to a protein of known function, with good sequence alignment along the entire length, and conserved active sites and motif sequences, is usually given the designation of the corresponding protein (and gene) of other bacteria (commonly *E. coli* or *Bacillus subtilis*).
- An ORF encoding a product that matches a characterized protein along most of its length but which has a low similarity (less than 40% amino acid identity) and no matches with protein motifs, is designated a putative protein sharing the putative function of the matched protein.
- An ORF encoding a product for which there are no matches with specific protein entries along the entire alignment of the ORF, but which shows strong similarity to a particular protein family (e.g., ABC transporters and transcription regulators with helix-turn-helix DNA-binding motifs) when analyzed by motif searches, cannot be assigned a specific gene and protein name. It is designated a probable member of the particular protein family.
- An ORF encoding a product with significant similarity to the conceptual translation of a gene, with conceptual orthologs in diverse bacterial species, but for which there is no experimental evidence of actual protein expression, and which does not belong to any known families, is a conserved hypothetical protein.
- An ORF encoding a product with no significant similarity to any sequence entries in any of the databases cannot be assigned features. There is no evidence to say whether it is actually a gene or not, and the product is designated a hypothetical protein.

In summary, there are no standard criteria for annotation procedures. Criteria for gene nomenclature have been proposed by the Institute for Genomic Research (TIGR, http://www.tigr.org/ER/db_assignmentextver2.shtml). The compilers of the SWISSPROT database have been verifying and attempting to standardize the annotation of microbial genomes through their HAMAP initiative (High-quality Automated Microbial Annotation of Proteomes, http://www.expasy.ch/sprot/hamap/). As the information derived from microbial genomes accumulates, the subsequent annotation procedures may be misled by comparisons with earlier annotated entries containing incorrect or surpassed information. Therefore, one should be vigilant about producing high-fidelity genome-sequence data, and must also constantly review information generated by transcriptome and proteome analyses.

## 4. Glossary

- **Active site:** a localized combination of amino acid side groups within the tertiary or quaternary structure of a protein that can interact with a chemically specific substrate and that provides the protein with biological activity.
- **Motif:** a conserved pattern of amino acids found in two or more proteins, which may have a similar biochemical activity (http://www.expasy.ch/prosite/) and often is near the active site of the protein.

- **Domain:** an extended sequence pattern that has a common evolutionary origin among the aligned sequences found by a homology search. A homologous domain is generally longer than a motif. The domain may include all of a given protein sequence or only a portion of the sequence. Some domains are complex and made up of several smaller homologous domains that became joined during evolution. The separate domains of a given protein may interact extensively.
- **Block:** an amino acid sequence pattern conserved within a family of proteins. The pattern includes a series of possible matches at each position in the represented sequences, but there are no inserted or deleted positions in the sequence.
- **Family:** a group of proteins with similar biochemical function that are more than 50% identical when aligned. This same cutoff is used by PIR. A family comprises proteins with the same function in different organisms (orthologous sequences), but may also include proteins in the same organism (paralogous sequences). If a multiple sequence alignment of a protein family reveals a common level of similarity throughout the lengths of the proteins, PIR refers to the family as homeomorphic. A homeomorphic domain may comprise several smaller homologous domains that are shared with other families. Families may be further subdivided into subfamilies or grouped into superfamilies based on respective higher or lower levels of sequence similarity.

## References

1. Ewing, B., Hillier, L., Wendl, M. C., and Green, P. (1998) Base-calling of automated sequencer traces using phred. I. Accuracy assessment. *Genome Res.* **8,** 175–185.
2. Ewing, B. and Green, P. (1998) Base-calling of automated sequencer traces using phred. II. Error probabilities. *Genome Res.* **8,** 186–194.
3. Maidak, B. L., Cole, J. R., Parker, C. T., Jr., Garrity, G. M., Larsen, N., Li, B., et al. (1999) A new version of the RDP (Ribosomal Database Project). *Nucleic Acids Res.* **27,** 171–173.
4. Gordon, D., Abajian, C., and Green, P. (1998) Consed: a graphical tool for sequence finishing. *Genome Res.* **8,** 195–202.
5. Lowe, T. M. and Eddy, S. R. (1997) tRNAscan-SE: a program for improved detection of transfer RNA genes in genomic sequence. *Nucleic Acids Res.* **25,** 955–964.
6. Nakamura, Y., Gojobori, T., and Ikemura, T. (2000) Codon usage tabulated from international DNA sequence databases: status for the year 2000. *Nucleic Acids Res.* **28,** 292.
7. Borodovsky, M., McIninch, J. D., Koonin, E. V., Rudd, K. E., Medigue, C., and Danchin, A. (1995) Detection of new genes in a bacterial genome using Markov models for three gene classes. *Nucleic Acids Res.* **23,** 3554–3562.
8. Delcher, A. L., Harmon, D., Kasif, S., White, O., and Salzberg, S. L. (1999) Improved microbial gene identification with GLIMMER. *Nucleic Acids Res.* **27,** 4636–4641.
9. Altschul, S. F., Gish, W., Miller, W., Myers, E. W., and Lipman, D. J. (1990) Basic local alignment search tool. *J. Mol. Biol.* **215,** 403–410.

10. Pearson, W. R. (1995) Comparison of methods for searching protein sequence databases. *Protein Sci.* **4,** 1145–1160.
11. Kyrpides, N. C. and Ouzounis, C. A. (1999) Whole-genome sequence annotation: "Going wrong with confidence." *Mol. Microbiol.* **32,** 886–887.
12. Tatusov, R. L., Koonin, E. V., and Lipman, D. J. (1997) A genomic perspective on protein families. *Science* **278,** 631–637.
13. Higgins, D. G., Thompson, J. D., and Gibson, T. J. (1996) Using CLUSTAL for multiple sequence alignments. *Methods Enzymol.* **266,** 383–402.
14. Thompson, J. D., Gibson, T. J., Plewniak, F., Jeanmougin, F., and Higgins, D. G. (1997) The CLUSTAL_X Windows interface: flexible strategies for multiple sequence alignment aided by quality analysis tools. *Nucleic Acids Res.* **25,** 4876–4882.
15. Apweiler, R., Attwood, T. K., Bairoch, A., Bateman, A., Birney, E., Biswas, M., et al. (2001) The InterPro database, an integrated documentation resource for protein families, domains and functional sites. *Nucleic Acids Res.* **29,** 37–40.
16. Williams, K. P. (1999) The tmRNA website. *Nucleic Acids Res.* **27,** 165–166.
17. Falquet, L., Pagni, M., Bucher, P., Hulo, N., Sigrist, C. J., Hofmann, K., et al. (2002) The PROSITE database, its status in 2002. *Nucleic Acids Res.* **30,** 235–238.
18. Kanehisa, M. and Goto, S. (2000) KEGG: Kyoto encyclopedia of genes and genomes. *Nucleic Acids Res.* **28,** 27–30.

# 4

## Comparative Genomics

*Digging for Data*

### Matthew B. Avison

#### Summary

Comparative genomics is a science in its infancy. It has been driven by a huge increase in freely available genome-sequence data, and the development of computer techniques to allow whole-genome sequence analyses. Other approaches, which use hybridization as a method for comparing the gene content of related organisms, are rising alongside these more bioinformatic methods. All these approaches have been pioneered using bacterial genomes because of their simplicity and the large number of complete genome sequences available. The aim of bacterial comparative genomics is to determine what genotypic differences are important for the expression of particular traits (e.g., antibiotic resistance, virulence, or host preference). The benefits of such studies will be a deeper understanding of these phenomena; the possibility of exposing novel drug targets, including those for antivirulence drugs; and the development of molecular techniques that reveal patients who are infected with virulent organisms so that health care resources can be allocated appropriately. With more and more genome sequences becoming available, the rise of comparative genomics continues apace.

**Key Words:** Comparative genomics; genome sequence; microarrays; proteomics; BLAST; genomic island; synteny; horizontal gene transfer; mutation; hybridization; bacteriology.

## 1. Introduction

Like so many "omics" techniques, comparative genomics can mean different things to different people. Indeed, anyone who has performed a BLAST search (*see* Chapter 2) for the sequence of their favorite protein using pre-existing protein databases is carrying out comparative genomics in its most basic sense. The hope, in this example, is that other researchers have studied similar proteins and that their results, together with the homology found, can

From: *Methods in Molecular Biology, vol. 266: Genomics, Proteomics, and Clinical Bacteriology: Methods and Reviews*
Edited by: N. Woodford and A. Johnson © Humana Press Inc., Totowa, NJ

be used to formulate a testable hypothesis that will keep one busy for a few months.

In recent years, with increased availability of bacterial genome sequences, it has become possible to compare entire genomes, in order to make large-scale observations about evolution, horizontal gene transfer, and genes that might be responsible for phenotypic differences between the comparator organisms. The aim remains the same as that of simple BLAST analysis, namely to formulate hypotheses that need to be tested.

It is also possible to compare genome sequences without large-scale sequencing by using various hybridization-based approaches, where the aim is to locate DNA sequences found in only one of two comparator organisms. This is proving a powerful and increasingly popular approach for locating genes involved in phenotypes specific to one particular isolate, but not shared with many better-characterized, or type, strains.

Comparative genomics has now moved into the functional genomics era. In this incarnation, the comparison is of the expression of a gene or group of genes in different organisms having different phenotypic characteristics.

This review discusses previous and ongoing comparative genomic projects, and will describe proteomic, functional genomic, and hybridization-based approaches to comparative genomics. The emphasis will be on the analysis of bacteria relevant to human disease.

## 2. Comparative Genome Sequencing Projects
### 2.1. Review of Large-Scale Genome Sequence Comparisons
#### 2.1.1. Escherichia coli

*Escherichia coli* K12 is the most widely studied and most completely understood bacterium. It is used as a model system for other organisms, and the publication of its genome sequence in 1997 was one of the highlights of bacteriology research in the last century *(1)*. Following publication of the genome sequence of a strain of enterohemorrhagic *E. coli* (EHEC) of serotype O157:H7, comparative genomics was used to reveal some of the determinants responsible for pathogenicity in the latter *(2)*. In fact, over 4 Mb of DNA sequence is shared by both strains, with very little variation. This common DNA makes *E. coli* what it is, while the remaining DNA is the key to strain-specific phenotypes. Approximately 0.5 Mb is unique to K12, and around 1.5 Mb is unique to EHEC O157:H7. Much of this unique DNA is located in strain-specific gene islands where the base composition (CG content) is not consistent with the background composition of the main part of the chromosome. This reflects horizontal transfer of regions from donor species with a chromosomal base composition different from that of *E. coli*. EHEC O157:H7-

specific islands encode many known and candidate virulence factors, including toxins, adhesion and invasion systems, and extracellular antigenic determinants. These findings are being exploited to develop sensitive typing systems aimed at identifying virulent *E. coli* strains. They have also fueled detailed analysis of the molecular basis of virulence in EHEC strains in general *(2)*.

Comparative genome sequence analysis facilitates the design of multilocus sequence typing (MLST) rationales (*see* Chapter 15). The comparison of K12 and O157:H7 strains of *E. coli* led to an MLST scheme that was designed to undertake an evolutionary analysis of a collection of pathogenic *E. coli* strains *(3)*. A mixture of O157:H7 strain-specific genes and those shared by K12 and O157:H7 were amplified using genomic DNA from members of the strain collection as template, and the amplicons were sequenced. Phylogenetic analysis was then performed; the results allowed the evolution of the various strains tested to be predicted, and led to the conclusion that the same virulence factors had been independently acquired by a number of different evolutionary lineages of *E. coli* at different times. This remarkable finding highlighted the prime importance of a few specific virulence factors in the way *E. coli* isolates infect patients, and will form the basis of further study *(3)*.

## 2.1.2. Pseudomonas aeruginosa

In August 2000, the genome sequence of *Pseudomonas aeruginosa* PA01 was reported *(4)* and represented the largest bacterial genome to have been completely sequenced at that time. Comparative analysis of the products of individual open reading frames (ORFs) with those from other complete genomes using the amino acid sequence alignment program blastp *(5)* showed that, at the time of the comparison, where homologs were present in other bacterial genome sequences, the closest matches were nearly always with *E. coli*. Furthermore, gene order, average gene size, and average gene spacing in *P. aeruginosa* were remarkably similar to those of *E. coli*. The larger genome of *P. aeruginosa,* therefore, reflects greater genetic complexity, and not simply genetic rearrangements and spacing out of genes, or the occurrence of more recent gene duplication and/or horizontal gene transfer events. This genetic complexity manifests itself as the presence of significantly more gene families than any other genome sequenced at that time, and not in simple duplications of gene families present in an ancestral bacillus. The conclusion drawn from these observations is that the diverse and unpredictable environment of *P. aeruginosa* has led to evolutionary pressure to generate new gene families that have specialized functions, but without losing pre-existing functions. It is far less likely that the opposite is true, i.e., relatively more "specialist" bacteria (like *E. coli*) have lost entire gene families; a few members of each would inevitably have been left behind by chance as a signature of their past existence. Understanding the forces that drive pseudomonads to evolve to cope with complex and varying environmental niches, is important in controlling the threat of these organisms in the clinical setting, where they are becoming increasingly prominent *(4)*.

### 2.1.3. Salmonella enterica

Members of the species *Salmonella enterica* represent prime targets for comparative genomic analysis. Isolates are generally very similar, with almost identical basic physiology, but different serovars can have very specific host ranges, and may cause infections characterized by very different symptoms *(6)*. Hence, determining which serovar-specific genetic elements are present will inevitably further our understanding of these specific disease processes. To date, however, only the genomes of strains of *S. enterica* sv. Typhi *(7)* and *S. enterica* sv. Typhimurium *(8)* have been published, together with a comparative analysis *(7)*.

*S. enterica* sv. Typhi is an important human pathogen, and is responsible for approximately half a million deaths worldwide per year. The genome sequence of the clinical isolate CT18 was compared first with the genomes of *E. coli* strains K12 and O157:H7 *(7)*, and subsequently with that of the laboratory strain of *S. enterica* sv Typhimurium LT2. *S. enterica* sv. Typhimurium causes gastroenteritis, and has a wide host range, though LT2 is not particularly pathogenic *(8)*. One striking feature to have emerged from the comparison was that strain CT18 has around 200 genes that are shared with the three comparator organisms, but which contain mutations such that no functional protein product can be produced *(7)*. These mutated genes are called *pseudogenes*. There is no obvious decrease in the ability of CT18 to survive outside the host compared to *S. enterica* sv. Typhimurium LT2 and the *E. coli* strains, but this reduced functionality might be one reason why *S. enterica* sv. Typhi strains infect or colonize only humans, whereas *S. enterica* sv. Typhimurium and *E. coli* strains show a much broader host range. These pseudogenes have been found in all *S. enterica* sv. Typhi isolates examined to date, so they are not a unique property of strain CT18. Comparative analysis of CT18, LT2, and *E. coli* K12 also revealed the locations of previously characterized pathogenicity islands (PIs). These genetic regions comprise the major differences between the two *S. enterica* serovars, with about 0.5 Mb of DNA being unique to each genome. These areas are undoubtedly responsible for host specificity and the differences in the types of disease caused by the two serovars, and will form the starting point for future work, particularly that aimed at the eradication of typhoid fever *(7)*.

An additional benefit of the *S. enterica* sv. Typhi CT18 genome-sequencing project was the complete sequencing of a 200 kb plasmid carried by this isolate. This plasmid is responsible for one of the most worrying features of modern *S. enterica* sv. Typhi strains, multidrug resistance, because it carries a large number of antibacterial, detergent, and heavy-metal-resistance determinants *(7)*. Indeed, this example highlights the potential of large-scale sequencing projects to determine a considerable amount of important information in the short time (and for a relatively small amount of money) it takes to sequence a large plasmid crammed with clinically relevant functions.

Recently, Hansen-Wester and Hensel analyzed the PIs of published *S. enterica* and *E. coli* genome sequences in great detail *(9)*. The rationale behind their analysis was that PIs tend to be located adjacent to tRNA gene loci that are anchors for bacteriophage integration, and so can be hot spots for the acquisition of genes by horizontal transfer *(10)*. There are around 80 tRNA loci in the *E. coli* K12 genome, and 2 kb on either side of each locus was aligned with 2 kb on either side of their homologs in the *E. coli* O157:H7, *S. enterica* sv. Typhi CT18, and *S. enterica* sv. Typhimurium LT2 genomes. This sort of analysis is very time consuming, requiring considerable user intervention, and as such it is not usually included in proteomic sequence blastp analyses similar to those described for *S. enterica* and *E. coli* isolates in **Subheadings 2.1.1.** and **2.1.3.** The analysis revealed four new regions of horizontal transfer specific to *S. enterica* strains. These regions have %GC contents that are far lower than the remainder of the *S. enterica* genome and contain a number of putative ORFs. Although deletion experiments did not reveal a role for these ORFs in the progression of infection in a mouse model, it is possible that they have a more subtle role that needs to be measured in a more targeted manner *(9)*. This is an excellent example of a focused comparative genomic project aimed at addressing a specific question, and represents the type of analysis that will be used increasingly in the future of comparative genomics as related to important human pathogens.

### 2.1.4. Yersinia pestis

The genomes of two *Yersinia pestis* strains, CO92 *(11)* and KIM *(12)*, have been sequenced. Both are responsible for plague pandemics, although CO92 is currently the most common strain. Its sequence appeared first, and has proved extremely important in our understanding of this highly virulent human pathogen. A large number of regions associated with horizontal gene transfer were located, together with regions showing the intriguing phenomenon of genome decay *(11)*. This occurs following evolution of an organism to grow in a new environment, via the acquisition of new genes by horizontal transfer. There is no longer selective pressure for the maintenance of genes that were important for its survival in its previous environment, and they are subsequently degraded by the accumulation of point and insertion mutations, until they appear as pseudogenes. This may be what is also happening to *S. enterica* sv. Typhi, albeit to a lesser extent *(7)*. The observation of gene decay in *Y. pestis* made by comparative genomics supports the previous proposal that *Y. pestis* evolved from an enteropathogenic *Yersinia pseudotuberculosis* isolate because, for the most part, it is the genes responsible for an enteric lifestyle that have decayed *(11)*. When the *Y. pseudotuberculosis* genome is sequenced, it will presumably then be possible to target genes responsible for enteric infection for further analysis. Genome decay is the opposite of the situation in *P. aeruginosa,* whose niche is a fluctuating environment, where the addition of new functions is sometimes good, but the loss of functions is mostly bad.

A comparative genomic analysis of the *Y. pestis* KIM and CO92 sequences *(12)* revealed that, while there is a large degree of conservation between the sequences of specific genes, their location in the genome is dramatically different. The reason for this is not clear, but the observation is at odds with comparisons of other members of the Enterobacteriaceae, where there is a remarkably consistent gene order (synteny). For example, the gene order is very similar in *E. coli* K12, *E. coli* O157:H7, *S. enterica* sv Typhimurium LT2, and *S. enterica* sv Typhi CT18, even though these organisms diverged over 100 million years ago *(7)*. This is in contrast to the two *Y. pestis* strains, which are far more closely related in time *(12)*. Putative virulence factor genes that differ between the two *Y. pestis* strains were observed following the comparative analysis, but the fact that these organisms are similarly pathogenic implies that these sequence differences are not particularly significant. Indeed this is an example where comparative genomics was not entirely appropriate, because the biology of the comparator organisms are so similar that little important information was obtained from the analysis. The observed lack of synteny between the isolates showed that the genome is structurally less stable than those of many other organisms, implying that *Yersinia* spp. are capable of rapidly altering their genomes to adapt to new environments if evolutionary pressure is applied *(12)*. The potential threat from this property should be taken very seriously indeed.

### 2.1.5. Staphylococcus aureus

*Staphylococcus aureus* is one of the most important nosocomial bacterial pathogens, and the variability of its virulence, antigenic determinants, and susceptibility to antibiotics makes it an ideal candidate for comparative genomics. A number of sequencing centers set out to complete the genomes of various *S. aureus* isolates, but the first to be published was that of methicillin-resistant (MRSA) isolate N315 from Japan *(13)*. In the same manuscript, the genome of the vancomycin-resistant (VRSA) strain, Mu50, was also reported. Both strains fall into clonetype II-A. The results showed that *S. aureus* is remarkably flexible in its ability to share genetic information with other organisms in the environment, including even eukaryotes. The number of transmissible elements present in the two genomes is large, and, despite close phenotypic similarity between the two organisms, there are strain-specific insertions, including relatively large amounts of extra DNA, in Mu50. Only about 50% of the genome could be traced back to a common ancestor *Bacillus/Staphylococcus*, which is less than might be predicted given the phylogenetic relatedness between the two groups as determined by rDNA sequencing. This is a consequence of large amounts of recent genetic movement in staphylococci *(13)*.

As with the other organisms, many of the mobile genetic elements in *S. aureus* encode important virulence determinants. A large proportion of the virulence genes have been previously reported to be important for soft-tissue infections. In addition, however, virulence factors were also found in the Mu50 and N315 genome sequences that have been associated with infection of skeletal tissue in other *S. aureus* strains. Because neither Mu50 nor N315 have been found in skeletal lesions, no one had looked for these specific virulence factors, but it appears from the genome sequences that these isolates have the potential to cause such infections.

Some of the enterotoxin genes located in Mu50 and N315 are strain specific, but most of the virulence factors identified are virtually identical in the two strains. Again, this is an example of two strains that are too phenotypically similar to make comparative genomics particularly revealing *(13)*. The exception to this was their antibiotic resistance, where the two strains displayed markedly different properties. As such, one major reason for the N315/Mu50 comparative genome analysis was to determine, en masse, the sequence of antibiotic resistance determinants present in both organisms. All of those located are present on mobile genetic elements; Mu50 carries one additional transposon, which accounts for its increased spectrum of resistance. However, no obvious single locus was identified that might account for vancomycin resistance in Mu50 *(13)*.

In a subsequent analysis, whole proteomic sequences from Mu50 were compared with those from N315 in an attempt to predict which genetic differences might lead to vancomycin resistance in Mu50 *(14)*. The rationale was that since vancomycin resistance can be selected from an MRSA population, it is most likely to be caused by mutation of an intrinsic determinant, or group of determinants, present in all MRSA isolates. This analysis was the first proteomic sequence analysis to be carried out, and the first comparative genome project to address a specific phenotypic difference between two organisms, and as expected the results obtained were complex *(14)*. The sequenced proteomes of Mu50 and N315 show significant differences in intrinsic gene products, as a large number of point mutations were observed. Importantly, none of these were in genes encoding proteins with a predicted role in cell wall biosynthesis. Given that the phenotype of vancomycin resistance is associated with overproduction and decreased crosslinking of the cell wall, only mutations in ORFs encoding proteins involved in this aspect of physiology were thought to be significant. A series of putative mutations likely to produce loss of function in cell wall and associated intermediary metabolism in Mu50, mainly resulting from frameshifts, were identified, and will provide the basis for future experiments to determine the genetic basis for vancomycin resistance in Mu50 *(14)*. However, subsequent analysis has revealed that some of the apparent mutations in Mu50 might be artifacts. At least one of the genes reported to be mutated has been checked by polymerase chain reaction (PCR) analysis and found to be intact *(15)*. Unfortunately, it seems that the Mu50 genome sequence was not properly completed prior to publication, and contains a high error frequency, due to incomplete coverage in some areas of the genome. This problem highlights what can happen when genome sequencing projects become too competitive. The conclusions of any comparative analysis aimed at addressing a specific question should be tempered with caution concerning the quality of the sequence. Furthermore, comparative proteomic sequence analysis is limited to mutations in ORFs leading to alteration/disruption in protein sequence. When addressing a specific phenotypic question, it is equally possible that mutations in noncoding DNA sequences (e.g., promoter sequences) might lead to changes in the expression of genes, leading to the phenotypic effect seen. This requires a hugely more time-consuming entire genome comparison, or a comparison of gene expression using one of the methods described in **Subheading 4.**

Recently, the genome sequence of the community-acquired *S. aureus* strain MW2 has been reported *(16)*. A comparison of this sequence with that of hospital-acquired N315 revealed that the majority of the genetic material was identical, and in the same gene order, but that there are a number of genomic islands that contain strain-specific genetic elements known to transfer horizontally. In some cases, these islands carry virulence genes, and there are far more in MW2 than in N315, though there are also a number of genes encoding functions less obviously associated with pathogenicity. It seems that these genomic islands are hot spots for transfer and rearrangement of genetic material, and may contribute to the high degree of variability seen among *S. aureus* isolates. It is possible that genomic island allotyping methods, which rely on determining the sequences of these hypervariable islands in groups of strains, will be a useful future method of determining the pathogenic potential of particular isolates *(16)*.

Interestingly, MW2 shows far less evidence of acquisition of resistance genes than N315, which is multiresistant, and MW2 contains relatively few mobile genetic elements. It might be that the community environment of MW2 is easier to survive in. Clearly, the use of antibiotics in hospitals is more intense than in the community, providing a stronger selective pressure for the acquisition of antibiotic resistance genes. In contrast, MW2 (but not N315) carries a bacteriocin determinant, which is believed to be important for community carriage. Bacteriocins kill competitor bacteria, thus clearing the way for colonization by organisms that produce them (and which are immune to their effects). In contrast, it is thought that N315 colonizes hospitalized patients because prior antibiotic therapy kills potential competitor bacteria *(16)*.

The first example of shotgun comparative genomics was performed with an *S. aureus* isolate. Herron et al. sequenced random clones from a gene library made from genomic DNA of an *S. aureus* isolate associated with bovine mastitis *(17)*. Approximately 10% of the genome was covered, and these pieces of sequence, scattered about the genome, were aligned with the complete genome sequence of *S. aureus* N315. Even this limited analysis revealed sequences that were part of genomic islands, according to background sequence, and were specific to the bovine strain. The ORFs present in these islands might encode proteins essential for infection of the bovine, rather than the human host. The analysis also showed a number of examples of horizontal gene transfer from gram-negative organisms not seen in previous *S. aureus* genome sequencing projects. Accordingly, shotgun genomic comparisons might have considerable value. If an ORF is located in a gene library from a particular species but is not found in the published genome of a member of the same species, it might have a strain-specific function. This function can be predicted and the prediction tested *(17)*.

### 2.1.6. Group A Streptococci

Acute rheumatic fever (ARF) and resultant heart disease is associated with infection by some strains of group A *Streptococcus* (GAS), but not others *(18)*. Recently, the complete genome sequences of two GAS strains were reported and compared. Strain MGAS8232 (serotype M18) is associated with ARF *(19)*, and strain SF370 (serotype M1) is not *(20)*. Approximately 10% of the genome is unique to each strain, with the remainder being identical *(19)*. The unique portions are spread around the genome in DNA islands, some of which (including at least five phage genomes in ARF strain MGAS8232) are clearly associated with recent horizontal gene transfer. Within the phage-associated DNA are 10 ORFs predicted to encode putative ARF-associated virulence factors; these are now the targets of research aimed at determining what causes ARF, and what might be done to prevent it in the future *(19)*. This is an excellent example of how targeted comparative genomics can yield important information about the specific disease associations of otherwise highly similar organisms.

## 2.1.7. Mycobacterium *spp.*

Tuberculosis (TB) is arguably the most important human infectious disease, with *Mycobacterium tuberculosis* killing approx 4 million people each year worldwide. Although the genome sequence of only one clinical isolate of *M. tuberculosis* (H37Rv) has been published so far *(21)*, the results have fueled a large number of comparative analyses *(22)*. The only direct genome sequence comparison among *Mycobacterium* spp. is between *M. tuberculosis* H37Rv and an animal-associated *Mycobacterium leprae* isolate. While *M. leprae* is recognized as an important human pathogen, it has never been possible to culture human *M. leprae* isolates in the laboratory in such a way that genomic sequence analysis can be performed; however, isolates that colonize certain animals (but do not cause disease) can be grown, albeit poorly, in the laboratory *(23)*. *M. leprae* has an unusual genome for a bacterium, because only approx 50% codes for proteins, with the remainder comprising large numbers of pseudogenes and regions of DNA that may have once coded for proteins, but which have mutated beyond recognition, possibly because of the loss of an important DNA polymerase proofreading activity. Furthermore, the genome of *M. leprae* is about 1.2 Mb smaller than that of other *Mycobacterium* spp., probably because of recombination events between distant randomly repetitive sequences. The overall result of mutation and deletion means that *M. leprae* has less than a third of the functional ORFs carried by *M. tuberculosis* H37Rv. This gene decay is predicted to impair severely the ability of *M. leprae* to grow, and may well explain the difficulty in culturing this bacterium in the laboratory *(23)*.

## 2.1.8. Helicobacter pylori

*Helicobacter pylori* causes a wide variety of different disease presentations, and it is possible, though not entirely clear, that different strains might have acquired different virulence factors to allow them to cause different symptoms. Comparison of the genome sequences of *H. pylori* strains 26695 and J99 suggests that there are around 50–100 genes unique to each strain *(24)*. This interstrain variation most probably reflects gene acquisition rather than loss, and this suggests that *H. pylori* has a genome that is in continual flux *(25)*. Evolutionary analysis of *H. pylori* strains has been considerably strengthened by comparative microarray analysis *(26)*.

## 2.2. Comparative Microarray Analysis

This is a hybridization-based approach to determining gene content, and the basic principle is similar to those described in **Subheading 3.**, though the use of microarrays allows a large amount of data to be collected at once (*see* Chapter 10). In this technique, the presence or absence of genes can be determined without further experiments being performed. However, despite its increasing popularity, comparative microarray experiments have major limitations. Firstly, to produce a complete microarray for any organism, the sequences of all the loci in the genome encoding ORFs need to be determined, in order to design PCR primers to generate amplicons to be spotted onto the microarray slide. Therefore, whole genome microarrays are usually constructed for strains whose genomes have been sequenced. Unfortunately, it is not entirely clear at present how representative of the natural population of a particular organism some genome sequence strains are. Second, only genes present in the genome sequence strain and absent in the comparator strain can be identified. This is because the result of hybridizing labeled DNA from the second comparator organism to the amplicons on the microarray is, in its most accurate form, binary; i.e., spots either become labeled, or they do not. Hence if genes are not present in the array (i.e., the genome sequence strain) but are present in the comparator organism, they will be missed. This can compound the problem of how representative genome sequence strains are. The third problem, which applies to all hybridization-based comparative approaches, is that a gene identified as being present in the comparator organism on the basis of hybridization on the array, may not actually be intact, and may not actually encode a functional protein. As seen in **Subheading 2.1.3.** with pseudogenes, a single nucleotide change (which would not affect hybridization to the array) can result in premature termination of a translation product, and the loss of the function encoded. Despite these problems, as more genome sequences become available and more arrays are produced, comparative microarray analysis is becoming increasingly valuable, and there have been several recent successes with this technique.

Comparative microarray analysis has been used to study the evolution of *E. coli* strains *(27)* using the 4300 ORF *E. coli* K12 microarray, which was developed following the release of the *E. coli* K12 genome sequence *(1)*, originally for gene expression (transcriptome) analysis *(28)*. In the comparative analysis, labeled total genomic DNAs from five *E. coli* strains of known evolutionary relatedness were individually washed over the array, and the number of positive hybridizations recorded in each case; 3782 ORFs were conserved in all strains. Analysis of these data allowed the authors to plot a complex predicted evolutionary path between the strains and, accordingly, to propose an evolutionary origin for them *(27)*. This information is not only of academic interest, as the rules discovered might be useful in predicting the future evolutionary potential of important pathogens.

Dziejman et al. recently published an analysis of the relatedness of *Vibrio cholerae* isolates responsible for epidemic and pandemic diseases *(29)*. The *V. cholerae* genome sequence strain, El Tor O1 N16961, was responsible for the seventh cholera pandemic, beginning in 1961 *(30)*. Prior to this time, however, El Tor O1 strains caused epidemic disease, and there are examples of El Tor O1 strains that do not cause disease at all. In order to determine some of the genes responsible for the pandemic nature of N16961, comparative microarrays were used to identify genes not present in epidemic El Tor O1 strains and/or environmental strains, but present in a N16961 microarray representing 3632 ORFs *(29)*. Among other discoveries about specific strains, the authors found 22 genes unique to N16961. Eleven of these genes were present as a pandemic-specific genomic island 1, and 8 more were found on pandemic-specific genomic island 2. The equivalent regions of the genomes of the other comparator strains were sequenced; this confirmed the absence of these specific gene clusters and revealed the exact points at which these mobile elements were inserted into the chromosome. Unfortunately, as is typical of genome-sequence analysis, the possible functions of the genes, and the specificity of these functions in pandemic disease, were not entirely obvious from the gene sequences and their homologies with other known proteins. However, the findings will stimulate a targeted analysis of gene function where no targets had been identified previously.

Microarray analysis has also been applied to studies of *Mycobacterium* spp. For example, Behr et al. used it to identify 14 regions that are absent from the avirulent *M. tuberculosis* strain BCG Pasteur, which is used as an attenuated vaccine, but that are present in the *M. tuberculosis* H37Rv (virulent) genome sequence strain. Most of these deletions were also found in other attenuated vaccine strains *(31)*. A review of the explosion of comparative microarray analysis in mycobacterial research has recently been published *(22)*.

## 3. Comparative Genomics Without the Need for Large-Scale Sequence Analysis

### 3.1. Bacterial Comparative Genomic Hybridization

It is reasonable to suppose that phenotypic variation among isolates of a bacterial species must be the result of genetic differences. Such variations often reflect horizontal gene transfer, as shown by the results from the large-scale genome-sequencing projects described in **Subheading 2.1.** It is important, if we are to understand which genotypes are responsible for specific phenotypes, that we be able to analyze differences in genetic complement between bacteria quickly and simply. However, it is not feasible to sequence the entire genome of strains representing every variant. A method that focuses specifically on

DNA sequences that are unique to one comparator organism, and so reduces the amount of information that needs to be sifted through, has the potential to enhance greatly our understanding of what makes bacteria express particular traits.

Comparative hybridization is a method by which one can determine which sections of DNA are present exclusively in one of two comparator bacteria, without having to sequence either strain's genome. The first part of this technique involves the separation of small DNA fragments generated by (frequent-cutting) restriction digestion of genomic DNA from the first comparator organism, using 2D DNA electrophoresis. In this procedure, molecules are separated by size in the first dimension and by base composition in the second. The DNA from this gel is then blotted onto a nitrocellulose membrane and the membrane washed with a solution containing radiolabeled DNA from the second comparator organism. The blot is then stripped and reprobed with radiolabeled DNA from the first comparator organism, where all the separated fragments become labeled. A comparison of the two autoradiographs using sophisticated image analysis software identifies the DNA fragments found only in the first comparator organism, because these do not hybridize with labeled DNA from the second comparator. In order to sequence these unique fragments, a second 2D gel is run in parallel, using identically fragmented genomic DNA from the first comparator organism, where each fragment has been radioactively labeled at its 3' end. The gel is then autoradiographed to locate organism one–specific DNA fragments, which can be eluted from the gel, ligated into a cloning vector, and sequenced (**Fig. 1**).

In the pioneering use of this method, Malloff et al. *(32)* compared two strains of *P. aeruginosa* and discovered that one of the strains carried an antibiotic-resistance cassette containing an aminoglycoside acetyltransferase gene. They highlighted one problem of the method, however, which is that many of the fragments can cluster in the center of the gel, so differences become impossible to resolve. One possible way of improving the method is to produce a number of pairs of gels, each using DNA digested with a different restriction enzyme, thus increasing the chance that every part of the genome appears in an uncluttered part of the gel in at least one experiment. Another major problem with any hybridization-based approach to checking gene content is that it does not give information about the ability of the gene to be translated into a functional protein. Pseudogenes (with frameshift mutations that stop the encoded protein from being expressed) still hybridize to their wild-type comparator, because there could be as few as one base difference in hundreds. To overcome these problems, comparative functional genomic approaches are required (*see* **Subheading 4.**).

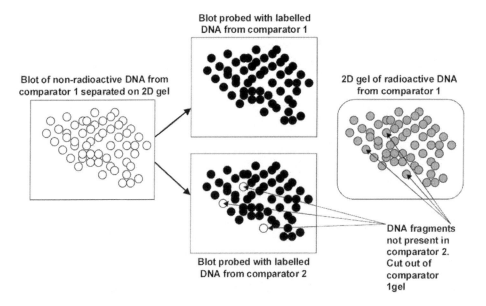

Fig. 1. Comparative genomic hybridization. DNA from comparator organism 1 is digested into many fragments using frequent cutting restriction enzymes, and divided into two equal aliquots. All fragments in one aliquot are radioactively labeled, and the fragments from both aliquots are separated by 2D gel electrophoresis in parallel, using identical conditions. DNA from the nonradioactive aliquot is blotted onto a nitrocellulose membrane and probed with radioactively labeled DNA from comparator 1 to locate all the fragments. The blot is then stripped and reprobed with radioactive DNA from comparator 2, revealing which fragments in comparator 2 do not have a homolog in comparator 1, seen as a lack of hybridization. It is not possible to remove DNA from nitrocellulose blots, so the equivalent positioned fragments on the 2D gel of the radioactive comparator 1 aliquot are visualized and cut from the gel.

## 3.2. Genomic Subtraction

This technique (**Fig. 2**) is now widely used to find sequences unique to one of two comparator genomes. The principle is to remove all sequences common to both genomes from a mixed solution of genomic DNA. This is achieved by digesting both genomes with the same restriction enzyme, and labeling with biotin all the double-stranded fragments from the first comparator organism at their 3' ends. The two sets of genomic fragments are heated to make the DNA single stranded, and then mixed. Cooling the mixture allows hybridization; on average, one biotinylated fragment from the first comparator DNA will hybridize with its nonbiotinylated complement in the DNA of the second comparator. If the biotinylated fragment does not have a complement in the second

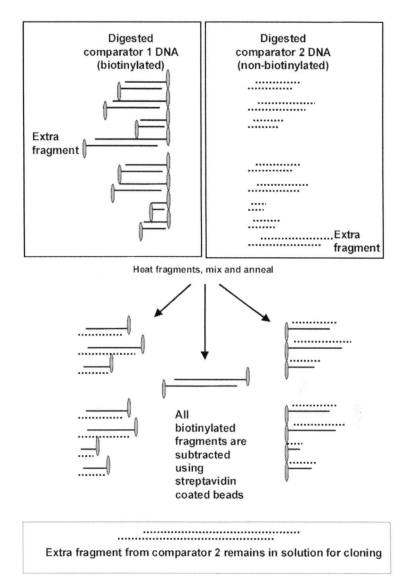

Fig. 2. Genomic subtraction. DNA from comparator organisms 1 and 2 is digested with the same restriction enzyme, and the 3' ends of all comparator 1 fragments are biotinylated. The digested DNA from both comparators is mixed and heated to make them single stranded, and then cooled to stimulate annealing. On average, one biotinylated fragment will anneal to its homolog in the nonbiotinylated DNA. Any resultant fragments that are biotinylated (singly or doubly) can be removed from solution using streptavidin-coated beads, which are pelleted by centrifugation. The remaining nonbiotinylated fragments can be ligated into a cloning vector. Problems come when, by chance, two complementary nonbiotinylated strands from comparator 2 reanneal, producing a false positive. To avoid this, biotinylated comparator 1 DNA can be used in excess.

comparator, it will anneal to its own biotinylated complement. Similarly, if a fragment in the second comparator does not have a complement in the first genome, it will hybridize with its own nonbiotinylated complement. Only the latter group (DNA sequences exclusively found in the second comparator organism) remain completely nonbiotinylated. All the biotinylated fragments can then be removed ("subtracted") by adding streptavidin-coated Sepharose beads or magnetic particles. Streptavidin binds to biotin, so when the beads are pelleted (by centrifugation or use of a magnetic bead separator), DNA fragments common to both comparators, or unique to the first, are pelleted. The fragments remaining in the supernatant can then be ligated to DNA linkers at both ends, and amplified using PCR with primers directed to the linker sequences. The PCR products can then be cloned and sequenced. The whole procedure can be repeated following biotinylation of the second comparator DNA, in order to identify genes that are uniquely present in the first comparator organism. However, this approach is not suitable for detecting the presence of pseudogenes, as discussed in **Subheading 2.2.**

Methods very closely related to this are becoming popular for addressing the genetic reasons why one strain of a bacterial species is pathogenic while another is not. Brown and Curtiss used it to find unique genomic regions associated with an avian-pathogenic *E. coli* isolate, using the non-pathogenic *E. coli* strain K12 as the comparator *(33)*. Pradel et al. used it to find genomic elements related to Shiga-toxin-producing *E. coli* strains *(34)*. The latter analysis used clinical isolates that do not produce Shiga toxin as comparators, and found some of the same genes that were previously observed in the O157:H7/K12 genomic sequence comparison of Perna et al. *(2)*, confirming their conclusion that some virulence factors have been acquired by different lineages of *E. coli* at different times, highlighting their significance and the selective pressure for their acquisition.

It is important to undertake proper follow-up genetic experiments to confirm involvement of DNA sequences in virulence. For example, when true virulence genes are deleted or disrupted, the result should be an avirulent phenotype. Accordingly, this is another example of where comparative genomics allows hypothesis generation.

### 3.3. Amplification Suppression (Subtractive Cloning)

In this method (**Fig. 3**), genomic DNA from comparator organism 1 is digested with a blunt-end-cutting restriction enzyme, and the mixture of products is separated into two fractions. In fraction 1, the 5' end of each DNA fragment is ligated to a single-stranded linker sequence 1, to which a PCR primer can anneal. Fraction 2 is similarly modified, but the linker is an unrelated sequence 2. The two tagged fractions are mixed together with a large excess of

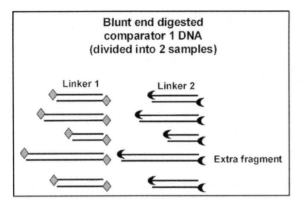

**Blunt end digested comparator 1 DNA (divided into 2 samples)**

Linker 1     Linker 2

Extra fragment

**Blunt end digested comparator 2 DNA (No linker added)**

Heat fragments, mix and anneal. Use of comparator 2 DNA in large excess minimises re-annealing of comparator 1 DNA if a complementary sequences is present in comparator 2 DNA.

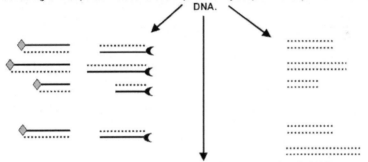

None of these products can be used as templates in PCR reactions primed with sequences complementary to the Linkers 1 and 2 because linkers are not present at both ends

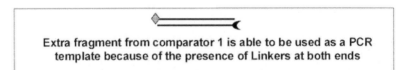

Extra fragment from comparator 1 is able to be used as a PCR template because of the presence of Linkers at both ends

Fig. 3. Amplification suppression. DNA from comparator organisms 1 and 2 is digested with the same restriction enzyme, and comparator 1 DNA is divided into two equal aliquots. The 5' ends of all fragments in both aliquots are ligated to single-stranded linker sequences, but a different linker sequence is used for each aliquot. The digested DNA from both comparators is mixed and heated to make it single stranded, and then cooled to stimulate annealing. On average, one fragment from comparator 1 will anneal to its homolog in comparator 2. Not all the possible arrangements of fragments are shown in the figure. Resultant fragments that contain linker sequences at one end or at nether end cannot be used in PCR reactions using primers complementary to the linker 1 and linker 2 sequences, and so are not amplified. Fragments found only in comparator 1 will always anneal in such a way that either the same linker is present at each end, or linker 1 is at one end and linker 2 is at the other. In either case, PCR using a mixture of primers complementary to linker 1 and linker 2 will be successful, and the PCR amplicon can be sequenced. False-positives occur when, by chance, complementary strands from comparator 1 reanneal, producing products with a linker at each end. To avoid this, comparator 2 DNA is used in excess.

genomic DNA from the second comparator organism, which has been digested with the same restriction enzyme. The mixture is heated to make all fragments single stranded, and then cooled to allow hybridization. The resultant mixture is subjected to PCR with primers that target linker sequences 1 and 2. The only hybridization products that result in a PCR amplicon are those with linker 1 at one end and linker 2 at the other. This occurs only when a DNA fragment from fraction 1 of the first comparator hybridizes with a complementary fragment in fraction 2 of the first comparator, but cannot happen if a similar complementary sequence is present in the second comparator, because the excess complementary DNA titrates out the linker-labeled DNA. Hence, only sequences found uniquely in the first comparator's genome are amplified, because no complementary sequence is present to titrate out the DNA. The resultant amplicons can then be cloned and sequenced, and the method repeated to find DNA unique to the second comparator organism.

Subtractive cloning was used by Ahmed et al. to estimate the level of variation between *Campylobacter jejuni* isolates that colonized birds and humans *(35)*. They successfully isolated 23 sequences found only in avian isolates, but could not rationalize a role for any of the sequences in pathogenesis. Their analysis did, however, confirm the usefulness of this technique for determination of sequence variation in a group of organisms from one species, though the methodology is not entirely robust and suffers from a high false-positive rate, as do most hybridization-based methodologies. DNA hybridization is very sensitive to the temperature and salt concentration of the reaction buffer used, and reproducibility relies on careful experimental technique and good-quality equipment.

Bahrani-Mougeot et al. used subtractive cloning to determine DNA sequences found in the uropathogenic *E. coli* strain CFT073, but not present in *E. coli* K12 MG1655 *(36)*. Their analysis provided evidence of some virulence factors previously identified by Zhang et al. using a similar methodology *(37)*, including fimbriae and hemolysins. They also identified some potential new virulence factors that need to be characterized further. It is interesting to note that genome sequencing of *E. coli* CFT073 is currently under way, and preliminary sequence analysis has already validated the subtractive cloning experiments, adding evidence that this is a useful approach to comparative genomic analysis when a genome sequence is not available *(36)*. With the commercialization of kits to undertake PCR subtractive cloning, this technique will undoubtedly become more commonplace. Again, it is not suitable for detecting the effect of pseudogenes on the complement of functional genes, and is likely to be most applicable to determining the location of genomic islands, which are very important in encoding strain-specific phenotypes.

## 4. Comparative Functional Genomics

Analysis of complete bacterial genomes, and comparisons of two related strains either by sequence or hybridization analysis, is increasingly useful in determining what makes comparator strains phenotypically distinct. However, there have been a number of problems with these approaches, as highlighted above. Moreover, they are generally too simplistic, as what functionally determines an organism's phenotype is its proteins, not its DNA. The activity of a protein is dictated both by its specific amino acid sequence and its level of expression. The problem of pseudogenes—where the gene is present in both organisms, but no product is produced by one, potentially because of a single base change—has already been discussed. A single base change might also result in a specific amino acid change (e.g., in an enzyme's active site), which will lead to a functionally different protein product. These complexities can be addressed by sequence proteome analysis using blastp or CLUSTAL W (*see* Chapter 2). However, we may also have to address the situation where an intact gene is present in two organisms, but its protein product is expressed in only one of them. This might be due to a mutation in the promoter, or in a transcription factor gene elsewhere on the chromosome. Analysis of the expression of genes, rather than just their sequence, is known as functional genomics.

### 4.1. Comparative Proteomics

In the functional genomics sense, a *proteome* is the complement of proteins expressed at any one time, while in the sequence sense, as used previously in this review, it is the complement of proteins with the potential to be expressed at some point in time. It is the former definition of the word proteome that will apply to this section. Clearly not all genes are expressed all the time, and subtle changes in gene expression from one organism to another might lead to dramatic phenotypic differences.

The method of proteomics is discussed elsewhere in this book (*see* Chapter 6), but, put simply, it is a way of separating as many of the proteins expressed by an organism as possible—as distinct spots on a gel or as peaks by liquid chromatography—and then identifying which protein each spot or peak corresponds to. As such, proteomic differences between strains are observed as spots/peaks that have different intensities in one organism as compared with another. The most dramatic result is when a spot/peak is entirely absent in one profile. The protein is then recovered from the spot/peak in the other comparator profile, and identified by proteolytic digestion and mass spectrometry, or by N-terminal sequencing. Care is needed when comparing profiles from organisms whose genome sequences are not known, because point mutations in a gene can lead to amino acid changes in the protein it encodes, leading to altered electrophoretic and chromatographic properties, and the observation of one spot/peak disappearing and one appearing in the comparator. Such false leads can result in a great deal of wasted time and resources *(38)*.

Difficulties with these techniques mean that they have been rarely used in comparative genomics (i.e., to compare different isolates). However, Betts et al. performed a comparative proteomic analysis of *M. tuberculosis* strain H37Rv, whose genome sequence has been published *(21)*, and strain CDC1551 (whose genome sequence has been determined, but not yet published). These strains share > 99% sequence identity, but have significant phenotypic differences, potentially explained by changes in gene expression *(39)*. The authors compared spots on 2D gels, and identified all the differentially expressed spots using mass spectrometry. Only 17 spot intensity differences were seen; the proteins representing 12 of them were identified. Four of the spot intensity differences were mobility variants caused by amino acid changes. Three of the proteomic differences were found to represent proteins encoded on a DNA element found in only one of the comparator organism genome sequences, validating the proteomic approach used. Unfortunately, this proteomic analysis did not provide a reason for the phenotypic differences seen between the organisms, though the (as yet unpublished) comparative genomic analysis did highlight a number of blastp differences that might explain the phenotypic differences. These important genetic changes were not reflected as proteomic changes; hence, either some of these genes were not expressed in the comparator encoding them under the conditions tested, or their respective spots were not resolved *(39)*. Poor resolution is a major problem with 2D electrophoresis, though it is less problematic when analyzing organisms expressing relatively few proteins, such as *M. tuberculosis (38)*.

## 5. Conclusions

The science of comparative genomics is still in its infancy. Because of the inherent complexity of even relatively simple organisms like bacteria, a large number of variables affect the expression of particular traits. Only when we are able to integrate all these comparative data, of sequence and of expression, will we be able to home in on what is important for a particular phenotype. Methodologies for data collection (and, more importantly, data analysis) are improving all the time, and soon we will begin to reap the rewards of genomic analysis. In the case of bacterial comparative genomics, the rewards are likely to be new vaccine or drug targets, and new diagnostic tests designed to help clinicians rapidly identify patients infected with virulent bacteria. In the meantime, bacterial genome sequencing continues apace, and comparative genomics is catching up.

## References

1. Blattner, F. R., Plunkett, G., 3rd, Bloch, C. A., Perna, N. T., Burland, V., Riley, M., et al. (1997) The complete genome sequence of *Escherichia coli* K-12. *Science* **277,** 1453–1474.
2. Perna, N. T., Plunkett, G., 3rd, Burland, V., Mau, B., Glasner, J. D., Rose, D. J., et al. (2001) Genome sequence of enterohaemorrhagic *Escherichia coli* O157:H7. *Nature* **409,** 529–533.
3. Reid, S. D., Herbelin, C. J., Bumbaugh, A. C., Selander, R. K., and Whittam, T. S. (2000) Parallel evolution of virulence in pathogenic *Escherichia coli*. *Nature* **406,** 64–67.
4. Stover, C. K., Pham, X. Q., Erwin, A. L., Mizoguchi, S. D., Warrener, P., Hickey, M. J., et al. (2000) Complete genome sequence of *Pseudomonas aeruginosa* PA01, an opportunistic pathogen. *Nature* **406,** 959–964.
5. Clarke, G. D., Beiko, R. G., Ragan, M. A., and Charlebois, R. L. (2002) Inferring genome trees by using a filter to eliminate phylogenetically discordant sequences and a distance matrix based on mean normalized BLASTP scores. *J. Bacteriol.* **184,** 2072–2080.
6. Edwards, R. A., Olsen, G. J., and Maloy, S. R. (2002) Comparative genomics of closely related salmonellae. *Trends Microbiol.* **10,** 94–99.
7. Parkhill, J., Dougan, G., James, K. D., Thomson, N. R., Pickard, D., Wain, J., et al. (2001) Complete genome sequence of a multiple drug resistant *Salmonella enterica* serovar Typhi CT18. *Nature* **413,** 848–852.
8. McClelland, M., Sanderson, K. E., Spieth, J., Clifton, S. W., Latreille, P., Courtney, L., et al. (2001) Complete genome sequence of *Salmonella enterica* serovar Typhimurium LT2. *Nature* **413,** 852–856.
9. Hansen-Wester, I. and Hensel, M. (2002) Genome-based identification of chromosomal regions specific for *Salmonella* spp. *Infect. Immun.* **70,** 2351–2360.
10. Hou, Y. M. (1999) Transfer RNAs and pathogenicity islands. *Trends Biochem. Sci.* **24,** 295–298.
11. Parkhill, J., Wren, B. W., Thomson, N. R., Titball, R. W., Holden, M. T., Prentice, M. B., et al. (2001) Genome sequence of *Yersinia pestis,* the causative agent of plague. *Nature* **413,** 523–527.
12. Deng, W., Burland, V., Plunkett, G., 3rd, Boutin, A., Mayhew, G. F., Liss, P., et al. (2002) Genome sequence of *Yersinia pestis* KIM. *J. Bacteriol.* **184,** 4601–4611.
13. Kuroda, M., Ohta, T., Uchiyama, I., Baba, T., Yuzawa, H., Kobayashi, I., et al. (2001) Whole genome sequencing of meticillin-resistant *Staphylococcus aureus*. *Lancet* **357,** 1225–1240.
14. Avison, M. B., Bennett, P. M., Howe, R. A., and Walsh, T. R. (2002) Preliminary analysis of the genetic basis for vancomycin resistance in *Staphylococcus aureus* strain Mu50. *J. Antimicrob. Chemother.* **49,** 255–260.
15. O'Neill, A. J. and Chopra, I. (2002) Insertional inactivation of mutS in *Staphylococcus aureus* reveals potential for elevated mutation frequencies, although the

prevalence of mutators in clinical isolates is low. *J. Antimicrob. Chemother.* **50,** 161–169.

16. Baba, T., Takeuchi, F., Kuroda, M., Yuzawa, H., Aoki, K., Oguchi, A., et al. (2002) Genome and virulence determinants of high virulence community-acquired MRSA. *Lancet* **359,** 1819–1827.

17. Herron, L. L., Chakravarty, R., Dwan, C., Fitzgerald, J. R., Musser, J. M., Retzel, E., et al. (2002) Genome sequence survey identifies unique sequences and key virulence genes with unusual rates of amino acid substitution in bovine *Staphylococcus aureus*. *Infect. Immun.* **70,** 3978-3981.

18. Cunningham, M. W. (2000) Pathogenesis of group A streptococcal infections. *Clin. Microbiol. Rev.* **13,** 470–511.

19. Smoot, J. C., Barbian, K. D., Van Gompel, J. J., Smoot, L. M., Chaussee, M. S., Sylva, G. L., et al. (2002) Genome sequence and comparative microarray analysis of serotype M18 group A *Streptococcus* strains associated with acute rheumatic fever outbreaks. *Proc. Natl. Acad. Sci. USA* **99,** 4668–4673.

20. Ferretti, J. J., McShan, W. M., Ajdic, D., Savic, D. J., Savic, G., Lyon, K., et al. (2001) Complete genome sequence of an M1 strain of *Streptococcus pyogenes*. *Proc. Natl. Acad. Sci. USA* **98,** 4658–4663.

21. Cole, S. T., Brosch, R., Parkhill, J., Garnier, T., Churcher, C., Harris, D., et al. (1998) Deciphering the biology of *Mycobacterium tuberculosis* from the complete genome sequence. *Nature* **393,** 537–544.

22. Cole, S. T. (2002) Comparative and functional genomics of the *Mycobacterium tuberculosis* complex. *Microbiology* **148,** 2919–2928.

23. Cole, S. T., Eiglmeier, K., Parkhill, J., James, K. D., Thomson, N. R., Wheeler, P. R., et al. (2001) Massive gene decay in the leprosy bacillus. *Nature* **409,** 1007–1011.

24. Janssen, P. J., Audit, B., and Ouzounis, C. A. (2001) Strain-specific genes of *Helicobacter pylori:* distribution, function and dynamics. *Nucleic Acids Res.* **29,** 4395–4404.

25. Garcia-Vallve, S., Janssen, P. J., and Ouzounis, C. A. (2002) Genetic variation between *Helicobacter pylori* strains: gene acquisition or loss? *Trends Microbiol.* **10,** 445–447.

26. Thompson, L. J. and de Reuse, H. (2002) Genomics of *Helicobacter pylori*. *Helicobacter* **7,** 1–7.

27. Ochman, H. and Jones, I. B. (2000) Evolutionary dynamics of full genome content in *Escherichia coli*. *EMBO J.* **19,** 6637–6643.

28. Richmond, C. S., Glasner, J. D., Mau, R., Jin, H., and Blattner, F. R. (1999) Genome-wide expression profiling in *Escherichia coli* K-12. *Nucleic Acids Res.* **27,** 3821–3835.

29. Dziejman, M., Balon, E., Boyd, D., Fraser, C. M., Heidelberg, J. F., and Mekalanos, J. J. (2002) Comparative genomic analysis of *Vibrio cholerae:* genes that correlate with cholera endemic and pandemic disease. *Proc. Natl. Acad. Sci. USA* **99,** 1556–1561.

30. Heidelberg, J. F., Eisen, J. A., Nelson, W. C., Clayton, R. A., Gwinn, M. L., Dodson, R. J., et al. (2000) DNA sequence of both chromosomes of the cholera pathogen *Vibrio cholerae. Nature* **406,** 477–483.

31. Behr, M. A., Wilson, M. A., Gill, W. P., Salamon, H., Schoolnik, G. K., Rane, S., et al. (1999) Comparative genomics of BCG vaccines by whole-genome DNA microarray. *Science* **284,** 1520–1523.

32. Malloff, C. A., Fernandez, R. C., and Lam, W. L. (2001) Bacterial comparative genomic hybridization: a method for directly identifying lateral gene transfer. *J. Mol. Biol.* **312,** 1–5.

33. Brown, P. K. and Curtiss, R., 3rd (1996) Unique chromosomal regions associated with virulence of an avian pathogenic *Escherichia coli* strain. *Proc. Natl. Acad. Sci. USA* **93,** 11,149–11,154.

34. Pradel, N., Leroy-Setrin, S., Joly, B., and Livrelli, V. (2002) Genomic subtraction to identify and characterize sequences of Shiga toxin–producing *Escherichia coli* O91:H21. *Appl. Environ. Microbiol.* **68,** 2316–2325.

35. Ahmed, I. H., Manning, G., Wassenaar, T. M., Cawthraw, S., and Newell, D. G. (2002) Identification of genetic differences between two *Campylobacter jejuni* strains with different colonization potentials. *Microbiology* **148,** 1203–1212.

36. Bahrani-Mougeot, F. K., Pancholi, S., Daoust, M., and Donnenberg, M. S. (2001) Identification of putative urovirulence genes by subtractive cloning. *J. Infect. Dis.* **183 (Suppl 1),** S21–S23.

37. Zhang, L., Foxman, B., Manning, S. D., Tallman, P., and Marrs, C. F. (2000) Molecular epidemiologic approaches to urinary tract infection gene discovery in uropathogenic *Escherichia coli. Infect. Immun.* **68,** 2009–2015.

38. Jungblut, P. R. (2001) Proteome analysis of bacterial pathogens. *Microbes Infect.* **3,** 831–840.

39. Betts, J. C., Dodson, P., Quan, S., Lewis, A. P., Thomas, P. J., Duncan, K. et al. (2000) Comparison of the proteome of *Mycobacterium tuberculosis* strain H37Rv with clinical isolate CDC 1551. *Microbiology* **146,** 3205–3216.

# 5

# Genome Plasticity

*Insertion Sequence Elements, Transposons and Integrons, and DNA Rearrangement*

## Peter M. Bennett

### Summary

Living organisms are defined by the genes they possess. Control of expression of this gene set, both temporally and in response to the environment, determines whether an organism can survive changing conditions and can compete for the resources it needs to reproduce. Bacteria are no exception; changes to the genome will, in general, threaten the ability of the microbe to survive, but acquisition of new genes may enhance its chances of survival by allowing growth in a previously hostile environment. For example, acquisition of an antibiotic resistance gene by a bacterial pathogen can permit it to thrive in the presence of an antibiotic that would otherwise kill it; this may compromise clinical treatments. Many forces, chemical and genetic, can alter the genetic content of DNA by locally changing its nucleotide sequence. Notable for genetic change in bacteria are transposable elements and site-specific recombination systems such as integrons. Many of the former can mobilize genes from one replicon to another, including chromosome-plasmid translocation, thus establishing conditions for interspecies gene transfer. Balancing this, transposition activity can result in loss or rearrangement of DNA sequences. This chapter discusses bacterial DNA transfer systems, transposable elements and integrons, and the contributions each makes towards the evolution of bacterial genomes, particularly in relation to bacterial pathogenesis. It highlights the variety of phylogenetically distinct transposable elements, the variety of transposition mechanisms, and some of the implications of rearranging DNA, and addresses the effects of genetic change on the fitness of the microbe.

**Key Words:** Insertion sequence; transposon; integron; gene cassette; plasmid; conjugation; transduction; transformation; pathogenicity island; transposition; site-specific recombination.

From: *Methods in Molecular Biology, vol. 266: Genomics, Proteomics, and Clinical Bacteriology: Methods and Reviews*
Edited by: N. Woodford and A. Johnson © Humana Press Inc., Totowa, NJ

## 1. Introduction

In the past, a bacterial species has been defined in part by the G+C content of its genome, and it was generally assumed that the overall G+C content reflected the base composition throughout the genome—i.e., it was more or less constant from one gene to the next, indicating a common line of descent for the majority, if not all, of the genes on the replicon. Underlying this assumption was the general belief that bacterial genomes change slowly with time, and that the bulk of genes on a particular genome will have coevolved over a considerable period and will therefore have been subject to the same selective pressures. In recent years, the sequences of several bacterial replicons (chromosomes and plasmids) have been reported (e.g., **refs. 1–4**) (*see also* www.tigr.org/tdb/mdb/mdbcomplete.html). Analyses of these data have indicated that the nucleotide compositions of individual replicons are not homogeneous; rather, the data indicate a significant degree of genetic flux involving acquisition of DNA sequences from diverse sources at different times (e.g., **ref. 2**). This leads to the conclusion that bacterial replicons, as gene collections, are dynamic rather than static. Over time, some genes are acquired and others are lost *(5)*. Those that confer selective advantage are preserved, while those that do not, atrophy and may be lost. Each replicon, whether it is a chromosome or a plasmid, appears to comprise a basic core that can be recognized because the DNA sequences have the same G+C content. In terms of the bacterial chromosome, this core encompasses the housekeeping genes that are essential for survival (i.e., growth and reproduction). For plasmids, the only essential genes and sequences are those needed for replication. The genes of the core structure display common codon usage (*see* Chapter 3), as would be expected, given that they will have been subject to prolonged coevolution. Inserted into the core are sequences that differ in nucleotide composition both from the core and from each other. This variation is indicative of different evolutionary descents before the sequences were fused, and implies a flow of DNA between different organisms (i.e., horizontal gene transfer and recombination). Lateral gene transfer involving mobile elements such as plasmids has been known for a long time; what had not been appreciated is the degree of sequence reassortment that has occurred among replicons, including bacterial genomes, in general. Among the genes in human pathogens that have been acquired in this way are those that confer pathogenicity and resistance to antibacterial agents. Although the sources of many of these determinants are not known, carriage of some exotoxin genes can be ascribed to a process termed *lysogenic conversion.* This refers to latent infection of a bacterial cell by a bacteriophage (phage) that integrates into the chromosome and becomes dormant. An example of an exotoxin encoded by a lysogenic phage is the cholera toxin of *Vibrio cholerae (6)*; others are the cytotoxin produced by disease-causing strains of *Corynebacterium diphtheriae;* the Shiga toxins (1 and 2), produced by *Shigella dysenteriae;* and streptococcal erythrogenic toxin *(7)*. Hence, to understand gene flux it is necessary to know something about how DNA is transferred between bacterial cells and rearranged by recombination.

## 1.1. Bacterial Genetic Exchange

DNA is transferred between bacterial cells by one of three mechanisms: conjugation, transformation, and transduction. Conjugation, mediated by plasmids *(8–10)* and conjugative transposons *(11–13)*, is a semiconservative replication process and normally involves transfer only of the element itself *(14,15)*. The persistence of a plasmid will usually reflect the ability to replicate autonomously in the new host, while the persistence of a conjugative transposon will reflect the element's ability to integrate into a DNA molecule in the new host cell (*see* **Subheading 5.**). Although plasmid acquisition can expand the gene set of a cell, it does not by itself involve gene reassortment of the type detected by sequence analysis. Nonetheless, conjugation can potentially mediate transfer of any DNA sequence. Indeed, it was precisely the ability to promote transfer of chromosomal DNA between *Escherichia coli* cells that first revealed the existence of the sex factor F, a conjugative plasmid *(10)*.

Transduction involves DNA transfer mediated by a phage *(16,17)*. In this case a segment of DNA that is not part of the phage genome is incorporated into a phage particle, either in place of or as an extension of the genome. This particle then injects its DNA load into another bacterial cell. There, nonphage DNA sequences may be rescued by recombination into one of the replicons in the new host. In contrast, transformation describes the genotypic alteration of a bacterial cell as a result of uptake of DNA from its environment *(18,19)*. Some if not all bacteria can scavenge DNA released by others and rescue the genetic content *(20)*. Again, incorporation of the foreign sequence requires integration into an existing replicon by some form of recombination. In both transduction and transformation, DNA sequences are usually rescued by RecA-dependent homologous recombination *(21)*. This can both remodel genes, as illustrated by the development of penicillin-resistant *Streptococcus pneumoniae (22)*, and increase the gene pool, as seen in the transduction experiments of Stanisich et al. *(23)*; homologous recombination is exploited in the gene knockout and allelic replacement strategies now frequently used in molecular genetics. However, as the name indicates, homologous recombination requires the participation of DNA sequences that have significant homology, and the recombination commonly results in DNA replacement rather than addition. Additions to a replicon are usually the result of transposition.

## 2. Transposition

Transposition is a genetic event in which a DNA sequence is translocated from one site to another usually unrelated site on the same or a different DNA molecule *(24)*. Transposition is, necessarily, a recombination event in the sense that DNA phosphodiester bonds are broken and reformed in the process. Most transposition events involve genetic structures called transposable elements, which comprise discrete DNA sequences with defined ends that generally mediate their own transposition. Although the term *transposable element* implies a defined genetic entity, the term *transposition* indicates only that there has been a genetic rearrangement by DNA relocation, which can occur by one of several mechanisms. Transposable elements are virtually ubiquitous in bacteria, and transposition is responsible for much of the DNA rearrangement seen among bacterial replicons. Genetic translocation may also involve DNA sequences not formally classified as transposable elements, such as pathogenicity islands and gene cassettes.

### 2.1. Transposable Elements: General Structure

Most reported DNA transpositions in bacteria involve transposable elements, either as the sole sequence transposed or as mediators of translocations of DNA sequences that are not normally part of a transposable element.

Most transposable elements conform to a simple model, encoding a protein or protein complex called a transposase, that mediates transposition, and individual elements may or may not encode other functions. The end sequences of the element are usually short inverted repeats (IRs) that are necessarily identical or nearly so, because they are binding sites for the transposase. The inverted configuration ensures transposase is positioned similarly at both ends of the element, i.e., facing outward, so it can cleave the phosphodiester bonds that anchor the element to its current carrier, both freeing it for transposition and guaranteeing that the element is conserved from one transposition to the next. These terminal IRs are element specific and usually are only 15–40 bp long. For a few transposable elements, the IRs are much longer (80–150 bp) and contain multiple transposase binding sites. In some cases, it has been shown that transposition involves formation of a protein–DNA complex called a transpososome *(25,26)*, in which the ends of the element are bound by transposase and aligned. Bringing the ends together presumably facilitates the subsequent insertion, where both ends have to be ligated essentially into the same site (*see* **Subheading 3.1.1.**). If the element encodes only the function needed for transposition and is 1–2 kb in size, it is called an insertion sequence (IS) element *(27,28)*. If the transposable element encodes functions other than those needed for transposition, e.g., a gene encoding antibiotic resistance, it is called a transposon. Such elements range in size from 3–4 kb (i.e., those that have only one or two additional genes) to elements more than 10 times this size, in some cases encompassing dozens of genes. In addition, certain phages, called transposing bacteriophages, use a form of transposition for replication. The best known of these is the *E. coli* mutator phage, Mu *(29)*. Related transposable elements encode related transposases and have related terminal IRs *(27, 30–32)*. However, not all transposable elements conform to the general description. Notable among these are the IS*91* family of elements, which lack IRs, indicating that these elements use a different transposition mechanism from most others *(33)*.

IS elements (*see* **Subheading 3.1.**) were the first transposable elements in bacteria to be recognized as such, although pride of place goes to phage Mu as the first bacterial transposon to be discovered *(34)*. The concept of transposable elements, however, was proposed a decade or more before the discovery of Mu. The idea was first put forward by Barbara McClintock in the late 1940s, before the structure of DNA was known, to explain results obtained from her elegant and detailed genetic analysis of maize kernel pigmentation *(35)*. She suggested correctly that variations in maize kernel pigmentation were the consequence of gene activation/inactivation resulting from translocation of one or more genetic elements from one chromosomal site to another. Physical evidence to support the concept came first from bacterial systems in the late 1960s *(36)*. These studies produced incontrovertible evidence for the existence of small, discrete, mobile DNA elements: IS*1*, IS*2*, and IS*3*. The importance of transposable elements in bacteria was subsequently illustrated by the discovery that clinically important antibiotic resistance genes were associated with bacterial transposons *(37–41)*.

Transposons can be described as either composite transposons (*see* **Subheading 3.2.**) or complex transposons (*see* **Subheading 3.3.**). The latter designation essentially indicates that the element is not a composite transposon. Composite transposons, for example Tn*5 (42)* and Tn*10 (43)*, have modular structures with a central sequence that encodes the function(s) that identifies the element phenotypically, but lacks transposition functions. This is flanked by two copies of the same IS element, forming either long terminal direct repeats or, more commonly, long terminal IRs. One or both of these IS elements provides the transposase for transposition, while the short IR sequences on the IS elements provide the sequences necessary for end recognition. When an IS element becomes a component of a composite transposon it does not immediately lose the ability to function as an independent element. Hence a composite transposon might comprise three distinct transposable sequences— two long terminal repeats and the entire composite structure. However, further genetic change and development of the composite element, a process that has been called *coherence,* can lead to fusion of the component parts and loss of the ability of one or both terminal IS elements to function independently. All that need be retained is a copy of the transposase gene(s) and the extreme short IR sequences. These minimal requirements have been exploited, for example, in the development of the mini-Tn5 series of transposons developed for mutation and gene analysis *(44)* (*see* Chapter 13).

Complex transposons lack both a modular structure and, in general, long terminal IRs. These elements are usually delimited by short IRs, of 15–40 bp depending on the element, as found on IS elements *(28)*. The IRs flank genes that encode both the functions needed for transposition and those that identify the element by conferring a particular phenotype on the host cell. No part of the structure can transpose independently of the rest. Some noncomposite transposons—e.g., Tn3 and related elements—can form composite transposable structures analogous to IS-based composite transposons *(45–48)*. Indeed, this is the basis of the in vivo cloning system that uses a deleted version of phage Mu, called mini-Mu, to mobilize chromosomal genes on to the broad host range plasmid RP4 *(29)*.

Transposing phages are typified by phage Mu, discovered in the early 1960s in a culture of *E. coli (34)*. Phage Mu is a temperate phage, in that it can lysogenize its host, and is so named because Taylor correctly surmised that the increased numbers of auxotrophic mutants in populations of *E. coli* lysogenized with Mu arose as a consequence of its insertion into genes encoding biosynthetic enzymes. Transposing phages like Mu use a form of transposition to replicate *(29)*, and differ from other transposable elements in that their lifecycles include an extracellular phase, in the form of phage particles.

## 2.2. Transposable Elements in Bacteria: Distribution

Transposable elements are widely distributed in nature among eubacteria and archeae, and are also found in lower and higher eukaryotes. IS elements, found in both prokaryotes and eukaryotes, are the most common, and hundreds have been identified in many different bacterial species, including Gram-positive and Gram-negative microbes and Archaebacteria, such as *Halobacterium* spp. and *Methanobrevibacter smithii (28)*. Wherever a serious search has been made for transposable elements, examples have been found. Indeed, the failure to find a transposable element is more likely to indicate a defect in the detection system than an absence of transposable elements. Furthermore, as the complete genome sequences of many different microbes are determined, new putative transposable elements are identified *(27)*. However, determination of whether these elements are real and active, as distinct from being pseudoelements or remnants of obsolete elements, requires experimental evidence that they can transpose.

Transposons have been found in many different bacterial species, often of clinical or veterinary origin. Our interest in things that directly affect the human condition, especially health, probably explains the preponderance of known transposons, both composite and complex, that carry antibiotic resistance genes, and the relative paucity of transposons carrying other markers. Most bacterial transposons have been found on plasmids rather than on other DNA molecules, but this probably reflects the fact that once a gene is located on a plasmid, its ability to spread to other bacteria is greatly increased. It also appears that, for some transposons such as Tn*3,* transposition to plasmids is much more frequent than transposition into the bacterial chromosome, despite transposition generally not being site specific and chromosomes being at least two orders of magnitude larger than most plasmids. The reason for this apparent paradox is unknown.

## 2.3. Designation of Transposable Elements

Insertion sequences and transposons are given the designations IS and Tn, respectively, and in each case, individual elements are identified by an italicized number (e.g., IS*10,* Tn*10*). To avoid the confusion that might arise from different elements being given the same identification numbers by different groups of researchers, it was suggested and widely accepted that numbers should be allocated from a central directory. This was administered by Esther Lederberg at the Department of Medical Microbiology, Stanford University, California *(49)*, who periodically published lists of new assignments. As the number of elements discovered increased, it became clear that this simple numbering system was not sufficiently informative, and authors began to elaborate. Accordingly, additional letters were included to indicate the source of the element, and the numbering became specific to the species. For example, IS*M1* is an element from *M. smithii,* while IS*Rm2* is an IS element from *Rhizobium meliloti,* and both are distinct from the elements IS*1* and IS*2,* respectively, which were the first transposable elements to be described. More recently, Chandler and Mahillion suggested that new IS elements be numbered as follows: IS followed by the first letter of the genus and then the first two letters of the species in which the element was discovered (namely, the convention used to name restriction endonucleases), plus a number chosen consecutively according to how many other new IS elements have been found in the species *(27,27a)*. Changes have also been instituted in transposon designations, in particular when numbering conjugative transposons (*see* **Subheading 5.**), and specifically those from *Bacteroides* spp., because the conventional numbering system fails to indicate the nature of the element. Such elements are now designated CTn plus a number. Nonetheless, the intent must remain to give each new transposable element its own distinct designation. A list of assigned IS numbers can be found at http://www-is.biotoul.fr.

## 2.4. Discovery of Transposable Elements in Bacteria

IS elements, and in particular IS*1,* IS*2,* and IS*3,* were discovered as a result of studying strongly polar mutations in the galactose and lactose operons of *E. coli,* i.e., mutations that prevented expression not only of the gene in which the mutation was located, but also of other genes in the operon located distal to (downstream from) the mutation *(50–52).* Such mutations represent a form of gene silencing. Although these mutations behaved like point mutations in that they could be mapped to a single site and could revert spontaneously at low frequency, the frequency of true reversion was not enhanced by chemical treatments known to enhance the generation of base substitution and frameshift mutations. To account for this anomaly, it was suggested that the mutations were the result of insertion of small segments of DNA at the point of mutation, a concept first proposed by Barbara McClintock to explain some of the genetic variation in maize *(35).* This idea was validated by a series of elegant genetic and biochemical experiments that used λ.*gal*-transducing phages and visualization of the IS element by electron microscopy of heteroduplex DNA *(36).* These experiments demonstrated the physical nature of transposable elements that, until then, had been simply a genetic concept. Further analysis established that these elements could also promote major DNA rearrangements, such as extended deletions *(53,54),* and could serve as mobile promoter elements, able to activate the expression of a gene adjacent to the point of insertion *(55).* This early work has been ably summarized by Starlinger *(56).*

## 3. Types of Transposable Elements

### 3.1. IS Elements

Bacterial IS elements are the simplest type of transposable element, and several hundred have been discovered and characterized to some extent, usually in terms of their sequences *(27).* They rarely exceed 2 kb in size and may be as small as 0.5 kb, which limits their genetic capacity in most cases to one or two genes. Most, as already stated, have short-terminal IRs that define the extremities of the element, and many have a single open-reading frame (ORF) that is presumed to encode the transposase. In a few cases, such as IS*1 (57),* IS*3 (58),* IS*10 (59),* IS*50 (60),* IS*91 (61),* and IS*903 (62),* this has been established experimentally by genetic and biochemical analysis. For many others, it is assumed that the putative genes encode transposition functions because the predicted gene products show significant homology to a known transposase. For putative IS elements discovered as a result of bacterial genome sequencing, the structural similarities to well-characterized IS elements are the sole evidence for IS status.

Some elements contain more than one ORF, with the smaller ORF overlapping the major one, wholly or in part. In one or two instances, the smaller ORFs have been shown to be real genes, encoding proteins that regulate transposition of the element *(27)*, although in most cases this remains to be confirmed experimentally. Such regulation is desirable, given that transposition can be mutagenic and is therefore potentially lethal to the cell.

An interesting gene arrangement is found in IS*1*. At 768 bp, this is one of the smallest IS elements, and it has two short ORFs, *insA* and *insB'*, which are adjacent but in different reading frames *(63,64)*. These sequences are cotranscribed, but while the former codes for a discrete peptide, the latter encodes the C-terminal segment of a fusion protein, InsAB', for which *insA* encodes the N-terminal segment. To create this hybrid protein, a proportion of the ribosomes that initiate translation of the *insA* region of the transcript undergo a –1 frameshift toward the end of this region before translation is halted by the termination codon that signals the end of the InsA sequence. This frameshift is directed by the nucleotide sequence at the point of changeover, and switches the translating ribosomes to the reading frame used to accommodate *insB (57,65)*. Translation then continues uninterrupted through the *insB'* region of the transcript to the translational termination codon that indicates the end of InsB'. It is the fusion product, InsAB', that is the transposase for the element. Both InsA and InsAB' are DNA-binding proteins and compete for the IRs of IS*1 (65)*. InsA competes with the transposase, denying it access to its sites of action and inhibiting transposition. InsA also represses expression of the insAB' transcript *(66,67)*. The transposition frequency is determined by the relative activities of these two proteins and is dependent on the ratio of the activities rather than the absolute activity of either. The transposases of IS*3* and like elements are also fusion proteins *(68)*.

From sequence comparisons it would appear that most bacterial IS elements belong to one of a small number of extended families. Approximately 20 apparently unrelated familial groups have been identified, and members of most of these are found in diverse eubacteria and archaebacteria *(27)*. The largest group comprises IS*3*-like and IS*5*-like elements, of which more than 70 of each are known; they are found in both eubacteria and archaebacteria, indicating, perhaps, an unusual degree of horizontal transfer and/or ancestral elements of considerable antiquity. The most comprehensive database on IS elements currently contains information on more than 800 elements from more than 196 bacterial species, representing 523 individual ISs (http://pc4.sisc.ucl.ac.be/is.html).

### 3.1.1. Transposition of IS Elements

Most of our information on the mechanisms of transposition used by different types of IS element has come from studies of IS*1*, IS*10*, IS*50*, IS*91*, IS*903*,
and IS*911*. An important early finding was that once an IS element has transposed to a new site, the element is bracketed by short direct repeats *(69,70)*.
Furthermore, while the same element at different sites of insertion is flanked
by repeats of the same size, the sequences differ from one insertion to another.
These repeats, 2–12 bp depending on the element, arise from a sequence found
at the target site prior to insertion of the element, and hence involve DNA
replication. It was quickly realized that this arrangement would result if, when
processing the target site for transposition, the DNA duplex was cleaved at
sites on each strand that are not directly opposite each other, but are displaced
by a few base pairs. Then, if the IS element, released from its donor site, is
joined to the single-strand extensions created by the staggered cut at the target
site, short, single-strand gaps are created at both ends of the element. DNA gap
repair synthesis would then produce the same short sequence (i.e., the short
direct repeats) at each end of the element. This is indeed what appears to happen with many transposable elements of all types—IS elements, composite
and noncomposite transposons, and transposing phages. Therefore, many different transposable elements create the same type of sequence arrangement by
transposing, namely a sequence delimited by terminal inverted repeats located
between short direct repeats. Hence, if such an arrangement is discovered when
analyzing new bacterial DNA sequences, it is reasonable to infer the presence
of a transposable element, particularly if there is a potential gene encoding a
transposase-like protein bracketed by the terminal IRs. Further experimental
work is then needed to determine whether the putative element is still
transpositionally active. As already pointed out, several examples of such discoveries have been reported as a result of bacterial genome sequence annotation. However, in the case of one or two IS elements, for example IS*91* and the
related elements IS*801* and IS*1294* *(71,72)*, transposition to a new site does not
generate target site duplications, which may indicate either that the mechanism
employs a flush cut at the target site, or that the element transposes by a mechanism fundamentally different from those used by most transposable elements,
as do IS*91*-like elements *(71,72)*.

The transposition mechanism of many IS elements, including IS*10* and IS*50*,
is conservative, in that the double-stranded form of the element is disconnected
from the donor DNA molecule and then inserted, in toto, into the recipient
DNA molecule. The simplest conservative mechanism is cut-and-paste, where
the ends of the sequence to be transposed are brought together (synapsed) in a
transposase–DNA complex (**Fig. 1**). The transposase cuts both DNA strands at

both ends of the element, releasing it from the donor DNA molecule, formally generating a free form of the element. True free forms of transposable elements are not normally detected, and probably usually remain as a component of a transposition complex, the transpososome, which attacks the target site. The 3'-OH groups at both ends of the element generated on release from the donor molecule, serve as nucleophiles that are used to attack nonopposed 5'-phosphate groups on opposite DNA strands at the target site to initiate two transesterification reactions *(73)*. These reactions require no additional energy input and connect the element to the target molecule by single-stranded DNA. The short single-strand gaps are believed to be filled by gap-repair DNA synthesis, generating the short direct-target-sequence repeats that bracket the element. The fate of the remainder of the donor DNA molecule, which is linearized when the IS element is excised, is unknown, but what little evidence there is suggests that it is probably degraded.

Some IS elements transpose conservatively, using a mechanism that involves a true circular intermediate (**Fig. 1**). The element is disconnected from the donor molecule, and the ends of the element are joined to produce a circular form. IS*3* and IS*911,* both of which are IS*3* group elements, use such a mechanism *(74,75)*. Production of the circular form of the element depends on two element-encoded proteins, OrfA and OrfAB, the latter being a fusion protein, the product of two open reading frames, *orfA* and *orfB*. As in the case of IS*1,* the two *orf*s are in different reading frames (0 and −1) and overlap to a small extent *(76)*. Again, production of OrfAB is the result of a directed ribosome frameshift. Both the OrfA and OrfAB proteins bind the terminal IRs of their cognate element, and both proteins are needed for high-frequency transposition (*cf,* the case of IS*1,* where the fusion protein InsAB is the transposase, while the second, smaller protein InsA inhibits transposition by competing with InsAB nonproductively for the IRs).

IS*911* circles have been isolated and shown in vitro to be efficient substrates for transposition *(74)*. When the element is disconnected from donor DNA and circularized, 3 bp of the donor, linked to one end of the element, are also removed and form a spacer between the terminal IRs of the element. This is lost when the element is inserted into a new site. One consequence of circularization of IS*911* (and IS*3*) is the creation of a new promoter for expression of the transposase genes from the temporary structure. This promoter is significantly stronger than that used to express the transposase genes on the linear structure *(77)*. Circularization places the terminal IRs beside each other. The −35 box of the new promoter is on one IR, while the −10 box is on the other, with the additional 3 bp ensuring correct spacing of the boxes in the new promoter. The generation of a powerful promoter ensures that the levels of the

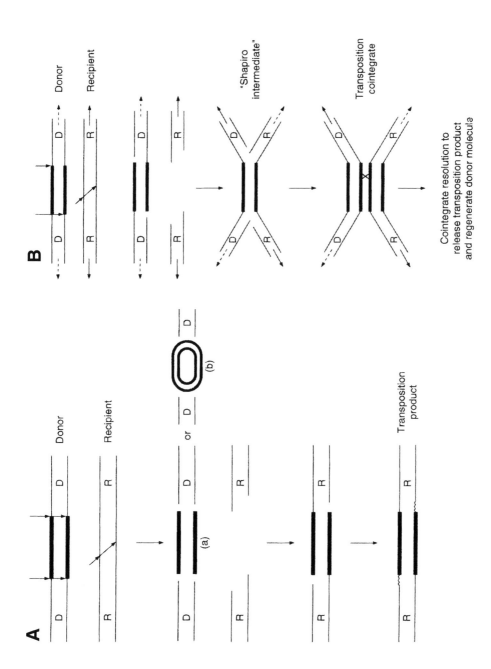

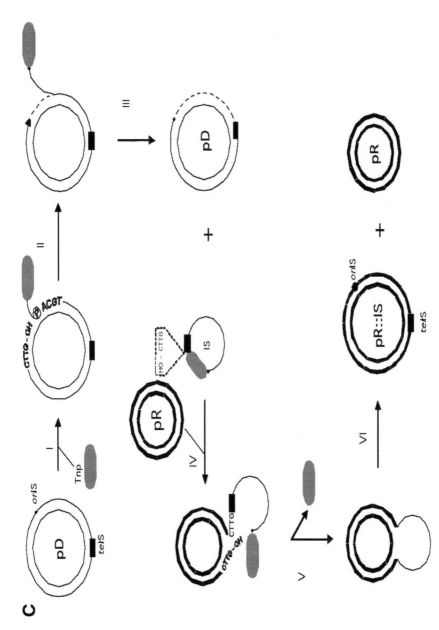

Fig. 1. Mechanisms of bacterial transposition. (A) Conservative transposition by "cut-and-paste" (*see* **Subheading 3.1.1.** for further explanation); (B) replicative transposition via "Shapiro intermediate" (*see* **Subheading 3.3.1.**); (C) RC transposition (*see* **Subheading 3.3.1.**).

transposase proteins are boosted when needed for the second step of the transposition, namely insertion of IS*911* into a new site.

Several IS elements that normally transpose in a conservative manner can also generate what are termed transposition cointegrates (**Fig. 1**). These structures are fusions of the transposition donor and the target molecule, and have two copies of the element in direct repeat, one at each replicon junction. They arise from an alternative, replicative transposition mechanism that generates what is termed a "Shapiro intermediate" (**Fig. 1**; *see* **Subheading 3.3.1.**). IS*1* can generate cointegrates and simple insertions. The cointegrates are relatively stable, particularly when formed and maintained in a *recA* background (i.e., in a host that is unable to carry out homologous recombination).

The formation of transposition cointegrates has been exploited to detect and isolate new transposable elements. An otherwise nonmobilizable probe plasmid carrying a resistance marker (e.g., a cloning vector) is introduced into the test strain, which carries a conjugative plasmid suspected of harboring a transposable element. The construct is then mated with an appropriately drug-sensitive recipient strain, and transconjugants expressing the resistance marker on the probe plasmid are selected. Such derivatives are recovered only if the probe plasmid is transferred to the recipient strain, which requires that it be joined to the conjugative plasmid. Cointegrate formation achieves this. The probe and conjugative plasmid may separate subsequent to transfer, a process termed *cointegrate resolution.* The products of the resolution process are the conjugative plasmid and a derivative of the probe plasmid carrying one of the two copies of the transposable element from the cointegrate. The copy of the element on the probe plasmid can then be further analyzed by standard molecular genetic analysis.

Transposition of IS*91*-like elements (IS*91,* IS*801,* and IS*1294*) in Gram-negative bacteria requires replication of these elements, but does not involve formation of Shapiro complexes. These elements are distinctive in that they lack terminal IRs, suggesting that the ends are processed differently *(71,72).* They display target-site specificity, and transposition does not create target-site duplications *(71,78,79),* although inserts may be flanked by short direct repeats. Each element has a highly conserved sequence at one end that resembles the leading-strand replication origins (ls-origin) of a number of plasmids found in Gram-positive bacteria that replicate by rolling circle (RC) replication *(33,71).* This mechanism generates a single-strand plasmid intermediate template, which is then converted to the more conventional double-stranded form. The other end of each element is also highly conserved, with a region of dyad symmetry preceding a tetranucleotide sequence identical to the preferred target site. The tetranucleotide delimits the end of the element. The sequence arrangement at this end of the element resembles a Rho-independent transcription termination signal. Transposition inserts the IS element on the 3' side of the target sequence, so for the next transposition the same short sequence is present at both ends (one originating from the carrier DNA, the other from the element itself). However, whereas the old target site is separated from the element on transposition, the "target-site sequence" that defines the "trailing" end is retained, and the element remains intact. Each IS91-like element has a single ORF that occupies most of the sequence and encodes a protein that clearly belongs to the family of RC-plasmid replication proteins (REPs) *(80)*. This and other findings suggest that transposition of IS91 and related elements involves RC replication and, probably, a transient single-stranded DNA intermediate *(71,72)* (**Fig. 1**). Single copies of these elements can translocate markers adjacent to the trailing end of the element and can mediate replicon fusions *(81)*.

### 3.2. Composite Transposons

As already discussed, a number of IS elements are able to function in pairs to mediate transposition of DNA segments caught between them (**Table 1**). These modular structures, which have a copy of the same IS element at each end, are called composite transposons, and have been found in Gram-positive and Gram-negative bacteria. In most cases, the two copies of the IS element form terminal IRs, although in a few composite transposons they form terminal direct repeats. The majority of known composite transposons encode resistance to one or more antibiotics (**Table 1**). Such composite structures transpose because each is seen by the transposition mechanism simply as a "stretched" version of the IS element. The IS element is recognized by its transposase via the short-terminal IRs that define the limits of the element. The composite structure, necessarily delimited by the same short IRs, is essentially indistinguishable from the element itself, except that it is larger. An interesting quirk of composite transposons is that, because transposition needs only a pair of IRs, when they are part of a circular molecule an event called "inside-out" transposition can occur. Instead of recognizing the outside IR of each terminal IS element of the transposon, the transposase recognizes the inside pair of IRs adjoining the central domain. Transposition using these IRs disconnects the IS elements from the central section that characterized the original transposon, which is replaced with the rest of the replicon. Although composite transposons are conveniently considered as a group because of their common construction, they may not be phylogenetically related, and it is important to remember that just as individual IS elements may transpose by different mechanisms (*see* **Subheading 3.1.1.**), transposition mechanisms of composite transposons may also differ.

**Table 1**
**Some Bacterial Composite Transposons**[a]

| Transposon | Size (kb) | Terminal elements | IS family | Target-site duplication | Marker(s)[b] |
|---|---|---|---|---|---|
| | | *Gram-negative bacteria* | | | |
| Tn5 | 5.7 | IS50 (IR) | IS4 | 9 | KmBlSm |
| Tn9 | 2.5 | IS1 (DR) | IS1 | 9 | Cm |
| Tn10 | 9.3 | IS10 (IR) | IS4 | 9 | Tc |
| Tn903 | 3.1 | IS903 (IR) | IS5 | 9 | Km |
| Tn1681 | 4.7 | IS1 (IR) | IS1 | 9 | HST |
| Tn2350 | 10.4 | IS1 (DR) | IS1 | 9 | Km |
| Tn2680 | 5.0 | IS26[c] | IS6 | 8 | Km |
| | | *Gram-positive bacteria* | | | |
| Tn3851 | 5.2 | NR | NR | NR | GmTbKm |
| Tn4001 | 4.7 | IS256 (IR) | IS256 | 8 | GmTbKm |
| Tn4003 | 3.6 | IS257 (DR) | IS6 | 8 | Tm |

[a]DR, direct repeat; IR, inverted repeat; NR, not reported; HST, heat-stable toxin.
[b]Resistance phenotype: Bl, bleomycin; Cm, chloramphenicol; Gm, gentamicin; Km, kanamycin; Sm, streptomycin; Tb, tobramycin; Tc, tetracycline; Tm, trimethoprim.
[c]IS26 and IS6 are synonyms.

One recently discovered manifestation of composite transposon formation concerns genes encoding CTX-M β-lactamases, which are found on a number of plasmids in some cefotaxime/ceftazidime-resistant clinical isolates of species such as *Klebsiella pneumoniae*. These plasmid-associated extended spectrum β-lactamase genes have emerged throughout the world in recent years as threats to chemotherapeutic efficacy, and are thought to have originated from the chromosomes of *Kluyvera* spp. They have been mobilized from *Kluyvera* chromosomes in the form of composite transposons by the IS1380-like element ISEcp1 (**27**). Unusually, however, these composite elements contain only one copy of ISEcp1 as a terminal sequence, which provides transposition function and the IR sequence at one end of the transposed sequence. The other terminal IR sequence needed for transposition originates not from a second copy of ISEcp1, but from a sequence that is found fortuitously on the *Kluyvera* chromosome, close to the $bla_{CTX-M}$ gene, where it is mistaken for an ISEcp1 IR sequence by the transposase (**82**). Accordingly, the mobilized $bla_{CTX-M}$ gene is flanked by an intact copy of ISEcp1 and a homolog of its terminal IR sequence, emphasizing the point that the necessary and sufficient components for transposition are a pair of appropriately oriented IR sequences and the cognate transposase.

Some IS elements, notably IS*10*, IS*50*, and IS*903*, were discovered because they form the terminal repeats of composite transposons encoding antibiotic resistance (**Table 1**). IS*10* provides the long terminal IRs of Tn*10*, a transposon that encodes inducible resistance to tetracycline *(83)*, whereas inverted copies of IS*50* constitute the terminal repeats of Tn*5*, which has an operon encoding resistance to kanamycin/neomycin, bleomycin, and streptomycin, although the last is not expressed in some hosts, including *E. coli (84)*. Resistance to kanamycin/neomycin is also encoded by Tn*903*, in which the single resistance gene is flanked by inverted copies of IS*903 (85,86)*. IS*10* and IS*50* both belong to the IS*4* family of elements, while IS*903* is an IS*5*-like element. It is likely that the gene for the SHV β-lactamase found on several R plasmids has been recruited on more than one occasion from the chromosome of *Klebsiella pneumoniae* on composite transposons generated by the IS*6*-like element IS*26* (Matthew Avison, University of Bristol, personal communication).

IS*50* encodes two proteins, Tnp (transposase) and Inh (inhibitor of transposition). These proteins are virtually identical and differ only in that Tnp is 55 amino acids longer than Inh at the N terminus. Tnp and Inh are both primary translation products, each being produced from its own transcript *(87)*. In Tn*10*, transposition is controlled by regulating production of IS*10* Tnp by means of a 69-nt antisense RNA that anneals to the *tnp* mRNA to block translation *(88)*. The *tnp* mRNA and antisense RNA are not equally stable; the former has a half-life of approx 40 s, whereas the antisense RNA has a half-life of approx 70 min. The molar ratio of these RNA species in steady state is approx 20:1 in favor of the antisense RNA, which means that IS*10* Tnp is synthesized at its highest rate immediately after the element is transferred to a new cell (a case of zygotic induction), with synthesis being markedly reduced thereafter. In contrast, IS*903* limits transposition by producing an unstable Tnp that does not accumulate *(89)*. Hence, the frequency of transposition is determined by the rate of transposase production at the time.

### 3.3. Complex Transposons

A complex transposon has a more complicated genetic structure than an IS element or a composite transposon, as genes that have no transposition functions (e.g., antibiotic resistance genes) have been recruited into the body of the element, rather than captured by "block assembly." However, it should be remembered that, just as the term *composite transposon* denotes only that the element has a modular structure with terminal IS repeats, so *complex transposon* implies only that the transposon is neither a composite transposon nor a transposing phage.

### 3.3.1. Tn3 and Related Transposons

The classic complex transposon is Tn*3*, which is derived from resistance (R) plasmid R1 *(70)*. It is essentially the same as Tn*1*, which originates from another R plasmid, RP4, and was the first transposon to be discovered *(41)*. Both elements were originally called Tn*A*, and their transposition functions are fully interchangeable. These transposons encode TEM1 and TEM2 β-lactamases, respectively, that differ by only a single amino acid (a Gln37Lys substitution) and have virtually identical substrate specificities, conferring resistance to some β-lactam antibiotics, notably ampicillin, carbenicillin, and early cephalosporins. Tn*1* and Tn*3* are widely distributed among Gram-negative bacteria from clinical and veterinary sources, notably the Enterobacteriaceae and *Pseudomonas aeruginosa,* and are usually found on plasmids. Tn*3* is more common. In the last few years, many variants (>100) have evolved from both enzymes, including the so-called extended-spectrum TEMs, which can hydrolyze third-generation cephalosporins such as ceftazidime and cefotaxime (which TEM1 and TEM2 do not hydrolyze), and TEM variants that are insensitive to β-lactamase inhibitors, such as clavulanic acid, that efficiently inhibit the parent enzymes (*see* http://www.lahey.org/studies/webt). The speed with which these variants have become established owes much to the widespread distribution of the parent genes on Tn*1* and Tn*3*, and also, perhaps, to the fact that acquisition of Tn*3* can confer a growth advantage both in vitro and in vivo in the absence of antibiotic selection (Enne, V. and Bennett, P., unpublished results).

Tn*3* is typical of an extended family of phylogenetically-related transposons from both Gram-negative and Gram-positive bacteria *(30)*. Among them they carry genes encoding resistance to many antibiotics, to mercuric ions, and catabolism of nutrients (**Table 2**). These elements have different, but clearly similar, short (35–40 bp)-terminal IR sequences and related transposition genes. Each encodes a transposase, TnpA, of approx 1000 amino acids. Most also encode a second recombination enzyme, TnpR, that is a site-specific recombinase of the resolvase/DNA invertase family, which mediates the second step in a two-step transposition mechanism, a reaction called cointegrate resolution. TnpR requires a specific site for action, called *res,* located beside *tnpR (90)*. A few elements, typified by Tn*4430,* encode an alternative cointegrate resolution enzyme, TnpI, that belongs to the phage λ integrase family *(91)*. One member of the Tn*3* family is only 3.2 kb long. IS*1071* looks like a typical IS element in that it consists only of the transposase gene and terminal IRs. The transposase is only distantly related to those encoded by other Tn*3*-like elements *(46)*.

**Table 2**
**Some Tn3-Like Transposons**

| Transposon | Size (kb) | Terminal IRs (bp) | Target (bp) | Marker(s)[a] | Subgroup |
|---|---|---|---|---|---|
| | | *Elements from Gram-negative bacteria* | | | |
| Tn*1* | 4.957 | 38/38 | 5 | Ap | Tn*3* |
| Tn*3* | 4.957 | 38/38 | 5 | Ap | Tn*3* |
| Tn*21* | 19.6 | 35/38 | 5 | HgSmSu | Tn*501* |
| Tn*501* | 8.2 | 35/38 | 5 | Hg | Tn*501* |
| Tn*1000* | 5.8 | 36/37 | 5 | None | Tn*3* |
| Tn*1721*[b] | 11.4 | 35/38 | 5 | Tc | Tn*501* |
| Tn*1722*[b] | 5.6 | 35/38 | 5 | None | Tn*501* |
| Tn*2501* | 6.3 | 45/48 | 5 | None | Tn*3* |
| Tn*3926* | 7.8 | 36/38 | 5 | Hg | Tn*501* |
| | | *Elements from Gram-positive bacteria* | | | |
| Tn*551* | 5.3 | 35 | 5 | Ery | Tn*501* |
| Tn*4430* | 4.2 | 38 | 5 | None | Tn*4430* |
| Tn*4556* | 6.8 | 38 | 5 | None | Tn*4652* |

[a]Ap, ampicillin; Cm, chloramphenicol; Ery, erythromycin; Hg, mercuric ions; Sm, streptomycin; Su, sulphonamide; Tc, tetracycline; xyl, xylose catabolism.
[b]Tn*1721* is a composite structure that utilizes Tn*1722* as its basis for transposition.

Most Tn*3*-related transposons belong to one of two main branches of the family. On each branch the transposition functions of the transposons are more closely related to each other than they are to those of elements of the other branch. The parent element from which Tn*3*-like elements evolved is unknown, but it has been speculated to be IS*1071* **(46)**. The relatively small (5 kb) element Tn*3* is the type element for one branch that includes Tn*1000* (also known as γδ), a cryptic element, and Tn*1546*, an element from *Enterococcus faecium* that encodes resistance to vancomycin **(30)**. The transposition functions on Tn*3*-like transposons occupy approx 4 kb, irrespective of the size of the transposon **(30)**. Tn*21*, the type element for the other main branch, is somewhat larger (20 kb), and is also a resistance transposon conferring resistance to streptomycin, spectinomycin, sulphonamides, and mercuric ions. This last feature is typical of Tn*21*-like elements **(31)**. The Tn*21*-like elements have essentially the same core structures and transposition functions, and differ primarily in the number and type of other resistance determinants they carry. The variation in the antibiotic resistance gene composition of Tn*21* and related transposons is believed to be of recent evolutionary origin, and reflects the activities of integrons and gene cassettes (*see* **Subheading 4.**). In addition to the two main family branches, there are some elements that are Tn*3*-like, but that are not closely related to Tn*3*, to Tn*21*, or to each other *(30)*.

There are two arrangements of transposition genes on Tn*3*-like transposons. On Tn*3* and its close relatives, *tnpA* and *tnpR* are adjacent, but have opposite orientations and are transcribed divergently; *res*, the specific site for cointegrate resolution, is located in the intergenic region. On Tn*501* and closely related elements, these features are arranged in the order *res, tnpR, tnpA,* with the two genes similarly oriented and transcribed away from *res*. In both arrangements the *tnpA* gene is at one end of the element and runs into one of the terminal IRs of the transposon. One consequence of the gene arrangement in Tn*3* is that TnpR, by binding to *res,* functions as a *tnpA* repressor *(92)* as well as a site-specific recombinase; Tn*3* derivatives with null mutations in *tnpR* transpose at markedly elevated frequencies and create cointegrates that are inefficiently resolved in a RecA-proficient host, and are essentially stable in a *recA*-defective background.

The mechanism of transposition of Tn*3*-like elements involves semi-conservative replication and recombination (**Fig. 1**), and generates cointegrates via Shapiro complexes (*see* below). Transpositions create target-site duplications, generally of 5 bp, which result from the target site being opened by staggered cuts across the duplex DNA. The mechanism differs from conservative mechanisms in that the element is only partly disconnected from the donor site. Single-strand cuts are made by the transposase at both ends of the element, on opposite strands of the duplex DNA, to expose 3'-OH groups. These nucleophiles are then used to attack phosphodiester bonds at the target site to execute two transesterification reactions on each strand of the DNA, 5 bp apart. The product is a "Shapiro intermediate," a structure proposed by J. A. Shapiro *(93)* (and independently, at the same time, by A. Arthur and D. Sherratt *[94]*). In this intermediate, the transposon is joined to the target, but remains attached to the donor. The transesterification reactions necessarily generate 3'-OH groups at both of the new junctions, which can be used to prime replication of the transposon by host-cell enzymes. When donor and target sites are on separate replicons, transposition results in replicon fusion producing a circular transposition cointegrate that contains the donor and recipient replicons joined by two copies of the transposon in direct repeat, one at each junction (**Fig. 1**). The cointegrate is normally processed to release two replicons—one a derivative of the recipient DNA molecule carrying a copy of the transposon, and the other indistinguishable from the transposon donor. This process, termed *cointegrate resolution,* is mediated by TnpR (or TnpI), the site-specific recombinase, using the pair of *res* sites on the cointegrate, one on each copy of the transposon (**Fig. 1**). Cointegrate resolution can also be effected by host RecA-dependent recombination, but is less efficient.

This replicative mechanism of transposition requires both ends of the transposon, which are processed in the same way, explaining why the transposon terminals are perfect or near-perfect IRs. The job of the sequences is to bind transposase and direct the cleavage reactions precisely to the ends of the element, preserving both sequences as parts of the transposon in the process. It is surprising, therefore, to find that transposition can still occur when one end of the transposon has been deleted, albeit at a much lower frequency. One-ended transposition is mediated by some but not all Tn*3*-like transposons, and requires one IR sequence and the appropriate transposase to generate what are essentially replicon fusions *(95–97)*. The entire donor replicon, or part thereof, is inserted into the recipient DNA molecule. One consequence of this is that the products of this form of transposition, unlike those with the intact element, are not of uniform size. Inserts in different transposition recombinants form a nested series. One end of all the inserts is the solitary IR sequence. Consistent with the mechanism being transposition, different insertions are flanked by different 5-bp direct repeats. The products are not generated by the replicative transposition mechanism described for Tn*3*-like elements, although components of that mechanism are clearly needed. Rather, the products have the hallmark of rolling-circle replication, as for IS*91*-like elements, where only one end is anchored (**Fig. 1**). How the other end is determined in these aberrant reactions is not known.

Some Tn*3*-like transposons, including Tn*3* but not Tn*21*, are subject to transposition immunity *(98)*. In such cases, the native element will not transpose at a detectable frequency to another DNA molecule that already has a copy of the element. The signal for inhibition is a single copy of the element's terminal IR sequence *(99)*. For Tn*3*, transposition immunity can be largely eliminated by increasing the level of transposase in the cell by a knockout mutation in *tnpR*. Transposition immunity is not restricted to Tn*3*-like elements and has been reported for Tn*7* *(100)* and phage Mu *(101)*, with the mechanism of transposition immunity in the latter having been well studied and characterized.

## 4. Integrons and Gene Cassettes

Tn*21* and related elements form a subgroup of closely related, relatively large (approx 20 kb) transposons that confer resistance to many antibiotics, including β-lactams, aminoglycosides, chloramphenicol, and trimethoprim. One characteristic of Tn*21*-like transposons is that antibiotic resistance genes cluster at essentially the same position in the different elements *(31)*, indicating nonrandom insertion. Given the variety of resistance genes involved (approx 40 to date), the fact that they migrate to the same point within an approx 20-kb sequence suggests that the different sequences have been captured by a site-specific recombination system. This involves elements called *integrons*

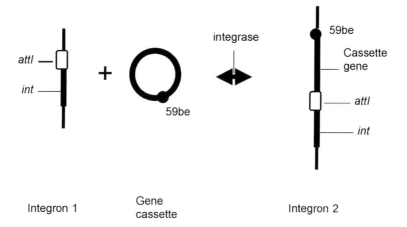

Fig. 2. Bacterial integrons: capture and release of gene cassettes (*see* **Subheading 4.** for further explanation).

*(102)* and small, mobile, gene-sized DNA sequences called *gene cassettes* *(103)*.

An integron is a genetic entity with a site used to capture other DNA sequences (the gene cassettes) by site-specific recombination *(104)*. The recombinase, called *integrase,* is encoded by the *int* gene, which is adjacent to the recombination site *attI*. Since the initial discovery of integrons on Tn*21*-like transposons, several different integron classes, classified according to their integrase genes, have been reported. Tn*21*-like transposons carry class 1 integrons, each of which encodes essentially the same integrase. Many of these integrons accommodate a gene for sulphonamide resistance, *sul1,* and a truncated gene, *qacEΔ1,* which in its intact state encodes resistance to quaternary ammonium compounds. Both genes are believed to be remnants of gene cassettes that have suffered partial deletion, locking what is left of each cassette into the integron, so that *qacEΔ1.sul1* now forms one end of the integron. Hence, the distinctive array of a class 1 integron is *int-attI*–gene cassette(s)–*qacEΔ1-sul1*.

A gene cassette is a genetic element that exists, formally at least, as a nonreplicating DNA circle when moving from one integron to another. It is usually a linear structure found in the chromosome, plasmids, or transposons such as Tn*21*. Gene cassettes are usually 500–1000 bp in size, which is large enough to accommodate a single gene, often an antibiotic-resistance gene *(103)*. In addition, each has what is called a 59 base element (59 be) or *attC* site that is paired with *attI* for integrase-mediated recombination *(105)* (**Fig. 2**). In general, cassette genes lack promoters, but insertion at *attI* places the gene in the correct orientation downstream of a promoter at the end of *int.* When a gene cassette has inserted into *attI,* the recombination site is reformed, and a second and then a third cassette can be inserted sequentially into the integron. Cassettes are separated by their 59 be *attC* sites, which creates new gene arrays that function as operons expressed from the integron promoter.

### 4.1. Classes of Integrons

In principle, there is no limit to the number of gene cassettes that can be captured, but functional constraints, such as reduced expression of genes as they are progressively moved away from the promoter, may limit the number in practice. However, class 1 integrons with eight gene cassettes have been reported *(106)*, and so-called super-integrons, with dozens of gene cassettes, first reported in *V. cholerae (107,108)*, have been found in a several other bacterial species *(109–111)*.

Apart from super-integrons, three other classes of integrons have been characterized. The basic class 1 integron (the class to which most characterized integrons belong) has no gene cassettes and is presumed to have the genetic form *int-attI.* New integrons are created when a gene cassette(s) is inserted at *attI,* or when a gene cassette(s) is deleted from an existing array. Although cassette insertion occurs at *attI,* individual cassettes can be removed from any part of the array by integrase-mediated excision using the 59-be regions that flank the cassette gene *(104).* An excised cassette can be reinserted into the residual integron or into another integron at the relevant *attI* site, or may simply be lost because it fails to reinsert and is unable to replicate autonomously. The movement of gene cassettes in and out of integrons is a random process, and whether a particular arrangement survives or is lost depends primarily on natural selection; when the gene combination is of selective advantage to the cell, it will be established in the microbial population by clonal expansion, at the expense of less fit systems.

The archetype class 2 integron is part of the transposable element Tn7 and includes genes conferring resistance to trimethoprim and streptomycin. The integrase shows less than 50% identity to the class 1 integrases and is nonfunctional owing to a stop codon within the reading frame *(112).* Only one example of a class 3 integron is known *(113).* While most of the genes discovered on gene cassettes in integrons of classes 1 to 3 are antibiotic resistance genes *(103),* it is likely that these represent only a small fraction of all possible cassette genes, reflecting only the keen interest of microbiologists in bacterial drug resistance. In this respect, it is worth noting that while super-integrons also contain gene cassettes carrying antibiotic resistance determinants, most cassettes in these elements do not encode antibiotic resistance.

Integrons and gene cassettes are undoubtedly responsible for much of the drug resistance shown today by clinical bacterial isolates, particularly by members of the Enterobacteriaceae and *P. aeruginosa.* These systems provide a genetic flexibility that has been exploited to the full in the evolution of bacterial drug resistance. That integrons have been found on the bacterial chromosome and on many plasmids, often as part of a transposon, simply underscores this flexibility. It is of interest, in this context, that the genes for the IMP and VIM metallo-β-lactamases, which can protect *P. aeruginosa* and some other Gram-negative bacteria against carbapenems such as imipenem and meropenem, are commonly found as gene cassettes on class 1 integrons *(104,114)*.

## 5. Conjugative Transposons

Members of one class of complex transposable elements, conjugative transposons, have the ability to move, not only from one DNA site to another, but also from one bacterial cell to another. This reflects the fact that these elements transpose by a conservative mechanism that utilizes a free circular intermediate *(12)*, similar to the transposition intermediates of some IS elements such as IS*911,* and that each element encodes a set of functions for cell-to-cell transfer. First described in Gram-positive bacteria *(115)*, elements of this type have also been found in Gram-negative anaerobes, such as *Bacteroides* spp. *(116)*. Elements of this type transpose and promote their own conjugal transfer from one cell to another. They are not plasmids because they lack replication functions and are replicated as part of the carrier replicon. However, they resemble plasmids because they can exist as circular extrachromosomal elements that can be transferred from one cell to another by conjugation.

Conjugative transposons are a subset of a larger group of mobile genetic elements that also includes *integrative "plasmids"* *(117,118)* and genomic islands (*see* **Subheading 8.**) and that use site-specific recombination systems to integrate into and excise from the carrier replicons upon which they rely for replication. Integrative "plasmids," such as R391 and R997 *(118)*, were originally thought to be true plasmids because they support conjugal transfer of their resistance genes, but numerous attempts to isolate them as plasmid DNA failed. This was because they exist primarily as elements integrated into the host cell chromosome and not as autonomous units. Self-transmissible genetic elements that must integrate into replicons to ensure replication, such as conjugative transposons, integrative "plasmids" and genomic islands, have been termed *integrative and conjugative elements,* or *ICEs* *(11)*. One notable ICE is SXT, a mobile genetic element from *Vibrio cholerae* that confers resistance to chloramphenicol, streptomycin, sulphamethoxazole and trimethoprim *(11,119,120)*.

## 5.1. *Tn*916

The type element Tn916 was found in a clinical isolate of *E. faecalis (121)* and confers resistance to tetracycline. Tn916 is 18.4 kb and is essentially identical to Tn918, Tn919, and Tn925, which were recovered independently from different clinical isolates. The related element Tn1545 is 23.5 kb and has genes encoding resistance to kanamycin and erythromycin, as well as tetracycline *(122)*. These transposons promote their own transfer from one bacterial cell to another at frequencies of $10^{-9}$ to $10^{-5}$.

Tn916 is a typical conjugative transposon. More than 50% of the structure is taken up by a block of conjugation functions at one end of the element *(123,124)*. Conjugal transfer of the transposon is consolidated in the recipient by transposition of the element into the chromosome, where it is replicated as part of the chromosome. Transposition genes map to the other end of the element. Knockout mutations in two genes, *xis-Tn* and *int-Tn,* which are needed for transposon excision *(125)*, prevent both transposition and conjugation. In contrast, conjugation functions are not needed for transposition, and elements with damaged transfer systems can still transpose from the chromosome onto a plasmid, although mutations in some conjugation genes do reduce the frequency of this type of event.

Transposition of Tn916-like elements occurs in two stages *(13,121)*. First, staggered double-strand cuts are made at the junctions of the element with the current carrier DNA, generating a DNA fragment with a 5' single-stranded hexanucleotide overhang at each end. The ends are then joined to create a circular form of the element with a 6-bp heteroduplex spacer separating the ends. This may then be reinserted into the host cell chromosome, be transposed onto a plasmid in the same cell, or be conjugally transferred to another cell and then transposed into the chromosome or a resident plasmid. The mechanism appears to be similar to the integration-excision system of phage λ and the transposition proteins are related to the Int and Xis proteins used by phage λ for integration into and excision from the *E. coli* chromosome *(126,127)*.

## 6. Site-Specific Transposition

Most transposable elements transpose into many sites on bacterial chromosomes or plasmids, although bias towards insertion into particular regions of replicons has been reported. For example, IS1 and Tn3 appear to favor insertion into AT-rich tracts, and Tn5, which usually displays very low target-site specificity, has nonetheless been reported to have insertion hotspots. This behavior reflects a bias on the part of the particular transposition system, but what governs the behavior is generally not known *(128)*. In contrast, a few transposable elements have a marked tendency to insert in one particular site. Notable among these is the complex transposon Tn7.

## 6.1. Tn7

Tn7, first identified on R plasmid R483 (37), has two resistance genes, dfrI and aadA, encoding dihydrofolate reductase I and aminoglycoside adenylyltransferase A, respectively, conferring resistance to trimethoprim, and to streptomycin and spectinomycin. These resistance genes comprise gene cassettes that are part of the class 2 integron carried on the transposon. The integrase gene has sustained a nonsense mutation that severely truncates the integrase protein, inactivating it. Accordingly, movement of the gene cassettes requires complementation of the damaged integrase. The transposon is 14 kb in size, generates 5-bp target-site duplications on transposition, and inserts at high frequency in only one orientation into single sites on several bacterial chromosomes, including those of E. coli, K. pneumoniae, and Pseudomonas and Vibrio spp. It also transposes onto plasmids with little or no site specificity, but at a markedly lower frequency (129). High-frequency, site-specific transposition occurs into a plasmid that has been engineered to have a copy of the chromosomal insertion site, which in E. coli is at 84', between phoS and glmS. The transposon inserts into the transcriptional terminator of glmS, about 30 nucleotides from the translation stop codon, which does not interfere with glmS operation (130). The target site where Tn7 is inserted, which determines the 5-bp duplication, and the attachment site, attTn7, which is necessary for locus specificity, are not coincident. The attTn7 sequence is approx 50 bp and includes a small part of glmS, and although the attTn7 sites in different bacterial species are homologs, the insertion sites are not, indicating that they are determined simply by their distance from attTn7. Hence, Tn7 uses sequence data stored in a highly conserved gene, glmS, to identify a suitable insertion site where transposon insertion will cause little or no damage.

Tn7 is also unusual in that it has five transposition genes, tnsA, B, C, D, and E, rather than the one or two seen with most other transposable elements. TnsA, B, and C are required for all transpositions. TnsAB is the transposase and is activated by TnsC. TnsD is needed for site-specific transposition (i.e., attTn7 recognition), but is replaced by TnsE for transposition into randomly chosen sites (131). Whereas for other elements the transposition proteins are multifunctional, those of Tn7 have discrete roles. Nonetheless, they function as a DNA–protein complex. In vitro analysis has revealed that transposition of Tn7 is conservative, with the transposon being disconnected from its carrier by double-stranded cuts at both ends of the element, creating trinucleotide ssDNA overhangs, which come from the carrier DNA, and recessed 3' OH groups, which mark the boundaries of the element. These nucleophiles are directed to attack the target, first causing breakage of phosphodiester bonds on both strands, with a 5-bp stagger, and then ligation of Tn7 into the gap via transesterification reactions. The resulting intermediate structure has the transposon flanked by very short single-strand gaps and 3-bp heteroduplexes formed from the trinucleotide overhangs attached to the transposon and one strand of the 5-bp target site. It is presumed that the short stretches of mismatched sequences are removed and that DNA duplex continuity is restored by the gap repair mechanism of the cell.

Site-specific Tn7 transposition in vitro involves a DNA–protein complex of transposon donor DNA, target DNA, transposition proteins TnsA, B, C, and D, and ATP. TnsC is an ATP-binding protein. Cell-free transposition is authentic in the sense that Tn7 transposes into a target plasmid with *att*Tn7 in a site- and orientation-specific manner *(132,133)*. The process is also subject to transposition immunity, as is Tn7 transposition in vivo *(100)*. While the terminal IRs of Tn7 are 30 bp, sequences 75 bp from one end and 150 bp from the other are needed for transposition. Each terminal sequence contains multiple imperfect repeats of a 22-bp sequence that has been shown to bind TnsB. These are thought to be nucleation centers for the assembly of the transposition complex.

### 6.2. Tn502

Other elements that show insertion specificity are Tn*502*, a 9.6-kb transposon found in Gram-negative bacteria, that encodes resistance to mercuric ions, and Tn*554*, a transposon from *Staphylococcus aureus* that encodes resistance to erythromycin and spectinomycin. Tn*502,* like Tn7, displays site and orientation specificity, but inserts into plasmid RP1 (RP4) rather than the bacterial chromosome *(134)*. When the insertion site is deleted, Tn*502* can transpose into many sites on the remainder of the plasmid at a significantly lower frequency, a behavior pattern that also resembles that of Tn7. Although it encodes mercury resistance, Tn*502* is not obviously related to Tn3-like transposons such as Tn*21* and Tn*501,* which also encode this resistance. Unfortunately, Tn*502* is still poorly characterized and there is insufficient information to say whether it is related to other site-specific elements.

### 6.3. Tn554

Tn*554* transposes primarily into a site designated *att*Tn*554* on the *S. aureus* chromosome in one orientation only. Five genes have been identified on the element: two resistance genes, *ery* and *spc,* and a three-gene transposition operon, *tnsABC (135)*. The products of transposition differ from those of most transposable elements in that a small part of the carrier DNA is transposed with the element *(136)*. These few additional base pairs are located at one end of the element when in the donor site, and are transferred to the other end of the element after transposition. Accordingly, a different short sequence is carried into the target site with each sequential transposition. The available data indicate a cut-and-paste mechanism via a circular intermediate, but further work is needed to consolidate this interpretation.

## 7. Transposing Bacteriophages

Transposable elements such as IS elements and transposons have no existence independent of the replicons that carry them, circular intermediates notwithstanding, since these are likely to be transient and remain complex bound. Some transposable elements, however, do have an independent state. For some phages, such as the *E. coli* temperate phage Mu, transposition is an essential aspect of their life cycles *(29)*. Bacteriophage Mu lysogenizes *E. coli* by inserting into the bacterial chromosome in an essentially random manner, using a conservative cut-and-paste transposition strategy. When induced to replicate, it does so by multiple rounds of replicative transposition, via Shapiro intermediates, into many sites on the chromosome. The former process was clearly established by exploiting the host-cell DNA methylation system. Phage particles were produced in a *dam⁺* host, where the phage DNA was fully methylated on both strands. These were then used to infect a *dam*-defective host in which there is no DNA methylation. The newly integrated phage DNA was found to be fully methylated, indicating that transposition had occurred without replication, rather than being hemimethylated, as would have resulted from a replicative mechanism *(137)*.

The linear genome of Mu is 37 kb, but phage particles have DNA molecules of 39 kb, with the extra length being derived from the host. The phage genome is packaged by a "head-full" mechanism, starting from one end. The end of the phage DNA that is specifically recognized and released by cutting the carrier DNA 150 bp from the end is the one nearest the phage's replication/transposition genes. Approximately 39 kb is packed into the phage head, and the DNA is then cut once more to release a copy of the phage genome flanked by two small regions of host DNA, 150 bp on one side and approx 2 kb on the other. The genome structure of Mu is unremarkable, showing the usual array of genes for phage construction (i.e., for head proteins, tail proteins, and so on). In addition, it has two replication genes, *A* and *B*, located close to one end (diagrammatically usually placed on the left). Several years after its discovery, it was realized that Mu replicates by transposition, and subsequent investigations into Mu replication in vitro have been invaluable to our understanding of the form of replicative transposition that involves Shapiro intermediates. Because replication of Mu, initially at least, involves intramolecular transposition events, gross molecular rearrangements result, which are most dramatically illustrated by the fragmentation of the chromosome into many smaller circles of DNA, each of which carries at least one copy of Mu. This kills the cell, and is an inevitable consequence of multiple rounds of indiscriminate intramolecular transposition. Unlike Tn*3*-like elements, which transpose by essentially the same mechanism, Mu lacks a cointegrate resolution system. Indeed, such a system would be superfluous, given that the purpose of replication is simply to create a bank of phage genomes that are processed and packaged into mature phage particles. It is of no consequence that cointegrates are not deliberately resolved, as the phage genome is released by other means, and lack of cointegrate resolution is no bar to further rounds of transposition *(29)*.

Mu was the first transposing bacteriophage to be discovered *(34)*. D108, which closely resembles Mu, is the only other transposing coliphage to have been described *(137)*. The two share 90% homology, with the principal regions of divergence being the ends of the genomes. One notable difference is at the 5'-ends of the A-genes, which results in poor complementation. Other transposing phages have been found in *Pseudomonas* spp., but although the sizes of their genomes are strikingly similar to that of Mu, there appears to be little or no homology with it. In *Vibrio* spp. the frequency of auxotrophs among certain phage lysogens is significantly raised over the normal frequency, findings reminiscent of the early results with *E. coli* and Mu.

## 8. Genomic Islands

Analysis of bacterial genome data indicates that bacterial chromosomes are not homogeneous, i.e., sequences on a particular chromosome do not all derive from the same phylogenetic source. Punctuating the genome core, which accommodates the housekeeping genes that are essential for growth and reproduction (these display common codon usage indicative of coevolution), are sequences with different G+C contents (indicating different evolutionary descents) *(5)*. These sequences, termed *genomic islands,* are most likely to have been acquired by horizontal transfer.

A subset of genomic islands, first detected in human pathogens and subsequently in animal and plant pathogens, contains genes encoding virulence determinants; these are called pathogenicity islands (PAIs) *(138)*. The origins of some PAIs are rather more evident than others, in that they are clearly lysogenic phages (such as those that encode the cholera and shiga toxins of *V. cholerae* and *S. dysenteriae,* respectively *[7]*) and transposable elements. PAIs accommodate genes encoding virulence determinants such as adherence factors, invasion factors, iron-uptake systems, protein-secretion apparatus, and bacterial toxins. They tend to be relatively large sequences of 10–200 kb and are often flanked by short direct repeats, suggesting involvement of transposable elements or site-specific recombination systems in their capture; they often house genes encoding proteins related to transposases and integrases, as well as IS elements or remnants of such elements. PAIs are defined by the following characteristics, although an individual PAI may not satisfy all of them *(139)*:

- Presence in pathogenic strains and absence/sporadic presence in less pathogenic strains of the same or a related species.
- Occupation of large chromosomal regions (often > 30 kb) and carriage of virulence genes (often many).
- tRNA genes, direct repeats, and/or insertion sequences at boundaries; (sometimes cryptic) mobility genes and sequences (e.g., insertion sequences, integrase and transposase genes, plasmid origins of replication) present within the element.
- Different G+C content from the remainder of the chromosomal DNA of the host bacteria.
- Found as compact, distinct units.
- Instability (prone to deletion).

A few PAIs are mobile, and in some instances movement is undoubtedly mediated by phage; examples would include the PAIs of *S. aureus (140)* and *V. cholerae (141)*. It is the possession of particular PAIs that distinguishes uropathogenic, enteropathogenic, and enterohemorrhagic strains of *E. coli* from each other and from avirulent ones *(138)*. PAIs are often located next to a tRNA gene, and the boundaries of the element may be marked by short direct repeats indicative of site-specific recombination. The presence of integrase genes on PAIs also points to acquisition by site-specific recombination, mediated perhaps by phage-like integrases. In some cases there is experimental evidence to support this view *(141,142)*.

Other genomic islands harbor antibiotic resistance genes. One important example is the methicillin resistance gene, *mecA,* in methicillin-resistant *S. aureus* (MRSA). It encodes a new penicillin-binding protein (PBP2') that shows considerably less affinity for most β-lactams than the normal complement of PBPs of *S. aureus.* The *mecA* gene is found in genomic islands of 21–67 kb *(143,144)* designated SCC*mec* (staphylococcal cassette chromosome *mec*) elements *(145)* that are inserted into specific sites on the *S. aureus* chromosome. Insertion and excision of these genomic islands is mediated by a site-specific recombination system encoded by two genes, *ccrA* and *ccrB*, closely linked to *mecA*. A mobile genetic element, closely related to the SCC*mec* elements, but lacking *mecA*, has been identified in a methicillin susceptible strain, *Staphylococcus hominis* GIFU12263. Designated SCC$_{12263}$, this element is a possible precursor of SCC*mec* elements and may serve as a vehicle for transfer of various genetic markers between staphylococcal species *(146)*.

## 9. Consequences of Transposition in Bacteria

The direct consequence of transposition is the insertion of one DNA sequence into another, which may disrupt a gene at the point of insertion, causing loss of gene function. Indeed, it was such a consequence that led to the discovery of transposable elements in bacteria *(50–52)*. The mutagenic activities of transposable elements have been successfully exploited in bacterial genetic analyses to locate genes of interest. A number of transposable elements have been used, and several have been engineered with considerable ingenuity to provide genetic tools that can be employed not only to discover genes of interest, but also to analyze their expression and to study gene product structure. Notable among these is the mini-Tn5 family of synthetic transposons *(44)* and the mini-Mu system, devised to enable in-vivo cloning *(29,147,148)*. Several ingenious "suicide systems" have also been devised to deliver transposons into cells, where they are established only if they transpose onto a resident DNA molecule, because the delivery vector itself fails to replicate after transfer. Since such transposition events are essentially random with respect to choice of target site, insertions may occur in many different genes and at many different sites within the same gene. The difficulty in practice is often that of identifying the desired insertion(s) among the many that are of little or no interest.

Transposable elements mediate DNA rearrangements other than straightforward self-insertion. A site at which an active transposable element is located may be a hotspot for deletions. These start precisely at one end of the element (either end is possible, but only one at a time) and extend for varying distances into the adjacent DNA. Inversions of DNA sequences adjacent to the element may also occur. Such rearrangements are usually accompanied by duplication of the transposable element on the DNA molecule. These events are random and infrequent, and the rearrangements reflect further rounds of transposition of the element into adjacent sites on the same DNA molecule. This ability of transposable elements to cause DNA rearrangement was detected early in the studies of the first bacterial transposable elements to be discovered *(54)*. For elements such as Tn3 and bacteriophage Mu, which transpose via Shapiro intermediates, intramolecular transposition automatically generates deletions and inversions, dependent only on how the 3'-OH ends of the element attack the linked target. If each initiates nucleophilic attack on a phosphodiester bond on the same strand as itself, then the transposition event will result in deletion of the DNA sequence between the "old" and "new" sites and the newly created copy of the transposon. The deleted DNA is released as a nonreplicating circular DNA species in which the ends of the lost DNA are joined to the copy of the transposon created by the transposition. In contrast, if each 3'-OH group attacks a phosphate group on the opposite DNA strand, then the transposition will create a duplicate but inverted copy of the transposon on the DNA molecule, and the two sections of the carrier DNA separated by the duplicate copies of the element will be inverted with respect to each other.

Many other transposable elements, including those thought to transpose by conservative cut-and-paste mechanisms, generate adjacent deletions and inversions. Precisely how this is achieved is poorly understood. In the case of some IS elements, some of these events may reflect homologous recombination across IS element duplications; single crossover events involving two directly repeated elements will release the intervening DNA sequence together with one copy equivalent of the element, in much the same way that bacteriophage λ is excised from the *E. coli* chromosome to initiate phage replication by site-specific recombination *(126,127)*. If, on the other hand, the recombination were to involve inverted copies of the element, then the DNA sequence between them will be inverted. An analogy for this would be the invertible segments of phages Mu and P1 that control host range *(149,150)* and the mechanism of flagella phase variation in *Salmonella* Typhimurium *(151)*, although once again these particular DNA inversions involve site-specific recombination systems and not homologous recombination. Hence, duplicate copies of a transposable element on a DNA molecule are loci of intrinsic instability and DNA rearrangement.

Transposable elements can also act as fusion points for DNA molecules. If two circular DNA molecules each carry a copy of the same transposable element, then homologous recombination between them can result in molecular fusion to create a single circular molecule essentially identical to a transposition cointegrate structure, i.e., one in which the two replicons are joined via a directly repeated copy of the element at each replicon junction. The first resistance plasmids isolated, typified by R1 and NR1 *(152,153)*, have such structures; however, it is impossible to determine retrospectively how these constructions were formed, i.e., by recombination between copies of the same element on independent replicons or by a transposition event. Suffice it to say, the arrangement does permit specific fragmentation of the molecules *(153,154)*. Replicon fusion, be it the result of a replicative transposition mechanism or of homologous recombination, offers the opportunity for conduction, that is, conjugal transfer of a nonconjugative DNA molecule by virtue of its attachment, possibly temporarily, to a conjugative element. Both sequences are transferred to the recipient as parts of a single DNA molecule, which may then fragment to release either the original two molecules (when the fusion involved recombination across homologous sequences) or one original and a derivative of the second now carrying a copy of the transposable element responsible for the fusion (when fusion reflected formation of a transposition cointegrate). Conduction systems have been used to study cryptic transposable elements, notably IS elements, using linked plasmid markers to follow the DNA rearrangements, e.g., IS*91 (61,72)*.

## 10. Concluding Remarks

The ability of transposable elements to insert into and to mediate gross molecular rearrangements of DNA can account for much of the macromolecular rearrangement that has been discovered among related bacterial plasmids, and provide the mechanism whereby some plasmids are able to integrate into other replicons, as when the F factor (and other conjugative plasmids) inserts into the chromosome of *E. coli* to create an Hfr strain *(155)*. Bacterial chromosomes are subject to the same rearrangements by the same means. It is now clear that many if not all bacterial chromosomes are molecular hybrids constructed from DNA segments that derive from diverse phylogenetic sources, giving these structures a mosaic appearance. Further, chromosomes are dynamic structures that are subject to continuous assault by all the mechanisms, physical and genetic, that can alter DNA molecules. Nonetheless, chromosomal rearrangement appears to be less frequent than the abundance of transposable elements in bacteria would lead us to expect. Indeed, if chromosome rearrangement were significantly more common than has been found, then it is likely that bacterial species boundaries would be somewhat more difficult to define than they are. However, this doesn't necessarily mean that the frequencies at which such rearrangements occur is less than expected; rather it may simply reflect a somewhat low incidence of survival of the products of such rearrangements. One possible reason for this is that many such rearrangements are either immediately lethal to the cell, because they involve loss of or damage to a vital function, or because the derivatives are less fit for survival in competition with their progenitor, so they fail to establish in the microbial flora. Survival fitness reflects not only the complement of genes possessed by the microbe but also the temporal expression of those genes and the balance of activities of the gene products. It is reasonable to believe that, over the prolonged period of time that constitutes the evolutionary history of a microbe, the balance of gene expression has been honed to optimize reproductive capability and survival characteristics in the environmental niche(s) occupied by the microbe. Accordingly, major genetic change is likely to upset this balance of gene expression to the detriment of the microbe. Although the change may appear small, the effect may be insidious, blunting the microbe's ability to compete in a variety of situations. Large deletion and inversion rearrangements are likely to be disadvantageous to the microbe in a competitive situation, while the consequences of gene acquisition by insertion will depend on the precise location of insertion and the balance of advantage/disadvantage imposed by the additional genetic information. New sequences offer the opportunity of new growth potential; e.g., acquisition of antibiotic resistance genes allows the microbe to survive, perhaps even thrive, in situations where the antibiotic-sensitive progenitor would perish; however, large-scale gene acquisition may not be without risk to the microbe, simply because the optimum balance of gene expression is inevitably altered by the introduction of the new genes. If they do not confer any immediate advantage, the nascent microbe may not be sufficiently competitive to survive long enough to exploit its new acquisitions. This aspect of microbial physiology/bacterial genetics deserves more attention than it has received to date. How the environmental fitness of a microbe is determined is poorly understood, yet it is clearly a question of major importance in relation to microbes that impact upon human health and well-being.

## References

1. Deng, W., Burland, V., Plunkett III, G., Boutin, A., Mayhew, G. F., Liss, P., et al. (2002) Genome sequence of *Yersinia pestis* KIM. *J. Bacteriol.* **184**, 4601–4611.
2. Welch, R. A., Burland, V., Plunkett III, G., Redford, P., Roesch, P., Rasko, D., et al. (2002) Extensive mosaic structure revealed by the complete genome sequence of uropathogenic *Escherichia coli. Proc. Nat. Acad. Sci. USA* **99**, 17,020–17,024.
3. Hayashi, T., Makino, K., Ohnishi, M., Kurokawa, K., Ishii, K., Yokoyama, K., et al. (2001) Complete genome sequence of enterohemorrhagic *Escherichia coli* O157:H7 and genomic comparison with a laboratory strain K12. *DNA Res.* **8**, 11–22.
4. Parkhill, J., Dougan, G., James, K. D., Thomson, N. R., Pickard, D., Wain, J., et al. (2001) Complete genome sequence of a multiple drug resistant *Salmonella enterica* serovar Typhi CT18. *Nature* **413**, 848–852.
5. Lawrence, J. G. and Roth, J. R. (1999) Genomic flux: genome evolution by gene loss and acquisition. In *Organization of the Prokaryotic Genome* (Charlebois, R. L., ed.), ASM Press, Washington, DC, pp. 263–289.
6. Waldor, M. and Mekalanos, J. (1996) Lysogenic conversion by a filamentous phage encoding cholera toxin. *Science* **272**, 1910–1914.
7. Waldor, M. K. (1998) Bacteriophage biology and bacterial virulence. *Trend Microbiol.* **6**, 295–297.
8. Thomas, C. M., ed. (2000) *The Horizontal Gene Pool: Bacterial Plasmids and Gene Spread*, Harwood Academic Publishers, The Netherlands.
9. Syvanen, M. and Kado, C. I., eds. (1998) *Horizontal Gene Transfer*, Chapman & Hall, London.
10. Broda, P., ed. (1979) *Plasmids*, W. H. Freeman & Co., Oxford.
11. Burrus, V., Pavlovic, G., Decaris, B., and Guédon, G. (2002) Conjugative transposons: the tip of the iceberg. *Mol. Microbiol.* **46**, 601–610.
12. Salyers, A. A., Shoemaker, N. B., Stevens, A. M., and Li, L. Y. (1995) Conjugative transposons: an unusual and diverse set of integrated gene transfer elements. *Microbiol. Rev.* **59**, 579–590.
13. Scott, J. R. and Churchward, G. G. (1995) Conjugative transposition. *Ann. Rev. Microbiol.* **49**, 367–397.

14. Zechner, E. L., de la Cruz, F., Eisenbrandt, R., Grahn, A. M., Koraimann, G., Lanka, E., et al. (2000) Conjugative-DNA transfer processes, in *The Horizontal Gene Pool: Bacterial Plasmids and Gene Spread* (Thomas, C. M., ed.), Harwood Academic Publishers, Amsterdam, pp. 87–174.

15. Wilkins, B. M. (1995) Gene transfer by bacterial conjugation: diversity of systems and functional specializations, in *Society for General Microbiology Symposium 52, Population Genetics of Bacteria* (Baumberg, S., Young, J. P. W., Wellington, E. M. H., and Saunders, J. R., eds.), Cambridge University Press, pp. 59–88.

16. Masters, M. (1996) Generalized transduction, in *Escherichia coli and Salmonella: Cellular and Molecular Biology, 2nd ed.* (Neidhardt, F. C., ed.), ASM Press, Washington, DC, pp. 2421–2441.

17. Weisberg, R. A. (1996) Specialized transduction, in *Escherichia coli and Salmonella: Cellular and Molecular Biology, 2nd ed.* (Neidhardt, F. C., ed.), ASM Press, Washington, DC, pp. 2442–2448.

18. Dubnau, D. (1999) DNA uptake in bacteria. *Ann. Rev. Microbiol.* **53,** 217–244.

19. Griffith, F. (1928) Significance of pneumococcal types. *J. Hyg.* **27,** 113–159.

20. Lorenz, M. G. and Wackernagel, W. (1994) Bacterial gene transfer by natural genetic transformation in the environment. *Microbiol. Rev.* **58,** 563–602.

21. Kowalczykowski, S. C., Dixon, D. A., Eggleston, A. K., Lauder, S. D., and Rehrauer, W. M. (1994) Biochemistry of homologous recombination in *Escherichia coli. Microbiol. Rev.* **58,** 401–465.

22. Dowson, C. G., Coffey, T. J., and Spratt, B. G. (1994) Origin and molecular epidemiology of penicillin-binding-protein-mediated resistance to β-lactam antibiotics. *Trend. Microbiol.* **2,** 361–366.

23. Stanisich, V. A., Bennett, P. M., and Ortiz, J. M. (1976) A molecular analysis of transductional marker rescue involving P-group plasmids in *Pseudomonas aeruginosa. Mol. Gen. Genet.* **143,** 333–337.

24. Craig, N. L. (1996) Transposition, in *Escherichia coli and Salmonella: Cellular and Molecular Biology, 2nd ed.* (Neidhardt, F. C., ed.), ASM Press, Washington, DC, pp. 2339–2362.

25. Davies, D. R., Goryshin, I. Y., Reznikoff, W. S., and Rayment, I. (2000) Three-dimensional structure of the Tn5 synaptic complex transposition intermediate. *Science* **289,** 77–85.

26. Savilahti, H., Rice, P. A., and Mizuuchi, K. (1995) The phage Mu transpososome core: DNA requirements for assembly and function. *EMBO J.* **14,** 4893–4903.

27. Chandler, M. and Mahillon, J. (2002) Insertion sequences revisited, in *Mobile DNA II* (Craig, N. L., Craigie, R., Gellert, M., and Lambowitz, A. M., eds.), ASM Press, Washington, DC, pp. 305–366.

27a. Chandler, M. and Mahillion, J. (2000) Insertion sequence nomenclature. *ASM News* **66,** 324.

28. Mahillon, J. and Chandler, M. (1998) Insertion sequences. *Microbiol. Mol. Biol. Rev.* **62,** 725–774.

29. Toussaint, A. and Résibois, A. (1983) Phage Mu: transposition as a life-style, in *Mobile Genetic Elements* (Shapiro, J. A., ed.), Academic Press, New York, pp. 103–158.

30. Grindley, N. D. F. (2002) The movement of Tn3-like elements: transposition and cointegrate resolution, in *Mobile DNA II* (Craig, N. L., Craigie, R., Gellert, M., and Lambowitz, A. M., eds.), ASM Press, Washington, DC, pp. 272–304.

31. Grinsted, J., de la Cruz, F., and Schmitt, R. (1990) The Tn21 subgroup of bacterial transposable elements. *Plasmid* **24,** 163–189.

32. Sherratt, D. (1989) Tn3 and related transposable elements: site-specific recombination and transposition, in *Mobile DNA* (Berg, D. and Howe, M., eds.), ASM Press, Washington, DC, pp. 163–184.

33. Garcillán-Barcia, M. P., Bernales, I., Mendiola, V., and de la Cruz, F. (2002) IS91 rolling-circle transposition, in *Mobile DNA II* (Craig, N. L., Craigie, R., Gellert, M., and Lambowitz, A. M., eds), ASM Press, Washington, DC, pp. 891–904.

34. Taylor, A. L. (1963) Bacteriophage-induced mutations in *E. coli. Proc. Nat. Acad. Sci. USA* **50,** 1043–1051.

35. McClintock, B. (1956) Controlling elements and the gene. *Cold Spring Harb. Symp. Quant. Biol.* **21,** 197–216.

36. Hirsch, H. J., Saedler, H., and Starlinger, P. (1972) Insertion mutations in the control region of the galactose operon of *E. coli* II. Physical characterization of the mutations. *Mol. Gen. Genet.* **115,** 266–276.

37. Barth, P. T., Datta, N., Hedges, R. W., and Grinter, N. J. (1976) Transposition of a deoxyribonucleic acid sequence encoding trimethoprim and streptomycin resistances from R483 to other replicons. *J. Bacteriol.* **125,** 800–810.

38. Berg, D. E., Davies, J., Allet, B., and Rochaix, J.-D. (1975) Transposition of R factor genes to bacteriophage λ. *Proc. Nat. Acad. Sci. USA* **72,** 3628–3632.

39. Foster, T. J., Howe, T. G. B., and Richmond, K. M. V. (1975) Translocation of the tetracycline resistance determinant from R100-1 to the *Escherichia coli* K12 chromosome. *J. Bacteriol.* **124,** 1153–1158.

40. Kleckner, N., Chan, R. K., Tye, B.-K., and Botstein, D. (1975) Mutagenesis by insertion of a drug-resistance element carrying an inverted repetition. *J. Mol. Biol.* **97,** 561–575.

41. Hedges, R. W. and Jacob, A. (1974) Transposition of ampicillin resistance from RP4 to other replicons. *Mol. Gen. Genet.* **132,** 31–40.

42. Reznikoff, W. S. (2002) Tn5 transposition, in *Mobile DNA II* (Craig, N. L., Craigie, R., Gellert, M., and Lambowitz, A. M., eds.), ASM Press, Washington, DC, pp. 403–422.

43. Haniford, D. B. (2002) Transposon Tn10, in *Mobile DNA II* (Craig, N. L., Craigie, R., Gellert, M., and Lambowitz, A. M., eds.), ASM Press, Washington, DC, pp. 457–483.

44. Lorenzo, V de, Herrero, M., Jakubzik, U., and Timmis, K. N. (1990) Mini-Tn5 transposon derivatives for insertion mutagenesis, promoter probing, and chromosomal insertion of cloned DNA in Gram-negative bacteria. *J. Bacteriol.* **172,** 6568-6572.

45. Allmeier, H., Cresnar, B., Greck, M., and Schmitt, R. (1992) Complete nucleotide sequence of Tn1721: gene organization and a novel gene product with features of a chemotaxis protein. *Gene* **111,** 11–20.

46. Nakatsu, C., Ng, J., Singh, R., Straus, N., and Wyndham, C. (1991) Chlorobenzoate catabolic transposon Tn*5271* is a composite class I element with flanking class II insertion sequences. *Proc. Nat. Acad. Sci. USA* **88,** 8312–8316.

47. Bennett, P. M. (1989) Bacterial transposons and transposition: flexibility and limitations, in *Genetic Transformation and Expression* (Butler, L. O., Harwood, C., and Moseley, B. E. B., eds.), Intercept, Andover, UK, pp. 283–303.

48. Dobritsa, A. P., Dobritsa, S. V., Popov, E. I., and Fedoseeva, V. B. (1981) Transposition of DNA fragment flanked by two inverted Tn*1* sequences. *Gene* **14,** 217–225.

49. Lederberg, E. M. (1981) Plasmid reference centre registry of transposon (Tn) allocations through July 1981. *Gene* **16,** 59–61.

50. Jordan, E., Saedler, H., and Starlinger, P. (1968) 0° and strong polar mutations in the gal operon are insertions. *Mol. Gen. Genet.* **102,** 353–365.

51. Shapiro, J. A. (1969) Mutations caused by the insertion of genetic material into the galactose operon of *E. coli. J. Mol. Biol.* **40,** 93–105.

52. Malamy, M. H. (1970) Some properties of insertion mutations in the *lac* operon, in *The Lactose Operon* (Beckwith, J. R. and Zipser, D., eds.), Cold Spring Harbor Laboratory, p. 359.

53. Habermann, P. and Starlinger, P. (1982) Bidirectional deletions associated with IS*4. Mol. Gen. Genet.* **185,** 216–222.

54. Reif, H. J. and Saedler, H. (1975) IS*1* is involved in deletion formation in the *gal* region of *E. coli* K12. *Mol. Gen. Genet.* **137,** 17–28.

55. Saedler, H., Reif, H. J., Hu, S., and Davidson, N. (1974) IS*2,* a genetic element for turn-off and turn-on of gene activity in *E. coli. Mol. Gen. Genet.* **132,** 265–289.

56. Starlinger, P. (1980) IS elements and transposons. *Plasmid* **3,** 241–259.

57. Sekine, Y. and Ohtsubo, E. (1989) Frameshifting is required for production of the transposase encoded by insertion sequence 1. *Proc. Nat. Acad. Sci. USA* **86,** 4609–4613.

58. Sekine, Y., Eisaki, N., and Ohtsubo, E. (1994) Translational control in production of transposase and in transposition of insertion sequence IS*3. J Mol. Biol.* **235,** 1406–1420.

59. Chalmers, R. M. and Kleckner, N. (1994) Tn*10*/IS*10* transposase purification, activation, and in vitro reaction. *J. Biol. Chem.* **269,** 8029–8035.

60. Weinreich, M. D., Mahnke-Braam, L., and Reznikoff, W. S. (1994) A functional analysis of the Tn*5* transposase: identification of domains required for DNA binding and multimerization. *J. Mol. Biol.* **241,** 166–177.

61. Mendiola, M. V., Jubete, Y., and de la Cruz, F. (1992) DNA sequence of IS*91* and identification of the transposase gene. *J. Bacteriol.* **174,** 1345–1351.

62. Derbyshire, K. M. and Grindley, N. D. (1992) Binding of the IS*903* transposase to its inverted repeat in vitro. *EMBO J.* **11,** 3449–3455.

63. Johnsrud, L. (1979) DNA sequence of the transposable element IS*1. Mol. Gen. Genet.* **169,** 213–218.

64. Ohtsubo, H. and Ohtsubo, E. (1978) Nucleotide sequence of an insertion element, IS*1. Proc. Nat. Acad. Sci. USA* **75,** 615–619.

65. Escoubas, J. M., Prère, M. F., Fayet, O., Salvignol, I., Galas, D., Zerbib, D., et al. (1991) Translational control of transposition activity of the bacterial insertion sequence IS*1*. *EMBO J.* **10**, 705–712.

66. Zerbib, D., Prentki, P., Gamas, P., Freund, E., Galas, D. J., and Chandler, M. (1990) Functional organization of the ends of IS*1:* specific binding site for an IS*1*-encoded protein. *Mol. Microbiol.* **4**, 1477–1486.

67. Machida, C. and Machida, Y. (1989) Regulation of IS*1* transposition by the *insA* gene product. *J. Mol. Biol.* **208**, 567–574.

68. Chandler, M. and Fayet, O. (1993) Translational frameshifting in the control of transposition in bacteria. *Mol. Microbiol.* **7**, 497–503.

69. Grindley, N. D. (1978) IS*1* insertion generates duplication of a nine base pair sequence at its target site. *Cell* **13**, 419–426.

70. Grindley, N. D. F. (1983) Transposition of Tn*3* and related transposons. *Cell* **32**, 3–5.

71. Tavakoli, N., Comanducci, A., Dodd, H. M., Lett, M. C., Albiger, B., and Bennett, P. M. (2000) IS*1294,* a DNA element that transposes by RC transposition. *Plasmid* **44**, 66–84.

72. Mendiola, M. V., Bernales, I., and de la Cruz, F. (1994) Differential roles of the transposon termini in IS*91* transposition. *Proc. Nat. Acad. Sci. USA* **91**, 1922–1926.

73. Craig, N. L. (1995) Unity in transposition reactions. *Science* **270**, 253–254.

74. Ton-Hoang, B., Polard, P., and Chandler, M. (1998) Efficient transposition of IS*911* circles in vitro. *EMBO J.* **17**, 1169–1181.

75. Polard, P., Prère, M.-F., Fayet, O., and Chandler, M. (1992) Transposase-induced excision and circularization of the bacterial insertion sequence IS*911.* *EMBO J.* **11**, 5079–5090.

76. Polard, P., Prère, M.-F., Chandler, M., and Fayet, O. (1991) Programmed translational frameshifting and initiation at an AUU codon in gene expression of bacterial insertion sequence IS*911.* *J. Mol. Biol.* **222**, 465–477.

77. Ton-Hoang, B., Betermier, M., Polard, P., and Chandler, M. (1997) Assembly of a strong promoter following IS*911* circularization and the role of circles in transposition. *EMBO J.* **16**, 3357–3371.

78. Richter, G. Y., Björklöf, K., Romantschuk, M., and Mills, D. (1998) Insertion specificity and trans-activation of IS*801. Mol. Gen. Genet.* **260**, 381–387.

79. Díaz-Aroca, E., Mendiola, M. V., Zabala, J. C., and de la Cruz, F. (1987) Transposition of IS*91* does not generate a target duplication. *J. Bacteriol.* **169**, 442–443.

80. Mendiola, M. V. and de la Cruz, F. (1992) IS*91* transposase is related to the rolling-circle-type replication proteins of the pUB110 family of plasmids. *Nucl. Acids Res.* **20**, 3521.

81. Comanducci, A., Dodd, H. M., and Bennett, P. M. (1989) pUB2380: an R plasmid encoding a unique, natural one-ended transposition system, in *Genetic Transformation and Expression* (Butler, L. O., Harwood, C., and Moseley, B. E. B., eds.), Intercept, Andover, UK, pp. 305–311.

82. Poirel, L., Decousser, J.-W., and Nordmann, P. (2003) Insertion sequence IS*Ecp1B* is involved in expression and mobilization of a *bla*$_{\text{CTX-M}}$ β-lactamase gene. *Antimicrob. Agents Chemother.* **47**, 2938–2945.

83. Chalmers, R., Sewitz, K., Lipkow, K., and Crellin, P. (2000) Complete nucleotide sequence of Tn*10. J. Bacteriol.* **182,** 2970–2972.

84. Reznikoff, W. S. (1993) The Tn*5* transposon. *Ann. Rev. Microbiol.* **47,** 945–963.

85. Oka, A., Sugisaki, H., and Takanami, M. (1981) Nucleotide sequence of the kanamycin resistance transposon Tn*903. J. Mol. Biol.* **147,** 217–226.

86. Grindley, N. D. and Joyce, C. M. (1980) Genetic and DNA sequence analysis of the kanamycin resistance transposon Tn*903. Proc. Nat. Acad. Sci. USA* **77,** 7176–7180.

87. Krebs, M. P. and Reznikoff, W. S. (1986) Transcriptional and translational sites of IS*50.* Control of transposase and inhibitor expression. *J. Mol. Biol.* **192,** 781–791.

88. Kleckner, N., Chalmers, R. M., Kwon, D., Sakai, J., and Bolland, S. (1996) Tn*10* and IS*10* transposition and chromosome rearrangements; mechanism and regulation in vivo and in vitro. *Curr. Topic. Microbiol. Immunol.* **204,** 49–82.

89. Derbyshire, K. M., Kramer, M., and Grindley, N. D. (1990) Role of instability in the *cis* action of the insertion sequence IS*903* transposase. *Proc. Nat. Acad. Sci. USA* **87,** 4048–4052.

90. Heffron, F., McCarthy, B. J., Ohtsubo, H., and Ohtsubo, E. (1979) DNA sequence analysis of the transposon Tn*3:* three genes and three sites involved in transposition of Tn*3. Cell* **18,** 1153–1163.

91. Mahillon, J. and Lereclus, D. (1988) Structural and functional analysis of Tn*4430:* identification of an integrase-like protein involved in the co-integrate resolution process. *EMBO J.* **7,** 1515–1526.

92. Gill, R. E., Heffron, F., and Falkow, S. (1979) Identification of the protein encoded by the transposable element Tn*3* which is required for its transposition. *Nature* **282,** 797–801.

93. Shapiro, J. A. (1979) Molecular model for the transposition and replication of bacteriophage Mu and other transposable elements. *Proc. Nat. Acad. Sci. USA* **76,** 1933–1937.

94. Arthur, A. and Sherratt, D. J. (1979) Dissection of the transposition process. *Mol. Gen. Genet.* **175,** 267–274.

95. Heritage, J. and Bennett, P. M. (1985) Plasmid fusions mediated by one end of Tn*A. J. Gen. Microbiol.* **131,** 1131–1140.

96. Avila, P., de la Cruz, F., Ward, E., and Grinsted, J. (1984) Plasmids containing one inverted repeat of Tn*21* can fuse with other plasmids in the presence of Tn*21* transposase. *Mol. Gen. Genet.* **195,** 288–293.

97. Mötsch, S. and Schmitt, R. (1984) Replicon fusion mediated by a single-ended derivative of transposon Tn*1721. Mol. Gen. Genet.* **195,** 281–287.

98. Robinson, M. K., Bennett, P. M., and Richmond, M. H. (1977) Inhibition of TnA translocation by TnA. *J. Bacteriol.* **129,** 407–414.

99. Lee, C.-H., Bhagwhat, A., and Heffron, F. (1983) Identification of a transposon Tn*3* sequence required for transposition immunity. *Proc. Nat. Acad. Sci. USA* **80,** 6765–6769.

100. Arciszewska, L. K., Drake, D., and Craig, N. L. (1989) Transposon Tn*7 cis*-acting sequences in transposition and transposition immunity. *J. Mol. Biol.* **207,** 35–52.

101. Mizuuchi, K. (1992) Transpositional recombination: mechanistic insights from studies of Mu and other elements. *Ann. Rev. Biochem.* **61,** 1011–1051.
102. Stokes, H. W. and Hall, R. M. (1989) A novel family of potentially mobile DNA elements encoding site-specific gene-integration functions: integrons. *Mol. Microbiol.* **3,** 1669–1683.
103. Recchia, G. D. and Hall, R. M. (1995) Gene cassettes: a new class of mobile element. *Microbiol.* **141,** 3015–3027.
104. Bennett, P. M. (1999) Integrons and gene cassettes: a genetic construction kit for bacteria. *J. Antimicrob. Chemother.* **43,** 1–4.
105. Stokes, H. W., Gorman, D. B., Recchia, G. D., Parsekhian, M., and Hall, R. M. (1997) Structure and function of 59-base element recombination sites associated with mobile gene cassettes. *Mol. Microbiol.* **26,** 731–745.
106. Naas, T., Mikami, Y., Imai, T., Poirel, L., and Nordmann, P. (2001) Characterization of In53, a class 1 plasmid- and composite-transposon-located integron of *Escherichia coli* which carries an unusual array of gene cassettes. *J. Bacteriol.* **183,** 235–249.
107. Rowe-Magnus, D. A., Guerout, A. M., and Mazel, D. (1999) Super-integrons. *Res. Microbiol.* **150,** 641–651.
108. Mazel, D., Dychinco, B., Webb, V. A., and Davies, J. (1998) A distinctive class of integron in the *Vibrio cholerae* genome. *Science* **280,** 605–608.
109. Nield, B. S., Holmes, A. J., Gillings, M. R., Recchia, G. D., Mabbutt, B. C., Nevalainen, et al. (2001) Recovery of new integron classes from environmental DNA. *FEMS Microbiol. Lett.* **195,** 59–65.
110. Rowe-Magnus, D. A., Guerout, A. M., Ploncard, P., Dychinco, B., Davies, J., and Mazel, D. (2001) The evolutionary history of chromosomal super-integrons provides an ancestry for multiresistant integrons. *Proc. Nat. Acad. Sci. USA* **98,** 652–657.
111. Vaisvila, R., Morgan, R. D., Posfai, J., and Raleigh, E. A. (2001) Discovery and distribution of super-integrons among pseudomonads. *Mol. Microbiol.* **42,** 587–601.
112. Hansson, K., Sundstrom, L., Pelletier, A., and Roy, P. H. (2002) IntI2 integron integrase in Tn7. *J. Bacteriol.* **184,** 1712–1721.
113. Collis, C. M., Kim, M. J., Partridge, S. R., Stokes, H. W., and Hall, R. M. (2002) Characterization of the class 3 integron and the site-specific recombination system it determines. *J. Bacteriol.* **184,** 3017–3026.
114. Walsh, T. R., Toleman, M. A., Hryniewicz, W., Bennett, P. M., and Jones, R. N. (2003) Evolution of an integron carrying $bla_{VIM-2}$ in Eastern Europe: report from the SENTRY antimicrobial surveillance program. *J. Antimicrob. Chemother.* **52,** 116–119.
115. Clewell, D. B. and Gawron-Burke, C. (1986) Conjugative transposons and the dissemination of antibiotic resistance. *Ann. Rev. Microbiol.* **40,** 635–659.
116. Salyers, A. A., Shoemaker, N. B., and Li, L. Y. (1995) In the driver's seat: the *Bacteroides* conjugative transposons and the elements they mobilize. *J. Bacteriol.* **177,** 5727–5731.

117. Osborn, A. M. and Böltner, D. (2002) When Phage, plasmids, and transposons collide: genomic islands, and conjugative- and mobilizable-transposons as a mosaic continuum. *Plasmid* **48**, 202–212.

118. Böltner, D. and Osborn, A. M. (2004) Structural comparison of the integrative and conjugative elements R391, pMERPH, R997, and SXT. *Plasmid* **51**, 12–23.

119. Hochhut, B., Lotfi, Y., Mazel, D., Faruque, S. M., Woodgate, R., and Waldor, M. K. (2001) Molecular analysis of antibiotic resistance gene clusters in *Vibrio cholerae* O139 and O1 SXT constins. *Antimicrob. Agents Chemother.* **45**, 2991–3000.

120. Beaber, J. W., Burrus, V., Hochhut, B., and Waldor, M. K. (2002) Comparison of SXT and R391, two conjugative integrating elements: definition of a genetic backbone for the mobilization of resistance determinants. *Cell. Molec. Life Sci.* **59**, 2065–2070.

121. Franke, A. E. and Clewell, D. B. (1981) Evidence for a chromosome-borne resistance transposon (Tn*916*) in *Streptococcus faecalis* that is capable of conjugative transfer in the absence of a conjugative plasmid. *J. Bacteriol.* **145**, 494–502.

122. Churchward, G. (2002) Conjugative transposons and related mobile elements, in *Mobile DNA II* (Craig, N. L., Craigie, R., Gellert, M., and Lambowitz, A. M, eds.), ASM Press, Washington, DC, pp. 177–191.

123. Flannagan, S. E., Zitzow, L. A., Su, Y. A., and Clewell, D. B. (1994) Nucleotide sequence of the 18-kb conjugative transposon Tn*916* from *Enterococcus faecalis*. *Plasmid* **32**, 350–354.

124. Senghas, E., Jones, J. M., Yamamoto, M., Gawron-Burke, C., and Clewell, D. B. (1988) Genetic organization of the bacterial conjugative transposon Tn*916*. *J. Bacteriol.* **170**, 245–249.

125. Storrs, M. J., Carlier, C., Poyart-Salmeron, C., Trieu-Cuot, P., and Courvalin, P. (1991) Conjugative transposition of Tn*916* requires the excisive and integrative activities of the transposon-encoded integrase. *J. Bacteriol.* **173**, 4347–4352.

126. Craig, N. L. (1988) The mechanism of conservative site-specific recombination. *Ann. Rev. Genet.* **22**, 77–105.

127. Nash, H. A. (1981) Integration and excision of bacteriophage lambda: the mechanism of conservative site specific recombination. *Ann. Rev. Genet.* **15**, 143–167.

128. Craig, N. L. (1997) Target site selection in transposition. *Ann. Rev. Biochem.* **66**, 437–474.

129. Craig, N. L. (1991) Tn*7*: a target site-specific transposon. *Mol. Microbiol.* **5**, 2569–2573.

130. Gay, N. J., Tybulewicz, V. L., and Walker, J. E. (1986) Insertion of transposon Tn*7* into the *Escherichia coli glmS* transcriptional terminator. *Biochem. J.* **234**, 111–117.

131. Craig, N. L. (2002) Tn*7*, in *Mobile DNA II* (Craig, N. L., Craigie, R., Gellert, M., and Lambowitz, A. M., eds.), ASM Press, Washington, DC, pp. 423–456.

132. Bainton, R. J., Kubo, K. M., Feng, J.-N., and Craig, N. L. (1993) Tn*7* transposition: target DNA recognition is mediated by multiple Tn*7*-encoded proteins in a purified in vitro system. *Cell* **72**, 931–943.

133. Bainton, R., Gamas, P., and Craig, N. L. (1991) Tn*7* transposition in vitro proceeds through an excised transposon intermediate generated by staggered breaks in DNA. *Cell* **65,** 805–816.

134. Stanisich, V. A., Arwas, R., Bennett, P. M., and de la Cruz, F. (1989) Characterization of *Pseudomonas* mercury-resistance transposon Tn*502,* which has a preferred insertion site in RP1. *J. Gen. Microbiol.* **135,** 2909–2915.

135. Carmo de Freire Bastos, M. D. and Murphy, E. (1988) Transposon Tn*544* encodes three products required for transposition. *EMBO J.* **7,** 2935–2941.

136. Murphy, E. and Lofdahl, S. (1984) Transposition of Tn*544* does not generate a target duplication. *Nature* **307,** 292–295.

137. Pato, M. L. (1989) Bacteriophage Mu, in *Mobile DNA* (Berg, D. E. and Howe, M. M., eds.), ASM Press, Washington, DC, pp. 23–52.

138. Hacker, J. and Kaper, J. B. (2000) Pathogenicity islands and the evolution of microbes. *Ann. Rev. Microbiol.* **54,** 641–679.

139. Davis, B. M. and Waldor, M. K. (2002) Mobile genetic elements and bacterial pathogenesis, in *Mobile DNA II* (Craig, N. L., Craigie, R., Gellert, M., and Lambowitz, A. M., eds.), ASM Press, Washington, DC, pp. 1040–1059.

140. Lindsay, J. A., Ruzin, A., Ross, H. F., Kurepina, N., and Novick, R. P. (1998) The gene for toxic shock toxin is carried by a family of mobile pathogenicity islands in *Staphylococcus aureus. Mol. Microbiol.* **29,** 527–543.

141. Karaolis, D. K. R., Somara, S., Maneval, D. R. Jr., Johnson, J. A., and Kaper, J. B. (1999) A bacteriophage encoding a pathogenicity island, a type IV pilus and a phage receptor in cholera bacteria. *Nature* **399,** 375–379.

142. Rankin, A., Schubert, S., Pelludat, C., Brem, D., and Hessemann, J. (1999) The high-pathogenicity island of *Yersinia,* in *Pathogenicity Islands and Other Mobile Virulence Elements* (Kaper, J. B. and Hacker, J., eds.), ASM Press, Washington, DC, pp. 77–90.

143. Hiramatsu, K., Cui, L., Kuroda, M., and Ito, T. (2001) The emergence and evolution of methicillin-resistant *Staphylococcus aureus. Trends Microbiol.* **9**, 486–493.

144. Ma, X. X., Ito, T., Tiensasitorn, C., Jamklang, M., Chongtrakool, P., Boyle-Vavra, S., et al. (2002) Novel type of staphylococcal cassette chromosome mec identified in community-acquired methicillin-resistant *Staphylococcus aureus* strains. *Antimicrob. Agents Chemother.* **46**, 1147–1152.

145. Katayama, Y., Ito, T., and Hiramatsu, K. (2000) A new class of genetic element, Staphylococcus Cassette Chromosome *mec*, encodes methicillin resistance in *Staphylococcus aureus. Antimicrob. Agents Chemother.* **44**, 1549–1555.

146. Katayama, Y., Takeuchi, F., Ito, T., Ma, X. X., Ui-Mizutani, Y., Kobayashi, I., and Hiramatsu, K. (2003) Identification in methicillin-susceptible *Staphylococcus hominis* of an active primordial mobile genetic element for the staphylococcal cassette chromosome *mec* of methicillin-resistant *Staphylococcus aureus. J. Bacteriol.* **185**, 2711–2722.

147. Groisman, E. A. and Casadaban, M. J. (1986) Mini-Mu bacteriophage with plasmid replicons for in vivo cloning and *lac* gene fusions. *J. Bacteriol.* **168,** 357–364.

148. Van Gijsegem, F. and Toussaint, A. (1982) Chromosome transfer and R-prime formation by an RP4::mini-Mu derivative in *Escherichia coli, Salmonella typhimurium, Klebsiella pneumoniae,* and *Proteus mirabilis. Plasmid* **7,** 30–44.
149. Koch, C., Mertens, G., Rudt, F., Kahmann, R., Kanaar, R., Plasterk, R. H., et al. (1987) The invertible G segment, in *Phage Mu* (Symonds, N., Toussaint, A., van de Putte, P., and Howe, M. M., eds.), Cold Spring Harbor Laboratory, Cold Spring Harbor, NY, pp. 75–91.
150. Toussaint, A., Lefebvre, N., Scott, J. R., Cowan, J. A., de Bruijn, F., and Bukhari, A. I. (1978) Relationships between temperate phages Mu and P1. *Virology* **89,** 146–161.
151. Zieg, J. and Simon, M. (1980) Analysis of the nucleotides sequence of an invertible controlling element. *Proc. Nat. Acad. Sci. USA* **77,** 4196–4200.
152. Sharp, P. A., Cohen, S. N., and Davidson, N. (1973) Electron microscope heteroduplex studies of sequence relations among plasmids of *Escherichia coli* II. Structure of drug resistance (R) factors and F factors. *J. Mol. Biol.* **75,** 235–255.
153. Rownd, R. and Mickel, S. (1971) Dissociation and reassociation of RTF and r-determinant of the R-factor NR1 in *Proteus mirabilis. Nature New Biol.* **234,** 40–43.
154. Clowes, R. C. (1972) Molecular structure of bacterial plasmids. *Bacteriol. Rev.* **36,** 361–405.
155. Bennett, P. M. and Richmond, M. H. (1978) Plasmids and their possible influence on bacterial evolution, in *The Bacteria: A Treatise on Structure and Function* (Gunsalus, I. C., ed.), Academic Press, NY, pp. 1–69.

# 6

## Exploring and Exploiting Bacterial Proteomes

### Stuart J. Cordwell

**Summary**

The plethora of data now available from bacterial genome sequencing has opened a wealth of new research opportunities. Many of these have been reviewed in preceding chapters. Genomics alone, however, cannot capture a biological snapshot from an organism at a given point in time. The genome itself is static, and it is the changes in expression of genes, leading to the production of functional proteins, which allows an organism to survive and adapt to a constantly changing environment. Proteomics is the term used to describe the global analysis of proteins involved in a particular biological process. Such processes may be analyzed via comparative studies that examine bacterial strain differences, both phenotypic and genetic, bacteria grown under nutrient limiting conditions, growth phase, temperature, or in the presence of chemical compounds, such as antibiotics. Proteomics also provides the researcher with a tool to begin characterizing the functions of the vast proportion of "hypothetical" or "unknown" proteins elucidated from genome sequencing and database comparisons. For example, study of protein–protein, protein–ligand, protein–substrate, and protein–nucleic acid interactions for a given target protein may all help to define the functions of previously unknown proteins. Furthermore, genetic manipulation combined with proteomics technologies can provide an understanding of how gene expression is regulated. This chapter examines the technologies used in proteome analysis and the applications of proteomics to microbiological research, with an emphasis on clinically-relevant bacteria.

**Key Words:** Mass spectrometry; microbial proteomics; posttranslational modifications; proteomic signatures; subcellular fractions; two-dimensional gel electrophoresis.

## 1. Introduction
### 1.1. The Proteome Approach

Proteomics as a science has depended on two major steps, namely, the separation and identification of individual components from a mixture or complex of proteins (**Fig. 1**). Historically, the separation of proteins has been performed

From: *Methods in Molecular Biology, vol. 266: Genomics, Proteomics, and Clinical Bacteriology: Methods and Reviews*
Edited by: N. Woodford and A. Johnson © Humana Press Inc., Totowa, NJ

using two-dimensional gel electrophoresis (2-DE), combining isoelectric focusing of proteins in the first dimension with SDS-PAGE separation in the second. While this technology has been available since 1975 *(1)*, tremendous advances have been made in the last 10 yr in the quality and reproducibility of protein separation, due to the advent of improved methodology such as immobilized pH gradients *(2,3)* and "narrow-range" 2-DE *(4,5)*, coupled with simplification and increased sensitivity of 2-D gel staining *(6)*. It is not the intention of this chapter to focus on 2-DE methodology, because many excellent reviews are available elsewhere *(7–9)*.

In more recent times, approaches to protein separation have included whole sample tryptic digestion followed by multiple rounds of liquid chromatography (LC) to separate peptides prior to their identification by mass spectrometry (MS) *(10)*. Such technology allows extremely rapid identification of individual proteins within a proteome. As such, it is rapidly becoming the method of choice for answering the question, "is this gene expressed?" Initial reports suggest that well over 50% of a simple bacterial proteome, or 33% of a yeast proteome, can be identified in this large-scale manner. Recent reports suggest that the technology can also be used to characterize posttranslational protein modifications *(11)*. For quantification or comparative analyses, peptides can be labeled with isotope tags (isotope-coded affinity tags, ICATs *[12]*) that allow comparisons of peptide "abundance" for a number of peptides from within each protein on a proteome-wide scale.

The second major step in proteomics is identification. Until recently, this required *N*-terminal Edman sequencing, an often slow and laborious process capable of identifying only a handful of proteins per day. While this information is undoubtedly useful, proteomics requires the capacity for high throughput, high sensitivity, and high resolution. The advent of protein MS for identification rather than characterization purposes, revolutionized protein chemistry *(13–15)*. The technology has advanced at such a rapid rate that many hundreds of proteins can now be identified in a single day and from femtomole $(10^{-15})$ amounts of peptide mixtures *(16)*. The combination of 2-DE or 2-DLC approaches with MS is now the accepted methodology for proteomics research. In this chapter, I will discuss how these technologies have been combined to yield new research discoveries in clinical and applied microbiology.

## 2. Bacterial Proteome Mapping

2-DE provides a visual map of expressed proteins from a given bacterium. Many groups have begun to identify protein spots from these 2-D gel maps to create virtual reference libraries of visible proteins. These reference maps generally involve high-throughput mass spectrometric identification of abundant proteins from a range of 2-DE gels covering various pH ranges. Such analyses

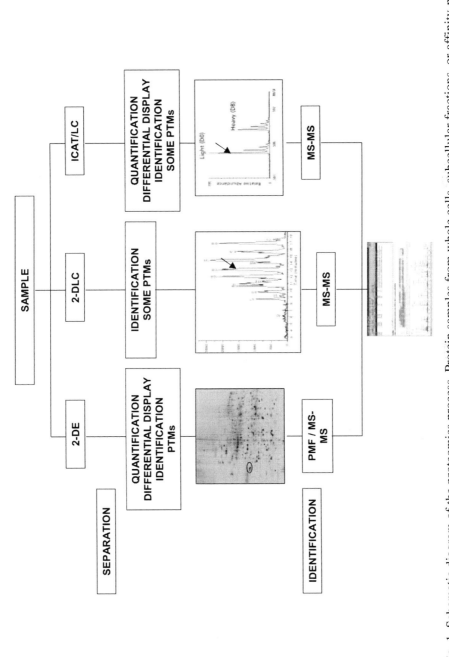

Fig. 1. Schematic diagram of the proteomics process. Protein samples from whole cells, subcellular fractions, or affinity-purified complexes are separated by 2-DE, 2-DLC, or ICAT prior to 2-DLC. Proteins and/or peptides are then identified by mass-spectrometric approaches including peptide mass fingerprinting (PMF) or tandem mass spectrometry (MS-MS).

can be broadened by taking a subcellular fractionation approach to map proteins specifically associated with bacterial membranes, or proteins secreted into the extracellular space. For whole cell proteomics, excellent reference maps have been established for a wide range of bacteria, including *Escherichia coli (17,18), Bacillus subtilis (19,20), Staphylococcus aureus (21,22), Mycobacterium tuberculosis (23), Mycoplasma pneumoniae (24)*, and *Helicobacter pylori (25)* (**Fig. 2**). Each of these maps contains the identifications of well over 100 protein spots, and they are useful for reconstructing physiological pathways under a chosen set of conditions. For example, an organism grown in rich, complex medium rarely expresses visible amounts of amino acid biosynthesis proteins, since amino acids are readily obtained from the medium. On the other hand, under specific nutrient starvation conditions, or in the presence of different sole sugar sources, the organism will respond by expressing those proteins necessary to utilize the substrates present in that particular environment. Whole-cell reference maps also form the starting point for more complex studies, including strain and species comparisons, the physiological response to altered environmental conditions, and the regulation of gene expression.

## 2.1. Subcellular Proteomes

### 2.1.1. Membrane and Extracellular Proteins

A great deal more useful information from bacterial proteomics can often be obtained by taking a "prefractionation" approach based on the separation of different subcellular fractions prior to 2-DE or peptide-LC separation. The stained 2-DE spot patterns are usually different, especially where the fractionation process leads to very pure fractions with little contamination from overlapping cellular compartments. A second method is the "sequential" extraction process *(26)*, where 2-DE is performed on solubility fractions, rather than sole cellular fractions. Subcellular fractionation has been used to examine bacterial membrane proteins from a wide variety of pathogenic species. The sodium carbonate enrichment-extraction method *(27)* provides an excellent separation of Gram-negative outer membrane proteins (OMPs) and has been used for mapping and species comparisons in *Pseudomonas aeruginosa (28,29)* and *Caulobacter crescentus (30)*. The method involves cell lysis by sonication or French Press and the removal of whole cells by centrifugation. Membranes are then precipitated in sodium carbonate, isolated by ultracentrifugation, and the proteins solubilized in 2-DE- or LC-compatible sample buffers. Separation of *P. aeruginosa* OMPs using 2-DE resulted in the identification of 189 proteins, where all except one (the ubiquitous chaperone GroEL) were predicted to maintain an outer membrane location by sequence prediction. Since GroEL is generally acknowledged to be the most abundant bacterial protein, any minor

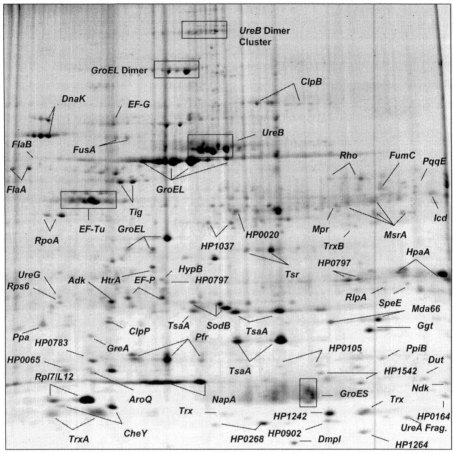

Fig. 2. 2-DE reference map of *H. pylori* cytosolic proteins. Proteins were identified by PMF.

cellular contamination would be detected by assaying the levels of this protein. On the other hand, GroEL may well shuttle to the membrane in its role as molecular chaperone and hence be detected at low levels in membrane-enriched fractions. Despite the success of this methodology, highly hydrophobic, membrane-associated proteins remain difficult to solubilize. In *P. aeruginosa,* only 7 of the 189 identified proteins had positive theoretical hydrophobicity values *(28),* indicating that such proteins remain insoluble for electrophoretic purposes. When OMPs from *Leptospira interrogans* were separated using 2-DE following detergent extraction in Triton X-114, the major OMPs included several lipoproteins and proteins not previously characterized in the organism, including an abundant protein of 21 kDa *(31).* A further enrichment step is to perform surface biotinylation and affinity selection prior to solubilization and 2-DE *(32)*. In this case, provided the biotin label does not pass through surface pores and label cytosolic proteins, the affinity purification should select for surface-accessible proteins, and should also allow for the elucidation of surface-exposed peptide epitopes.

While Gram-negative membrane proteins are generally simple to enrich and separate, Gram-positive membrane proteins have been highly problematic. This appears to be a solubility issue, given that the thick peptidoglycan layer remains difficult to fractionate or remove entirely, and that the proteins themselves tend to be highly hydrophobic. In whole-cell analyses of the *S. aureus* proteome *(21)*, envelope proteins were poorly represented, while subcellular fractionation of these organisms routinely resulted in fractions containing abundant cellular proteins. For example, recent work by Hughes et al. characterized several glycolytic enzymes as being surface associated in *Streptococcus agalactiae (33)*. However, it remains unclear whether this is a true association or merely an artifact associated with poor enrichment of Gram-positive envelope proteins.

Investigations of bacterial surface proteins, particularly interstrain comparisons, provide one means of appraising potential vaccine and diagnostic candidates. A second fractionation can be used to examine proteins secreted by bacteria into culture supernatants. Extracellular proteins are often involved in pathogenic processes and include toxins, proteases, and other proteins involved in breaking down host defenses or the acquisition of nutrients. Bacterial secretion is generally growth-phase dependent; therefore, the standard methodology is to grow cells in culture until the stationary phase is reached. The cells are then removed via centrifugation and filtration, and the free proteins in the culture supernatant are precipitated with ice-cold trichloroacetic acid, acetone, or methanol, alone or in combination. With regard to the characterization of surface-associated proteins, fractions derived from culture supernatants are often a mixture of cytosolic proteins (present owing to cell autolysis and turnover during growth), membrane proteins (particularly those sheared off due to physical interactions in culture), and truly secreted proteins. The presence or absence of a signal sequence may aid in the identification of secreted proteins. While this approach can be used for a variety of bacteria, it has been technically difficult to prepare extracellular fractions from organisms requiring protein additives such as serum in their growth medium, since these added proteins will also be detected using proteomics methodology. Recently, two groups have described studies of secreted proteins from *H. pylori,* an organism that is usually grown in the presence of serum *(34,35)*. Separation and identification of culture supernatant proteins have also been performed for a variety of bacteria including *M. tuberculosis (23), B. subtilis (36), S. aureus (37),* and *P.*

*aeruginosa (29).* In *S. aureus,* culture supernatants were examined to detect virulence factors regulated by the SarA regulatory protein (a global regulator of *S. aureus* virulence factors), whereas in *P. aeruginosa* such comparisons were performed to examine differences between two strains with different modes of pathogenicity *(29).*

### 2.1.2. Immunogenic Proteins and Vaccine Discovery

The characterization of expressed proteins using proteomic technologies can be applied to the discovery of novel vaccine targets, antimicrobials, and diagnostics, and to determine what cellular effects newly synthesized drug compounds may induce. Affinity approaches, as outlined later in this chapter, are also useful for determining which proteins directly interact with a compound of interest. Furthermore, proteomics is an excellent approach for monitoring the cellular response of bacteria to a given antimicrobial, first to provide a database of responses to known compounds, and then to begin screening the functions of new compounds in an attempt to characterize their mode of action (*see* Chapter 11).

As described above, particularly useful information about a microorganism can be gained by examining proteins associated with a chosen subcellular compartment. This becomes even more relevant when attempting to characterize immunogenic proteins from within a complex mixture. 2-DE combined with mass spectrometry and Western blotting using patient sera is a highly effective method of predicting potential vaccine targets *(38,39)* (**Fig. 3**). However, this approach often detects highly abundant cytosolic proteins as immunogenic constituents, yet their role in pathogenicity remains controversial. It seems likely that proteins such as the chaperone GroEL and enzymes involved in glycolysis become immunogenic in the host, perhaps only following cell autolysis. However, at least one study has shown convincing evidence that the "housekeeping" proteins elongation factor Tu, and pyruvate dehydrogenase beta subunit, are present as fibronectin-binding surface proteins in *M. pneumoniae* *(40).* This suggests that, at least in this microorganism, such proteins are capable of performing more than one function in vivo.

Prefractionation of surface-exposed and secreted proteins, combined with 2-DE and Western blotting, may provide much more focused information on potential vaccine targets. Furthermore, new methodology is also becoming available that utilizes surface display of genomic peptide libraries fused to *E. coli* OMPs, for screening against sera with high antibody titers *(41).* This method also allows the identification of specific epitopes that interact with antibody. Interestingly, 60 antigenic proteins were characterized in *S. aureus,* and all were predicted to be surface or secreted proteins, suggesting that immu-

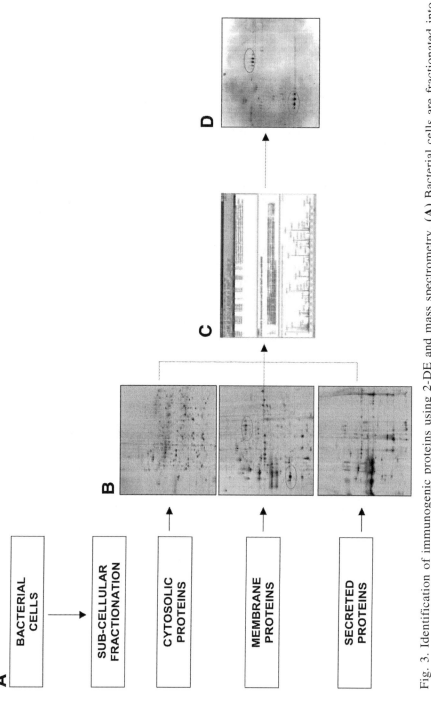

Fig. 3. Identification of immunogenic proteins using 2-DE and mass spectrometry. (**A**) Bacterial cells are fractionated into subcellular compartments. (**B**) Fractions are separated by 2-DE. (**C**) Proteins are identified by mass spectrometry. (**D**) Western blotting of membrane fraction and immunogenic proteins are mapped back to those identified in (**C**).

noreactive housekeeping proteins may well be an artifact when detected using the traditional Western blotting process *(41)*.

## 3. Microbiological Applications of Proteomics

### 3.1. Proteomic "Signatures"

A proteomic signature is defined as "...the subset of proteins whose alteration in expression is characteristic of a response to a defined condition or genetic change" *(42)*. A signature may contain proteins involved in the same metabolic pathway, proteins with similar functions, and/or those with shared regulation. For example, proteins whose expression is up- or downregulated in response to oxidative stress include those responding directly to the environmental stimulus, as well as others (including general stress proteins and those with housekeeping functions) that are "diagnostic" or "typical" of the response, even though their functions are unrelated. This approach can also be used to compare the responses of bacteria to antimicrobial compounds with both known and unknown modes of action. The patterns of protein expression following exposure to the antimicrobials can be used to create a database of responses. Each signature is then compared with similar signatures, and used to predict the targets and mode of action of given compounds, both alone and by comparison to those with known targets (**Fig. 4**). For example, the proteomic signatures for the *S. aureus* response to cell wall–active antibiotics *(43)*, show that this group of antibiotics induce or repress similar proteins, including peptide methionine sulfoxide reductase (which is indicative of oxidative stress) and GroES (which is indicative of protein synthesis stress), while antibiotics with different mechanisms cause wholly different protein expression profiles. For large-scale screening, determination of the expression patterns produced by hundreds or thousands of compounds allows their grouping in terms of related mechanisms of action, based upon proteomic signatures.

### 3.2. Applications

The widest use of microbial proteomics has been in the generation of maps of expressed proteins, and the application of these maps to the study of protein expression differences induced by changes in environment (e.g., heat/cold shock, oxidative stress) or genetic background (e.g., gene knockout and genetic overexpression). It is beyond the scope of this chapter to deal with every study that has used proteomics technologies in the study of bacterial physiology. However, interested readers should consult the pioneering work of Neidhardt and VanBogelen in *E. coli (44–46)*, and the work of Hecker in *B. subtilis (47–49)*. These studies have elucidated the bacterial proteins involved in heat and

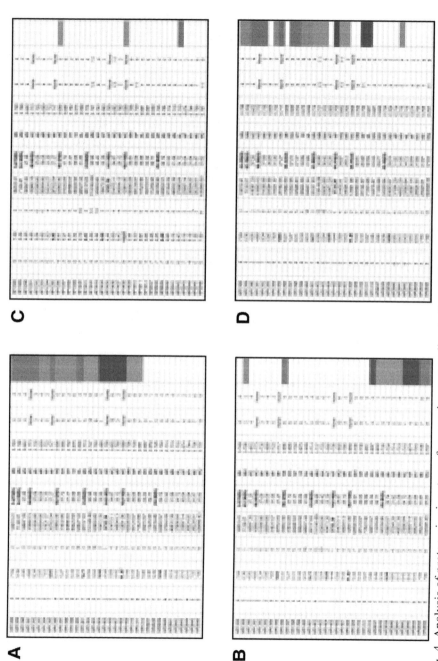

Fig. 4. Analysis of proteomic signatures for proteins responding to four different chemical compounds. (**A**) Known compound; (**B–D**) unknown compound. Pattern of protein up- or downregulation (as indicated by color) shows that (D) is likely to have a similar mode of action to (A).

cold shock, oxidative stress, the SOS response, nutrient starvation, the effect of antibiotics, and many other settings.

Proteomics can also be used to define *regulons,* or sets of genes and their corresponding proteins that are under the control or influence of a given regulatory system. The approach has been used to define proteins under the control of global regulators such as SigB in *B. subtilis* and *S. aureus (50)*, the global virulence regulator SarA in *S. aureus (37)*, Vfr in *P. aeruginosa (51)*, and many others. Proteomics technologies, especially in combination with "transcriptomics," are therefore an indispensable resource for examining new regulatory systems in microbial genomes.

In recent times, proteomics has been used as a tool for analyzing molecular differences between strains or species of bacteria with different phenotypes, e.g., virulent and avirulent strains of a given organism. Excellent examples of this type of study include comparison of different clinical isolates of *H. pylori,* which was carried out in conjunction with Western blotting to detect immunogenic proteins *(25)*, and comparisons of strains of *M. tuberculosis* and vaccine strains of *M. bovis* BCG *(23)*. A similar study was performed to examine the protein differences between methicillin-susceptible and methicillin-resistant strains of *S. aureus (21)*.

## 4. Protein Complexes and Posttranslational Modifications

### 4.1. Posttranslational Modifications

The major reason proteomics has established itself as an important technology in the postgenome era is its ability to focus on protein interactions and posttranslational modifications (PTMs). While DNA microarray technology allows a comprehensive view of changes in gene expression induced by changes in environmental or genetic stimuli, by monitoring levels of mRNA (*see* Chapter 10), it cannot monitor changes in protein modification induced by the same stimulus. For example, although the levels of a particular mRNA may remain constant, a vast change in the phosphorylation state of the coded protein may occur. While it is true that there is a poor correlation between mRNA and protein abundance in higher organisms *(52,53)*, and that mRNA is thus a poor predictor of protein levels in these organisms, this association has not yet been well tested in bacteria. Microarray and proteomics studies of the *ctrA-cckA* two-component regulatory system in *C. crescentus (54)* suggest that the correlation may be much better in prokaryotes. However, for a significant proportion of proteins changes in abundance under different biological conditions may be due to altered PTM state, rather than increased or decreased genetic expression.

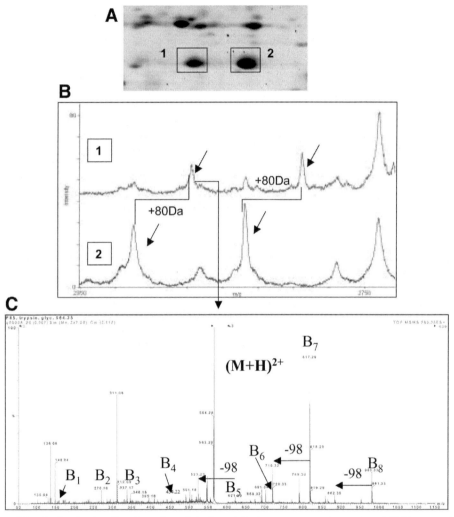

Fig. 5. Analysis of a phosphorylated protein by mass spectrometry. (**A**) Two spot isoforms apparent on 2-D gels and identified as the same protein. (**B**) Comparative mass spectrometric analysis shows spot 1 contains two phosphorylated peptides indicated by the addition of 80 Da. (**C**) Sequence analysis using MS-MS to identify the phosphorylated amino acid.

This chapter concentrates on major PTMs such as phosphorylation, glycosylation, and fragmentation; however, analysis of PTMs in bacteria has not routinely been performed on a global scale. Several methods exist to examine phosphorylation, including radiolabeling, antiphospho antibodies, and affinity enrichment prior to LC-MS *(55–57)*, while the elucidation of specific phosphorylation sites relies heavily on comparative MS (**Fig. 5**). "Isoforms" of spots visualized on 2-DE gels rarely provide enough information to determine the PTM state of a protein species, unless combined with one of the above techniques. For 2-DE approaches, incorporating radiolabel into biologically active material provides one approach to analyze phosphorylated proteins globally. However, the sensitivity of such labeling often means there is insufficient material remaining to allow identification and further characterization of the protein spots using MS, unless highly specialized sample concentration and desalting approaches are taken. Furthermore, the sensitivity of this approach also makes gel image comparisons with visible stains prone to errors. A second method involves the use of overlapping antiphospho antibodies (including antiphosphoserine, phosphothreonine, and phosphotyrosine), but these approaches suffer from many of the same disadvantages as radiolabeling. Western blotting is not compatible with postseparation analyses such as MS, and triplicate 2-DE gels need to be performed prior to blotting and treatment with each of the available antibodies. Furthermore, it is clear that the resolution and specificity of each antiphospho antibody is different, leading to some being more useful or reliable than others. Finally, no commercially available antibodies exist for phosphohistidine, which is the crucial phosphorylation in bacterial two-component regulatory systems. This, combined with the poor stability of the phosphate-histidine bond, makes the study of bacterial phosphorylation highly complex.

The attachment of various carbohydrates has long been known to modify protein structure and function in eukaryotes, but until recently, prokaryotes were not thought to modify their proteins in this way *(58)*. In the last few years, a handful of bacterial glycoproteins have been identified, mainly in Gram-negative bacteria, although the initial discovery of such proteins stemmed from the discovery of *S*-layer glycoproteins in Gram-positive archaeal species. Known prokaryotic glycoproteins are mainly those involved in outer membrane structures (such as pilins in the pili of *Neisseria meningitidis, Neisseria gonorrhoeae,* and *P. aeruginosa,* and the flagellins in the flagellae of *P. aeruginosa* and *Campylobacter jejuni*), and proteins confined to the outer membrane that are thought to be involved in bacterial adhesion and pathogenicity (e.g., OspAB in *Borrelia burgdorferi*). Such glycosylated proteins are generally detected by their unique spot appearance on 2-D gels, combined with the use of specific stains such as periodic acid–Schiff reagent, or one of the glyco-specific fluorescent dyes that are now available *(59)*. The elucidation of the structure of glycosylated peptides, particularly identification of the sugars and their specific sites of attachment, are best generated by comparative MS.

The list of potential PTMs is vast (*see* DeltaMass at http://www.abrf.org for a full list). One series of modifications that is often neglected, yet is proving to have significant biological importance, is fragmentation, or specific cleavage of translated proteins into functional subunits. These modifications can be found only by using a proteomics approach with 2-DE as the separation technology. However, it should be noted that many fragments of abundant proteins detected on 2-DE gels are products of protein turnover and nonspecific degradation. With *H. pylori,* for example, nearly 20 variants of the abundant chaperone GroEL, differing in p*I* and mass, have been found (unpublished data). Nonetheless, known functional cleavage products of *H. pylori* include the gamma-glutamyltranspeptidase *(60)*, where the alpha and beta subunits are cleaved posttranslationally. It appears likely that bacteria with very small genomes may use this approach to increase their genetic capacity through multifunctional proteins or protein subunits.

### 4.2. Protein–Protein and Protein–Ligand Interactions

The study of changes in protein expression associated with alterations to biological conditions can be achieved through differential display proteomics, although this does not necessarily provide much truly functional information. Functional proteomics relies heavily on methodology to elucidate intracellular protein–protein interactions, or the individual components of complexes of proteins that perform a given function. Protein interactions are generally studied using one of two approaches: the yeast two-hybrid system *(61)*, or affinity capture combined with mass spectrometry of the individual components *(13)*. As with global analyses of PTMs, little work has been performed to characterize protein complexes or interactions in bacteria, although a genome-wide survey of interactions was performed using the yeast two-hybrid system in *H. pylori (62)*. The two approaches contrast in their utility for bacterial analyses. The yeast two-hybrid system has remarkably high-throughput and can assay an entire genome of "baits" and "preys" for potential interactions, but it remains highly predictive, and further assays must be carried out to verify whether the proposed interactions occur in vivo. The affinity-capture approach begins with a target "protein of interest" and uses antibodies to capture that target in vitro, and those proteins to which the target is also bound in the living cell (**Fig. 6**). In more recent times, this approach has allowed high-throughput characterization of protein complexes in yeast *(63,64)*.

In bacteria, protein interactions have generally been studied by using host extracellular matrix proteins such as fibronectin, fibrinogen, and laminin as the affinity bait, and then determining which bacterial proteins are capable of binding, and hence which proteins are potential pathogenicity factors in host–pathogen interactions. Generally, this utilizes an affinity chromatography approach using antifibronectin (or other host extracellular matrix protein)

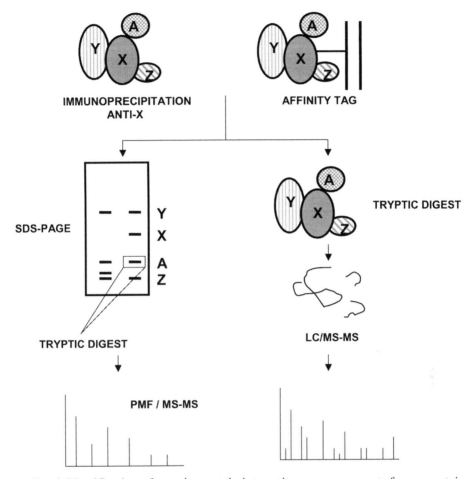

Fig. 6. Identification of protein–protein interactions or components from a protein complex. Proteins are either immunoprecipitated with an antibody against a protein of interest, or purified using an affinity column. Complexes can be separated with SDS-PAGE or digested immediately with trypsin prior to 2-DLC/MS-MS.

affinity chromatography columns. The bound proteins are then eluted from the column and either separated by SDS-PAGE prior to MS, or examined directly by MS posttryptic digestion. Similarly, a vast array of different ligands, including ATP, GTP, calcium, sugars, DNA, and RNA can be used as the affinity bait prior to washing through whole-cell proteins. This provides further functional information about the bound proteins. Such interactions are particularly amenable to high-throughput chip-based assays, as shown for a genome-wide screen of proteins with kinase function in yeast (*65*). The list of potential ligands also includes antibiotics, enabling direct detection of drug–protein interactions.

It is particularly difficult to use standard proteomics tools, such as 2-DE and MS, to determine the direct interactions of drugs with their specific microbial targets. Some predictions, as described above, can be made depending on the physiological response of the organism, as viewed via expression changes in metabolic and regulatory pathway proteins. A more targeted approach may be to use the drug itself as an affinity ligand for molecular "fishing" of any interacting proteins prior to MS. For example, one group has used vancomycin derivatives bound to sepharose as the basis of an affinity experiment to select for vancomycin-binding proteins in an *E. coli* model *(66)*. These researchers determined that at least six enzymes involved in peptidoglycan assembly were bound by this compound.

## 5. Conclusions

The many technologies encompassed under the term *proteomics,* especially combined with molecular biological methods such as mRNA expression analysis, are a powerful approach for understanding biological function in bacteria. This approach is a global method for characterizing the response of bacteria to environmental or genetic changes, as well as characterizing protein function through protein–protein, protein–ligand, and protein–nucleic acid interactions. Furthermore, proteins involved in host–pathogen interactions can be rapidly identified from within a complex mixture. Finally, proteomic studies are essential for comprehending posttranslational modifications of proteins that may be an influence in biological function, and that cannot be determined using a molecular biological approach. In the microbial postgenome era, proteomics has become, and will remain, an important tool for understanding the biology of microbial pathogens.

## References

1. O'Farrell, P. H. (1975) High-resolution two-dimensional electrophoresis of proteins. *J. Biol. Chem.* **250,** 4007–4021.
2. Görg, A., Obermaier, C., Boguth, G., Harder, A., Schiebe, B., Wildgruber, R., et al. (2000) The current state of two-dimensional electrophoresis with immobilized pH gradients. *Electrophoresis* **21,** 1037–1053.
3. Görg, A., Postel, W., and Gunther, S. (1988) The current state of two-dimensional electrophoresis with immobilized pH gradients. *Electrophoresis* **9,** 531–546.
4. Cordwell, S. J., Nouwens, A. S., Verrills, N. M., Basseal, D. J., and Walsh, B. J. (2000) Subproteomics based upon protein cellular location and relative solubilities in conjunction with composite two-dimensional electrophoresis gels. *Electrophoresis* **21,** 1094–1103.

5. Wildgruber, R., Harder, A., Obermaier, C., Boguth, G., Weiss, W., Fey, S. J., et al. (2000) Towards higher resolution: two-dimensional electrophoresis of *Saccharomyces cerevisiae* proteins using overlapping narrow immobilized pH gradients. *Electrophoresis* **21**, 2610–2616.

6. Patton, W. F. (2002) Detection technologies in proteome analysis. *J. Chromatogr. B* **771**, 3–31.

7. Rabilloud, T. (2001) Two-dimensional gel electrophoresis in proteomics: old, old fashioned, but it still climbs up the mountains. *Proteomics* **2**, 3–10.

8. Dunn, M. J. and Görg, A. (2001) Two-dimensional polyacrylamide gel electrophoresis for proteome analysis, in *Proteomics: From Protein Sequence to Function* (Pennington, S. R. and Dunn, M. J., eds.), BIOS Scientific, Oxford, UK, pp. 43–63.

9. Cordwell, S. J. (2002) Acquisition and archiving of information for bacterial proteomics: from sample preparation to database. *Methods Enzymol.* **358**, 207–227.

10. Washburn, M. P., Wolters, D., and Yates III, J. R. (2001) Large-scale analysis of the yeast proteome by multidimensional protein identification technology. *Nature Biotechnol.* **19**, 242–247.

11. MacCoss, M. J., McDonald, W. H., Saraf, A., Sadygov, R., Clark, J. M., Tasto, J. J., et al. (2002) Shotgun identification of protein modifications from protein complexes and lens tissue. *Proc. Natl. Acad. Sci. USA* **99**, 7900–7905.

12. Gygi, S. P., Rist, B., Gerber, S. A., Turecek, F., Gelb, M. H., and Aebersold, R. (1999) Quantitative analysis of complex protein mixtures using isotope-coded affinity tags. *Nature Biotechnol.* **17**, 994–999.

13. Pandey, A. and Mann, M. (2000) Proteomics to study genes and genomes. *Nature* **405**, 837–846.

14. Mann, M., Hendrickson, R. C., and Pandey, A. (2001) Analysis of proteins and proteomes by mass spectrometry. *Annu. Rev. Biochem.* **70**, 437–473.

15. Wilm, M. (2000) Mass spectrometric analysis of proteins. *Adv. Protein Chem.* **54**, 1–30.

16. Wilm, M., Shevchenko, A., Houthaeve, T., Breit, S., Schweigerer, L., Fotsis, T., et al. (1996) Femtomole sequencing of proteins from polyacrylamide gels by nanoelectrospray mass spectrometry. *Nature* **379**, 466–469.

17. VanBogelen, R. A, Abshire, K. Z., Moldover, B., Olson, E. R., and Neidhardt, F. C. (1997) *Escherichia coli* proteome analysis using the gene-protein database. *Electrophoresis* **18**, 1243–1251.

18. Tonella, L., Hoogland, C., Binz, P. A., Appel, R. D., Hochstrasser, D. F., and Sanchez, J.-C. (2001) New perspectives in *Escherichia coli* proteome investigation. *Proteomics* **1**, 409–423.

19. Hecker, M. and Völker, U. (2001) General stress response of *Bacillus subtilis* and other bacteria. *Adv. Microb. Physiol.* **44**, 35–91.

20. Buttner, K., Bernhardt, J., Scharf, C., Schmid, R., Mader, U., Eymann, C., et al. (2001) A comprehensive two-dimensional map of cytosolic proteins of *Bacillus subtilis*. *Electrophoresis* **22**, 2908–2935.

21. Cordwell, S. J., Larsen, M. R., Cole, R. T., and Walsh, B. J. (2002) Comparative proteomics of *Staphylococcus aureus* and the response of methicillin-resistant and methicillin-sensitive strains to Triton X-100. *Microbiology* **148,** 2765–2781.
22. Hecker, M., Engelmann, S., and Cordwell, S. J. (2003) Proteomics of *Staphylococcus aureus*—current state and future challenges. *J. Chromatogr. B* **787,** 179–195.
23. Jungblut, P. R., Schaible, U. E., Mollenkopf, H. J., Zimny-Arndt, U., Raupach, B., Mattow, J., et al. (1999) Comparative proteome analysis of *Mycobacterium tuberculosis* and *Mycobacterium bovis* BCG strains: towards functional genomics of microbial pathogens. *Mol. Microbiol.* **33,** 1103–1117.
24. Regula, J. T., Ueberle, B., Boguth, G., Görg, A., Schnolzer, M., Herrmann, R., et al. (2000) Towards a two-dimensional proteome map of *Mycoplasma pneumoniae. Electrophoresis* **21,** 3765–3780.
25. Jungblut, P. R., Bumann, D., Haas, G., Zimny-Arndt, U., Holland, P., Lamer, S., et al. (2000) Comparative proteome analysis of *Helicobacter pylori. Mol. Microbiol.* **36,** 710–725.
26. Molloy, M. P., Herbert, B. R., Walsh, B. J., Tyler, M. I., Traini, M., Sanchez, J.-C., et al. (1998) Extraction of membrane proteins by differential solubilization for separation using two-dimensional gel electrophoresis. *Electrophoresis* **19,** 837–844.
27. Molloy, M. P. (2000) Two-dimensional electrophoresis of membrane proteins using immobilized pH gradients. *Anal. Biochem.* **280,** 1–10.
28. Nouwens, A. S., Cordwell, S. J., Larsen, M. R., Molloy, M. P., Gillings, M., Willcox, M. D. P., et al. (2000) Complementing genomics with proteomics: the membrane subproteome of *Pseudomonas aeruginosa* PAO1. *Electrophoresis* **21,** 3797–3809.
29. Nouwens, A. S., Willcox, M. D. P., Walsh, B. J., and Cordwell, S. J. (2002) Proteomic comparison of membrane and extracellular proteins from invasive (PAO1) and cytotoxic (6206) strains of *Pseudomonas aeruginosa. Proteomics* **2,** 1325–1346.
30. Phadke, N. D., Molloy, M. P., Steinhoff, S. A., Ulintz, P. J., Andrews, P. C., and Maddock, J. R. (2001) Analysis of the outer membrane proteome of *Caulobacter crescentus* by two-dimensional electrophoresis and mass spectrometry. *Proteomics* **1,** 705–720.
31. Cullen, P. A., Cordwell, S. J., Bulach, D. M., Haake, D. A., and Adler, B. (2002) Global analysis of outer membrane proteins from *Leptospira interrogans* serovar Lai. *Infect. Immun.* **70,** 2311–2318.
32. Sabarth, N., Lamer, S., Zimny-Arndt, U., Jungblut, P. R., Meyer, T. F., and Bumann, D. (2002) Identification of surface proteins of *Helicobacter pylori* by selective biotinylation, affinity purification, and two-dimensional gel electrophoresis. *J. Biol. Chem.* **277,** 27,896–27,902.
33. Hughes, M. J. G., Moore, J. C., Lane, J. D., Wilson, R., Pribul, P. K., Younes, Z. N., et al. (2002) Identification of major outer surface proteins of *Streptococcus agalactiae. Infect. Immun.* **70,** 1254–1259.

34. Bumann, D., Aksu, S., Wendland, M., Janek, K., Zimny-Arndt, U., Sabarth, N., et al. (2002) Proteome analysis of secreted proteins of the gastric pathogen *Helicobacter pylori. Infect. Immun.* **70,** 3396–3403.
35. Kim, N., Weeks, D. L., Shin, J. M., Scott, D. R., Young, M. K., and Sachs, G. (2002) Proteins released by *Helicobacter pylori* in vitro. *J. Bacteriol.* **184,** 6155–6162.
36. Hirose, I., Sano, K., Shioda, I., Kumano, M., Nakamura, K., and Yamane, K. (2000) Proteome analysis of *Bacillus subtilis* extracellular proteins: a two-dimensional protein electrophoretic study. *Microbiology* **146,** 65–75.
37. Ziebandt, A. K., Weber, H., Rudolph, J., Schmid, R., Hoper, D., Engelmann, S., et al. (2001) Extracellular proteins of *Staphylococcus aureus* and the role of SarA and SigmaB. *Proteomics* **1,** 480–493.
38. Grandi, G. (2001) Antibacterial vaccine design using genomics and proteomics. *Trends Biotechnol.* **19,** 181–188.
39. Vytvytska, O., Nagy, E., Bluggel, M., Meyer, H. E., Kurzbauer, R., Huber, L. A., et al. (2002) Identification of vaccine candidates of *Staphylococcus aureus* by serological proteome analysis. *Proteomics* **2,** 580–590.
40. Dallo, S. F., Kannan, T. R., Blaylock, M. W., and Baseman, J. B. (2002) Elongation factor Tu and E1 β subunit of pyruvate dehydrogenase complex act as fibronectin binding proteins in *Mycoplasma pneumoniae. Mol. Microbiol.* **46,** 1041–1051.
41. Etz, H., Minh, D. B., Henics, T., Dryla, A., Winkler, B., Triska, C., et al. (2002) Identification of in vivo expressed vaccine candidate antigens from *Staphylococcus aureus. Proc. Natl. Acad. Sci. USA* **99,** 6573–6578.
42. VanBogelen, R. A., Schiller, E. E., Thomas, J. D., and Neidhardt, F. C. (1999) Diagnosis of cellular states of microbial organisms using proteomics. *Electrophoresis* **20,** 2149–2159.
43. Singh, V. K., Jayaswal, R. K., and Wilkinson, B. J. (2001) Cell wall-active antibiotic induced proteins of *Staphylococcus aureus* identified using a proteomic approach. *FEMS Microbiol. Lett.* **199,** 79–84.
44. Herendeen, S. L., VanBogelen, R. A., and Neidhardt, F. C. (1979) Levels of major proteins of *Escherichia coli* during growth at different temperatures. *J. Bacteriol.* **139,** 185–194.
45. VanBogelen, R. A., Kelley, P. M., and Neidhardt, F. C. (1987) Differential induction of heat shock, SOS, and oxidation stress regulons and accumulation of nucleotides in *Escherichia coli. J. Bacteriol.* **169,** 26–32.
46. VanBogelen, R. A., Olson, E. R., Wanner, B. L., and Neidhardt, F. C. (1996) Global analysis of proteins synthesized during phosphorus restriction in *Escherichia coli. J. Bacteriol.* **178,** 4344–4366.
47. Hecker, M., Schumann, W., and Völker, U. (1996) Heat-shock and general stress response in *Bacillus subtilis. Mol. Microbiol.* **19,** 417–428.
48. Antelmann, H., Scharf, C., and Hecker, M. (2000) Phosphate starvation-inducible proteins of *Bacillus subtilis:* proteomics and transcriptional analysis. *J. Bacteriol.* **182,** 4478–4490.

49. Eymann, C., Homuth, G., Scharf, C., and Hecker, M. (2002) *Bacillus subtilis* functional genomics: global characterization of the stringent response by proteome and transcriptome analysis. *J. Bacteriol.* **184**, 2500–2520.

50. Gertz, S., Engelmann, S., Schmid, R., Ziebandt, A. K., Tischer, K., Scharf, C., et al. (2000) Characterization of the sigma(B) regulon in *Staphylococcus aureus. J. Bacteriol.* **182**, 6983–6991.

51. Suh, S. J., Runyen-Janecky, L. J., Maleniak, T. C., Hager, P., MacGregor, C. H., Zielinski-Mozny, N. A., et al. (2002) Effect of *vfr* mutation on global gene expression and catabolite repression control in *Pseudomonas aeruginosa. Microbiology* **148**, 1561–1569.

52. Anderson, L. and Seilhamer, J. (1997) A comparison of selected mRNA and protein abundances in human liver. *Electrophoresis* **18**, 533–537.

53. Gygi, S. P., Rochon, Y., Franza, B. R., and Aebersold, R. (1999) Correlation between protein and mRNA abundance in yeast. *Mol. Cell. Biol.* **19**, 1720–1730.

54. Jacobs, C., Ausmees, N., Cordwell, S. J., Shapiro, L., and Laub, M. T. (2003) The functions of the CckA histidine kinase in *Caulobacter* cell cycle control. *Mol. Microbiol.* **47**, 1279–1290.

55. Oda, Y., Nagasu, T., and Chait, B. T. (2001) Enrichment analysis of phosphorylated proteins as a tool for probing the phosphoproteome. *Nature Biotechnol.* **19**, 379–382.

56. Mann, M., Ong, S. E., Gronborg, M., Steen, H., Jensen, O. N., and Pandey, A. (2002) Analysis of protein phosphorylation using mass spectrometry—deciphering the phosphoproteome. *Trends Biotechnol.* **20**, 261–268.

57. Kaufmann, H., Bailey, J. E., and Fussenegger, M. (2001) Use of antibodies for detection of phosphorylated proteins separated by two-dimensional gel electrophoresis. *Proteomics* **1**, 194–199.

58. Benz, I. and Schmidt, M. A. (2002) Never say never again: protein glycosylation in pathogenic bacteria. *Mol. Microbiol.* **45**, 267–276.

59. Steinberg, T. H., Pretty On Top, K., Berggren, K. N., Kemper, C., Jones, L., Diwu, Z., et al. (2001) Rapid and simple nanogram detection of glycoproteins in polyacrylamide gels and on electroblots. *Proteomics* **1**, 841–855.

60. Chevalier, C., Thiberge, J. M., Ferrero, R. L., and Labigne, A. (1999) Essential role of *Helicobacter pylori* gamma-glutamyltranspeptidase for the colonization of the gastric mucosa of mice. *Mol. Microbiol.* **31**, 1359–1372.

61. Fromont-Racine, M., Rain, J. C., and Legrain, P. (2002) Building protein-protein networks by two-hybrid mating strategy. *Methods Enzymol.* **350**, 513–524.

62. Rain, J. C., Selig, L., De Reuse, H., Battaglia, V., Reverdy, C., Simon, S., et al. (2001) The protein–protein interaction map of *Helicobacter pylori. Nature* **409**, 211–215.

63. Gavin, A.-C., Bösche, M., Krause, R., Grandi, P., Marzloch, M., Bauer, A., et al. (2002) Functional organization of the yeast proteome by systematic analysis of protein complexes. *Nature* **415**, 141–147.

64. Ho, Y., Gruhler, A., Heilbut, A., Bader, G. D., Moore, L., Adams, S. L., et al.

(2002) Systematic identification of protein complexes in *Saccharomyces cerevisiae* by mass spectrometry. *Nature* **415,** 180–183.

65. Zhu, H., Bilgin, M., Bangham, R., Hall, D., Casamayor, A., Bertone, P., et al. (2001) Global analysis of protein activities using proteome chips. *Science* **293,** 2101–2105.

66. Sinha Roy, R., Yang, P., Kodali, S., Xiong, Y., Kim, R. M., Griffin, P. R., et al. (2001) Direct interaction of a vancomycin derivative with bacterial enzymes involved in cell wall biosynthesis. *Chem. Biol.* **8,** 1095–1106.

# II

## APPLICATION OF GENOMICS TO DIAGNOSTIC BACTERIOLOGY

# 7

## Molecular Diagnostics

*Current Options*

### B. Cherie Millar and John E. Moore

### Summary

Diagnostic medical bacteriology consists of two main components: identification and typing. Molecular biology has the potential to revolutionize the way in which diagnostic tests are delivered in order to optimize the care of infected patients, whether they are in hospital or in the community. Since the discovery of the polymerase chain reaction (PCR) in the late 1980s, an enormous amount of research has enabled the introduction of molecular tests into several areas of routine clinical microbiology. Molecular biology techniques continue to evolve rapidly, and many laboratories have been reluctant to introduce these new methods due to concerns that the technology would become outdated. In consequence, the vast majority of clinical bacteriology laboratories do not currently use any molecular diagnostics, although such technology is becoming more widespread in specialized regional laboratories, as well as in national reference laboratories. Presently, molecular biology offers a wide repertoire of techniques and permutations. This chapter is intended to explore the application of these in the diagnostic laboratory setting.

**Key Words:** Bacteria; broad-range; *Burkholderia cepacia*; cystic fibrosis; DNA extraction; identification; microarray; nucleic acid sequence-based amplification (NASBA); PCR, real-time PCR; sequencing; 16S rRNA.

## 1. Introduction

The last decade of the twentieth century witnessed an exponential increase in molecular biological techniques, following the cellular and protein era of

From: *Methods in Molecular Biology, vol. 266: Genomics, Proteomics, and Clinical Bacteriology: Methods and Reviews*
Edited by: N. Woodford and A. Johnson © Humana Press Inc., Totowa, NJ

the 1970s and 1980s. This molecular explosion has allowed significant developments in many areas of the life sciences, including bacteriology. This chapter reviews the current situation with regard to the application of molecular biology in the area of medical bacteriology, particularly for the identification of causal agents of infection. The chapter also highlights the diversity of techniques that are available, either as research tools or for use in a routine setting. The impact of adopting molecular diagnostics on management, cost, labor, and space is also discussed. The overall aim is to provide an appreciation of the role that molecular diagnostics may play in routine clinical microbiology, and how these techniques can best be integrated to enhance the health care system.

## 2. Applications of Molecular Diagnostic Identification

Molecular identification should be considered in three scenarios: (1) the identification of organisms isolated in pure culture, (2) the rapid identification of organisms in clinical specimens, and (3) the identification of organisms from nonculturable specimens (e.g., culture-negative endocarditis).

### 2.1. Difficult-to-Identify Organisms

Most diagnostic clinical microbiology laboratories rely on a combination of colonial morphology, physiology, and biochemical/serological markers for identification, either to the genus level or, more frequently, to species level. It is important that organisms be correctly identified for a number of reasons, including optimal empirical treatment, the correct epidemiological reporting of causal agents in a given disease state, and infection control purposes. Sometimes it is argued that physicians may simply accept the Gram stain result and corresponding antibiogram in order to determine the optimum management of an infected patient. This lack of identification requires caution, for the reasons stated above. Consequently, there is a need to identify organisms of clinical significance reliably in a cost-effective and timely manner. Most laboratories currently rely on identification through biochemical profiling with the API-identification schema, the BBL-Crystal system, or other commercially available phenotypic systems. However, there are some organisms for which such methods are unable to give a reliable identification (e.g., nonfermenting Gram-negative rods).

### 2.2. Rapid Identification From Clinical Specimens

Traditional culture may take several days to generate sufficient growth of an organism to allow identification to be undertaken. Such delays mean that patients are treated empirically until the culture result is known, which may result in suboptimal management of some patients. There is thus a clear role for molecular identification techniques, where results can be made available rapidly.

## 2.3. Identification of Organisms in Culture-Negative Specimens

Since its origins in the late nineteenth century, bacteriology has largely been based on the ability to culture organisms in vitro. The forefathers of bacteriology, including Pasteur and Koch, were ardent exponents of bacteriological culture, and the affinity between the bacteriologist and laboratory culture has remained strong for the past 100 yr. Indeed, the ability to culture bacteria in vitro still remains the cornerstone of this discipline. However, there are some clinical situations where conventional culture fails to identify the causal organism. Reasons for this failure are numerous and include prior antibiotic therapy (e.g., treatment of acute meningitis with i.v. benzylpenicillin), involvement of fastidious organisms (e.g., HACEK organisms in cases of endocarditis), or those requiring specialized cell culture techniques for isolation (e.g., *Chlamydia* spp. and *Coxiella burnetti*), or slow-growing organisms (e.g., *Mycobacterium* spp.). Molecular methods may be included in the laboratory's diagnostic algorithm to enable rapid and reliable identification.

## 3. Gene Targets

Unlike diagnostic virology, which targets both DNA and RNA, the majority of diagnostic assays in bacteriology have been based around the amplification of target DNA. This offers the advantage that DNA is a stable molecule compared with mRNA, which has a short half-life *(1–3)*. Generally molecular diagnosis in clinical bacteriology is not concerned with the viability status of organisms, but with their presence or absence in patients with particular clinical conditions (e.g., the detection of meningococcal DNA in the cerebrospinal fluid (CSF) of a patient with suspected meningitis). The scenario is different, however, where medical bacteriology interfaces with food/public health microbiology. In this situation, qualitative detection of DNA from pathogenic foodborne bacteria is usually insufficient, and can even be misleading; evidence of viable organisms is required. For example, if *Salmonella* is suspected in a sample of dried milk powder, it is insufficient to detect *Salmonella* DNA, which may reflect archival DNA from dead cells killed during the drying process. It is important to give careful consideration to what one wishes to achieve from a molecular assay.

## 3.1. Universal Gene Targets

### 3.1.1. Ribosomal RNA Targets

Where there is no indication of the identity of a bacterial organism, amplification of DNA encoding ribosomal RNA (rRNA) in conjunction with sequencing of the amplicon has proven to be valuable *(4,5)*. In bacteria, there are three genes encoding the three types of RNA (5S, 16S, and 23S rRNA) found in the bacterial ribosome. The 16S rRNA gene has been most commonly employed for identification purposes (*see* **Table 1** and Chapter 14), due to it being highly conserved and commonly having several copies in the bacterial ribosome. 16S rRNA genes are found in all bacteria and accumulate mutations at a slow, constant rate over time; hence they may be used as "molecular clocks" *(30)* (*see* Chapter 16). Since 16S rRNA has regions that are highly conserved in all known bacteria, "broad-range" PCR primers may be designed to amplify intervening regions, even without phylogenetic information about an isolate. In addition to these conserved regions, highly variable regions of the 16S rRNA sequence can provide unique signatures for any bacterium.

Recently, sequencing of 16S–23S rRNA intergenic spacer regions has become popular due to its high copy number and, more importantly, its high sequence variability *(31,32)*. Primers, which are directed towards highly conserved regions of the 16S and 23S rRNA genes, may either be universal or may target specific genera, including *Bartonella* spp. *(33)*, *Chlamydia* spp. *(34)*, *Tropheryma whippelii (35)*, *Mycobacterium* spp. *(36)*, and *Salmonella* spp. *(37)*. The sequence of 23S rRNA genes has also been used for bacterial species identification. Anthony et al. *(38)* reported that the 23S rRNA locus shows more variation between species of medical importance than the 16S rRNA locus. Using universal 23S rRNA primers together with a hybridization assay with specific oligonucleotide probes, they were able to detect and identify the bacteria from 158 positive blood cultures. They concluded that the accuracy, range and discriminatory power of their assay could be continually extended by adding further oligonucleotides to their panel, without significantly increasing complexity and cost. The 23S rRNA gene locus has also been used successfully to detect *Stenotrophomonas maltophilia* from patients with cystic fibrosis *(39)*.

Overall, 16S–23S rRNA and 23S rRNA assays have not been used as widely as those targeting only the 16S rRNA gene, and this probably reflects the relative availability of sequence information. Presently, the only universal bacterial sequence-based identification scheme available commercially (the MicroSeq 500 16S ribosomal DNA [rDNA] bacterial sequencing kit from Applied Biosystems, Foster City, CA) is based on the 16S rRNA gene *(40)*.

Sequence-based identification methods employing rRNA gene loci require the use of appropriate analytical software. BLASTn and FASTA (*see* Chapter 2) software tools are commonly employed to make comparisons between the determined query sequence and those deposited in sequence databases (**Table 2**). Interpretative criteria should be used in order to ascertain the identification of the unknown sequence against its most closely related neighbor *(12)*.

**Table 1**
**Applications of 16S rDNA PCR to Identify Causal Agents of Bacterial Infection**

| Infection | Clinical specimen | Reference |
|---|---|---|
| Bacterial endophthalmitis | Vitreous fluid and aqueous humor specimens | *(6)* |
| Blood-borne sepsis | Blood-EDTA | *(7)* |
| Chronic prosthetic hip infection | | *(8)* |
| Detection of tick-infecting bacteria | | *(9)* |
| Endocarditis | Isolate | *(10)* |
| | Heart valve | *(11,12)* |
| | Blood culture | *(13)* |
| | Blood | *(14)* |
| | Arterio-embolic tissue | *(15)* |
| Endodontic infection | | *(16)* |
| Febrile episodes in leukemic patients | | *(17)* |
| *Helicobacter* spp. osteomyelitis in an immunocompetent child | Biopsy of bone lesion | *(18)* |
| Intra-amniotic infection | Amniotic fluid | *(19)* |
| Intra-ocular infection | | *(20)* |
| Maxillary sinus samples from ICU patients | | *(21)* |
| Meningitis | CSF | *(22)* |
| | Blood-EDTA | *(23)* |
| Nasal polyps. chronic sinusitis | | *(24)* |
| Peritonitis | CAPD fluid from culture-negative peritonitis | *(25)* |
| Rat bite fever | Blister fluid | *(26)* |
| Reactive arthritis | Synovial fluid/tissue | *(27)* |
| Septic arthritis | | *(28)* |
| Wound infection | Wound tissue from venous leg ulcer | *(29)* |

### 3.1.2. Other Universal Targets

Although 16S rRNA gene sequences may be employed successfully to identify many bacterial species, they lack sufficient discrimination to identify isolates of some genera to species level (e.g., *Burkholderia cenocepacia* and *B. multivorans*). In this circumstance, sequences of other essential genes, such as those encoding heat shock proteins (HSP), which enable cells to survive a

**Table 2**
**Commonly Employed Sequence Alignment Software Tools Used
in Conjunction With Nucleotide Sequence Databases**

| Sequence identification tools | Nucleotide sequence databases |
|---|---|
| BLASTn (Basic Local Alignment Sequence Tool) | http://www.ebi.ac.uk/embl/ (UK) |
|  | http://www.ddbj.nig.ac.jp/ (Japan) |
| http://www.ncbi.nlm.nih.gov/blast/ (USA) | http://www.ncbi.nlm.nih.gov/Genbank/ GenbankSearch.html |
| http://dove.embl-heidelberg.de/ Blast2/ (Germany) | |
| http://www.ebi.ac.uk/blast/index.html (UK) | |
| http://www-btls.jst.go.jp/ (Japan) | |
| | |
| FASTA | |
| | |
| http://www.ebi.ac.uk/fasta33/ | |
| | |
| Other | |
| | |
| Ribosomal database project (www.cme.meu.edu/RDP/html/index.html) | |
| MicroSeq (Commercial) (www.appliedbiosystems.com) | |
| *SmartGene* IDNA (Commercial) (www.smartgene.ch) | |

variety of environmental stresses (HSP60, HSP65, *groEL, groER,* and so on), have been shown to be useful *(41,42)*. HSPs, or *chaperonins,* appear to be constituents of the cellular machinery for protein folding, degradation and repair *(43)*.

Other "universal" gene loci have also been targeted, including the *recA* locus *(44)* and cold-shock proteins *(45)*. However, a major disadvantage of these targets is the relatively limited sequence data available in public databases against which to compare and identify a query sequence. For this reason, these targets are not commonly used, or are applied within well-defined populations, e.g., organisms belonging to the *Burkholderia cepacia* complex *(46)*.

### 3.2. Specific Targets

Molecular identification of bacteria based on specific gene targets has the obvious advantage of conferring a higher degree of specificity than identification based on use of universal or broad-range primers. This approach requires extensive knowledge of the genome sequence of the target species, to allow the selection of suitable targets and development of a specific assay. Examples of such specific targets include the hippuricase gene for the differentiation of hippurate-hydrolyzing campylobacters *(Campylobacter jejuni)* from non-hydrolyzing campylobacters *(47)*, and the *ctrA* gene of meningococci for the diagnosis of meningococcal meningitis *(48)*. The specific target does not necessarily have to be associated with a PCR assay, but may be used in combination with several other nucleic acid amplification/analysis techniques *(see* **Table 3**).

Presently, there are several hundred specific assays available for the identification of a diverse variety of bacteria that are too numerous to detail in this section. However, any published assay is potentially troubled with pitfalls associated with poor design, and these may lead to poor specificity and/or low sensitivity. Therefore, before any assay is adopted into a routine diagnostic service, the published method must be optimized empirically in the user's laboratory, and the user must acquire through experience knowledge of the strengths and weaknesses of the assay. This is essential for reliable, reproducible, and accurate interpretation of end results.

### 3.3. Antibiotic Resistance Markers

Antibiotic resistance in bacterial pathogens has become an important topic both nationally and internationally. Some scientists are forecasting the emergence of the "postantibiotic era," where it will be difficult to control common infections owing to the emergence of high-level multiresistance in most medically-important bacterial pathogens. Consequently, there has been great interest in rapid molecular detection of antibiotic resistance genes *(109)*, particularly in causal agents that are fastidious or nonculturable. Currently, most hospitals are concerned with the occurrence of methicillin-resistant *Staphylococcus aureus* (MRSA) and glycopeptide-resistant enterococci (GRE) on their wards, particularly surgical wards. Several workers have published molec-ular methods to detect MRSA using simple PCR assays that target the *mec*A gene locus *(110,111)*. Screening patients in intensive-care units (ICU) for MRSA carriage by *mec*A PCR is a useful tool, and allows infection-control teams to segregate MRSA-positive patients from noncolonized patients, thereby minimizing the opportunity for nosocomial spread.

### 3.4. Genomovar Analysis of Burkholderia cepacia in Cystic Fibrosis

Infection with the *B. cepacia* complex (BCC) is an important cause of increased morbidity and reduced survival in patients with cystic fibrosis (CF) *(112)*. Certain members of the BCC are transmissible, and epidemics have been described in a number of CF centers *(113)*. Several factors have been identified as markers of strain transmissibility, including the *B. cepacia* epidemic strain marker (BCESM) *(114)* and the cable pilus gene *(115)*. Most units in the UK now segregate patients infected with such strains from other CF patients, and

**Table 3**
**Description of Components, Work Flow, and Nucleic Acid Amplification/Analysis Techniques Involved in the Molecular Diagnosis of Medically Important Bacteria**

| Specimen categories | Nucleic acid amplification/analysis | Characterization/identification of amplicon |
|---|---|---|
| Amniotic fluid (19) | Block-based PCR (75) | Automated sequence analysis (4) |
| Arterio-embolic tissue (15) | - single round (13) | DNA-DNA hybridization (16) |
| Ascitic fluid (49) | - semi-nested (76) | Enzyme immunoassay (PCR-EIA) (97) |
| Atheroma (50) | - nested (77) | Restriction enzyme analysis (PCR-REA) (98) |
| Blood culture (13) | - multiplex (78) | Restriction fragment length polymorphism |
| Blood-EDTA (23) | Branched DNA signal amplification | (RFLP) (67) |
| Bone (18) | (75,79) | Single-strand conformational polymorphism |
| Bone marrow (51) | DNA-hybridization/probe assay (75) | (SSCP) (99,100) |
| Breast milk (52) | In situ PCR/RT-PCR (80) | |
| Bronchioelar lavage (BAL) (53) | Ligase chain reaction (LCR) (81–87) | *Genotyping/molecular epidemiology* |
| Cerebral spinal fluid (CSF) (22) | Microarrays (88) | Amplified fragment length polymorphism |
| Cervical specimen/tissue (54) | Nucleic acid sequence-based | (AFLP) (101) |
| Culture isolate (pure) (13) | amplification (NASBA) (89,90) | Arbitrary-primed - PCR (AP-PCR) (102) |
| Feces (55) | PCR-target capture/hybrid capture (91) | Multilocus sequence typing (MLST) (103) |
| Fixed tissue sample (56) | Q β replicase system (75,92) | Pulsed field gel electrophoresis (PFGE) (104) |
| Heart valve (57) | Real-time PCR (61) | Random amplification of polymorphic DNA |
| Lymph node tissue (58) | - Light cycler (93) | (RAPD) (105) |
| Middle ear fluid (59) | - Taqman (94) | - BOX PCR (106) |
| Nasal polyps/sinus/lavage (24) | Reverse transcriptase PCR | - ERIC PCR (107) |
| Paraffin-embedded tissue (58) | (RT-PCR) (75) | - rep PCR (108) |
| Pus (wound & blister) (26,60) | Strand displacement | Ribotyping (74) |
| Plasma (61) | amplification (SDA) (95) | Single-strand conformational polymorphism |
| Pleural effusion/fluid (62) | Transcription mediated | (SSCP) (99,100) |
| Prosthetic device (8) | amplification (TMA) (96) | |
| Saliva (63) | | |
| Semen/sperm (64) | | |
| Serum (49) | | |
| Skin (65,66) | | |
| Sputum (67) | | |
| Swabs (65) | | |
| Synovial fluid (28) | | |
| Tissue (wound) (29) | | |
| Urine (68) | | |
| Vaginal fluid (69) | | |
| Vitreous humor (70) | | |

*Nucleic acid extraction*

| | | |
|---|---|---|
| Alkali/heat lysis (71) | Commercial kits (71,73) | |
| | (e.g., Qiagen/Roche) | |
| Automated DNA extraction (72) | "In house methods" (71) | |
| (e.g., MagNAPure) | | |
| Boil (73) | Phenol/chloroform (74) | |
| Centrifugation/Chelex-100/boil (52) | Silica capture/guanidine | |
| | hydrochloride treatment/ | |
| | Proteinase K/Lysozyme (71) | |

infection-control guidelines have recently been published by the UK CF Trust to help reduce the potential for cross-infection *(116)*.

Nine genomovars of the BCC have been formally described, and initial studies indicate that genomovar II has less clinical impact than genomovar III *(117)*. Presently, *B. cenocepacia* (formerly genomovar III) is found in the majority of patients infected with BCC, followed by *B. multivorans* (formerly *B. cepacia* genomovar II). Some centers believe that complete segregation of patients, irrespective of the genomovar of their infecting strain, is the best policy to reduce the risk of cross-infection between patients with BCC; others feel that it is important to segregate patients with genomovar III strains only from patients with other BCC types (e.g., Belfast and Vancouver).

Early accurate identification of BCC is therefore of critical importance so that patients can be segregated, if necessary, and the potential for further epidemics can be reduced. BCC organisms present the clinical microbiologist with a diagnostic dilemma, as there are extremely few or, in some cases, no biochemical or growth-related phenotypic tests that reliably distinguish between these organisms. Hence, there has been a variety of molecular tests to differentiate BCC genomovars *(118,119)*. For example, RecA is a multifunctional, ubiquitous protein involved in general genetic recombination events and in DNA repair. The *recA* locus may contribute significantly to the overall genomic plasticity of the BCC, and is a strong candidate target for rapid molecular assays *(120)*. Molecular methods may aid CF centers to determine whether patients are colonized with this pathogen chronically, transiently, or not at all.

## 4. Selection of Molecular Diagnostic Procedures

Molecular assays rely on three basic components: (1) nucleic acid extraction, (2) amplification/analysis, and (3) detection of the amplified product (**Table 3**). Most assays allow a wide variety of permutations and combinations of methods, depending on the ultimate goal, and almost all types of clinical specimen have been subjected to a variety of DNA extraction protocols. There are myriad options available for nucleic acid amplification/analysis, and several considerations will help determine the most appropriate type of assay (**Table 4**). If speed is an important factor, as it is, for example, in the detection of meningococcal DNA in CSF from patients with suspected meningitis, then real-time assays (*see* Chapter 9), such as the Roche LightCycler or the ABI Taqman 7700 systems, should be adopted. They allow detection in a relatively short time, are gel-less systems, and also allow quantification of copy numbers. When detection of numerous different targets is important, a multiplex PCR format is more appropriate.

**Table 4**
**Criteria Used in the Determination of the Most Appropriate Molecular Method to Use**

| Criteria | Appropriate molecular method of choice | Comment/example |
|---|---|---|
| Time/speed to detection | Real-time applications (LightCycler/TaqMan 7700 system) | Detection of meningococcal DNA from clinical specimens in children with suspected meningitis *(48)* |
| Quantification | LightCycler/TaqMan 7700 system/NASBA | Determination of effect of antibiotic intervention |
| Multiple targets | Multiplex PCR/ microarrays/ hybridization probe assay | Determination of multiple respiratory pathogens in sputa |
| Viability | NASBA/RT-PCR | Determination of viable pathogens in foodstuffs or detection of viable but noncultural (VNC) organisms *(89)* |
| Commercial availability | MicroSeq (ABI Ltd.), LCR (Abbott) | Identification by comparison of query organism with high-quality database *(4)*. Detection of Chlamydia in genital specimens *(82,85,87)* |
| Throughput | High real-time PCR | Automated DNA extraction followed by real-time PCR |
|  | Low block-based PCR |  |

## 5. Laboratory Management of Molecular Assays

All molecular assays, including broad-range and specific PCR, are prone to contamination problems due to their high sensitivity. This has the potential to lead to false-positives *(121,122)*. Problems may arise at any stage between specimen reception and final molecular analysis, and these should be identified and appropriate control measures established to minimize each risk (*see* **Table 5** for a working example). To ensure the minimum contamination risk, including PCR amplicon carryover, a key element is design of a successful workflow through geographically separated areas, as detailed in **Fig. 1**. Although many hospital laboratories have limited work space, which makes separation of pre- and post-PCR areas difficult to achieve, it is important that adequate space be allocated to ensure compliance with standard for accreditation of molecular diagnosis *(123)*.

**Table 5**
**Sources and Controls of DNA Contamination in Molecular Diagnostic Analyses**

| | Specimen collection | Nucleic acid extraction | PCR analysis |
|---|---|---|---|
| | *Type:* Blood <br> *Vial:* EDTA <br> *Location:* Hospital ward <br> *Personnel:* Nurse, junior doctor | *Method:* Commercial kit <br> *Additional reagents:* Llysozyme, water, Tris-HCl <br> *Location:* Class II biological safety cabinet <br> *Personnel:* MLSO | *Reagents:* Water, buffer, MgCl$_2$, dNTPs, *Taq*, primers <br> *Location:* PCR cabinet in different location to nucleic acid extraction and post-pCR <br> *Personnel:* MLSO |
| **Contamination risk** | 1. Commensal flora on patient's skin <br> 2. Commensal flora on staff's skin <br> 3. Archival DNA in specimen vial <br> 4. Archival DNA in EDTA solution <br> 5. Inappropriate collection of specimen by personnel | 1. DNA from kit reagents <br> 2. DNA from additional reagents <br> 3. Location <br> 4. Class II biological safety cabinet <br> 5. Pipets <br> 6. Equipment e.g., centrifuge, heating block, vortex <br> 7. Reaction vials (1.5 mL) <br> 8. Personnel | 1. DNA from reagents <br> 2. Location <br> 3. PCR cabinet <br> 4. Pipets <br> 5. Equipment, e.g., vortex <br> 6. Reaction vials (1.5 mL) and PCR tubes <br> 7. Personnel |
| **Control actions to eliminate exogenous contamination** | 1. Iodine scrub of puncture site <br> 2. Wearing of sterile gloves and protective clothing <br> 3/4. Employment of laboratory prepared and quality-controlled DNA-free blood-EDTA vials | 1/2. All reagent purchased should be of molecular grade. All reagents should be screened prior to use. Contaminating DNA should be monitored using negative DNA extraction controls. | 1. All reagents purchased should be of molecular grade. All reagents should be screened prior to use. PCR master mix minus the primers and template DNA should be UV irradiated for 15 min prior to amplification. Contaminating DNA should be monitored using negative PCR setup controls. |

*(continued)*

**Table 5 (*Continued*)**

| Specimen collection | Nucleic acid extraction | PCR analysis |
|---|---|---|
| 5. Education of personnel | 3. A dedicated pre-PCR room should be used, ideally under positive pressure. Unidirectional work flow. | 2. A dedicated pre-PCR room should be used, ideally under positive pressure and separate from DNA extraction procedures. Unidirectional work flow. |
| | 4. Cabinet should be cleaned thoroughly and UV irradiated for a minimum period of 2 h prior to use. Cabinet should be serviced regularly. | 3. Cabinet should be cleaned thoroughly and UV irradiated for a minimum period of 2 h prior to use. Cabinet should be serviced regularly. |
| | 5. Dedicated pipets should be used and should not be removed from the class II cabinet. Plugged sterile tips should be employed throughout. Pipets should be cleaned and the barrels UV irradiated for a minimum of 2 h prior to use. | 4. Dedicated pipets should be used and should not be removed from the PCR setup cabinet. Plugged sterile tips should be employed throughout. Pipets should be cleaned and the barrels UV irradiated for a minimum of 2 h prior to use. |
| | 6. Equipment should be dedicated to extraction purposes only. | 5. Equipment should be dedicated to PCR setup purposes only |
| | 7. Reaction vials should be pre-autoclaved and DNA-free. | 6. Reaction vials and PCR tubes should be preautoclaved and DNA-free. |
| | 8. Education of personnel. Use of sterile gloves and dedicated laboratory clothing for DNA extraction purposes only. | 7. Education of personnel. Use of sterile gloves and dedicated laboratory clothing for PCR setup purposes only. |
| | 9. Avoidance of reusable laboratory glassware. | 8. Avoidance of reusable laboratory glassware. |

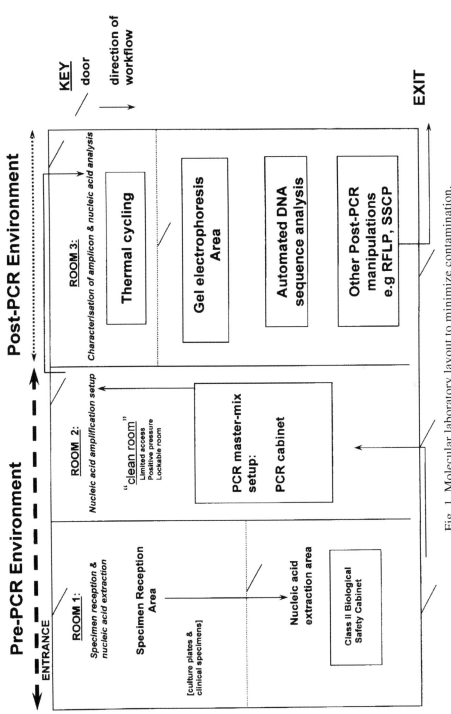

Fig. 1. Molecular laboratory layout to minimize contamination.

Successful employment of PCR in the detection of causal agents of infectious disease is critically dependent on both the quantity and quality of controls used in assays. **Table 6** lists reasons for false-positive and false-negative findings. It should be noted that for each PCR-based test, sensitivity should be evaluated for each specimen type, prior to routine implementation. It is vital that several negative and positive controls be set up during each diagnostic run. Negative and positive controls should include (1) DNA extraction control, (2) PCR set up control, and (3) PCR amplification control. For DNA extraction purposes, the positive control should include clinical material artificially spiked with organisms (e.g., blood culture spiked with *Escherichia coli*). For clinical tissue specimens, where a true positive specimen may be difficult to mimic, internal positive controls may be employed, such as following DNA extraction and amplification of the β-globin gene *(13)*. With respect to PCR controls, the positive control should be bacterial DNA extracted from a pure culture. Ideally, the positive control should include two components: (1) a specimen generating a weak signal, due to low copy numbers of target, and (2) a specimen generating a strong signal, as the result of high copy numbers of target. By employing these controls, especially the negative controls, it is easier to identify the point of contamination within the diagnostic assay—e.g., DNA extraction contamination-free, but contaminated at PCR set-up stage. Positive controls are also important, particularly those included in the DNA extraction procedures, as they serve to identify possible inhibition of the PCR reaction owing to inhibitory agents in the biological specimen, which coelute with extracted DNA, e.g., sodium polyanetholesulfonate in blood-culture material *(13)*. For a comprehensive review on PCR inhibition with respect to biological specimens, *see* Wilson *(124)*.

### 5.1. Standardization and Harmonization

Numerous molecular methodologies exist for the identification and genotyping of both culturable and nonculturable bacteria (*see* Chapter 14). Various commercial approaches have been described, such as the MicroSeq PCR-DNA sequencing identification system, but the majority of clinical microbiology laboratories use in-house methods. To date, there have been few attempts to standardize bacterial detection protocols among laboratories at a local, national, and international level, although several European centers have attempted this for specific organisms *(125–130)*. Lack of standardization (e.g., laboratories may use a published method, but with in-house variations and/or with different reagent suppliers) may lead to apparent anomalies in the epidemiology of a variety of infectious diseases. There are limited studies detailing the effect of such variation on qualitative reporting of results, and this area requires urgent attention.

**Table 6**
**Reasons for Obtaining False-Positive and False-Negative Results in Molecular Diagnostic Assays**

| Reasons for false-positive results | Reasons for false-negative results |
|---|---|
| Carry over contamination (amplicons) from previously amplified products | Inhibition of PCR reaction |
| Presence of exogenous target DNA in reagents, water, kits, sterile blood culture material | Inadvertent loss of template nucleic acid target owing to poor extraction, handling and storage protocols |
| Poor primer design (nonspecificity) | Digestion of nucleic acid template with endogenous DNAses and RNAses |
| Inadequate amplification conditions | Poor primer design (nonconserved regions at primer site[s] in variants) |
| Contamination from laboratory personnel | Poor intrinsic sensitivity of nucleic acid amplification/analysis detection system |
| | Poor sensitivity of nucleic acid amplification/analysis reaction |
| | Poor specificity |
| | Inadequate amplification conditions |

More recently, attempts have been made to standardize bacterial subtyping techniques, through the actions of PulseNet (http://www.cdc.gov/pulsenet/), which is primarily for bacterial food-borne disease surveillance in the United States, and the ESF Network for Exchange of Microbial Typing Information European Network (ENEMTI) (http://lists.nottingham.ac.uk/mailman/listinfo/ enemti). ENEMTI is a network of European laboratories that aims to standardize methods and data-exchange protocols for Internet-based comparison of microbial fingerprinting data. This project aims to develop an Internet-based database system for DNA fingerprints that is readily accessible and user-friendly for microbiologists with only limited computer expertise. In addition, the European Society for Clinical Microbiology and Infectious Disease (ESCMID) has a specific working group, the ESCMID Study Group on Epidemiological Markers (ESGEM), whose objectives are to evaluate critically microbiological typing systems and make recommendations for their appropriate use. They also aim to promote collaborative research into microbiological typing systems, to develop standardized methodology for specific pathogens, and to provide a forum for the exchange of ideas and the development of consensus strategies. To this end they may work with individuals and companies active in this research area to foster the development of further technological advances in microbial typing (http://www.escmid.org).

**Table 7**
**Advantages and Disadvantages of the Adoption of Molecular Assays into Clinical Bacteriology**

| Diagnostic criterion | Advantages | Disadvantages |
|---|---|---|
| *Accuracy of identification* | Aid in identifying etiological agents of infections, which are difficult to culture, including:<br><br>• Negative cultures<br>• Expensive cultures<br>• Slow-growing organisms<br>• Fastidious organisms<br>• Cell-dependent organisms<br>• Category three cultures where a designated secure cell culture laboratory is required<br>• Difficult, specific culture requirements where limited serological tests exist | • Problems associated with contaminant organisms, however, these problems may be aided by inclusion of appropriate DNA extraction and PCR controls (*121*)<br><br>• The agent identified should be considered with respect to the patient's medical and general history |
| *Time to detection*<br>Where specimens are:<br>(a) Culture-positive and/or serology-positive<br>(b) Culture-negative and/or serology-positive<br>(c) Culture-negative and serology-negative<br>*Impact on therapy* | Identification of causal agent following antibiotic therapy<br>• Confirmation of conventional detection result<br>• More rapid detection than conventional culture for fastidious and cell-dependent organisms<br>• Confirmation of serology result. Detect and highlight nonspecific serological false-positive results<br>• Identification of causal agent when all conventional diagnostic assays are negative<br>• Appropriate antibiotic therapy can be commenced sooner or modified earlier in the presence of a molecular identification from a culture-negative/serology-negative specimen<br>• Provision for PCR detection of antimicrobial resistance gene determinants in culture-negative PCR-positive specimen | • Longer time required for molecular PCR and sequence analysis than culture and serology for nonfastidious or cell-dependent organisms<br>• Longer time required for molecular PCR and sequence analysis than serology |
| *Cost-effectiveness* | • Cost-effective particularly with culture-negative/serology-negative specimens to avoid extended analysis for several potential pathogens either by specific culture and serological testing<br>• Economic and early use of most appropriate cost-effective antibiotic treatment regimen<br>• Economic and optimized in-patient stay | • Not cost-effective when conventional culture and serology give quality and early identification result<br>*Specialized equipment*<br>Necessary to purchase/lease specialized equipment usually with costly maintenance contracts |

**Table 7** (*Continued*)

| Diagnostic criterion | Advantages | Disadvantages |
| --- | --- | --- |
| | | *Space allocation* |
| | | *Lack of education in modern molecular-based technologies* Medical laboratory scientific officers, clinical scientists and medical microbiologists all must understand the principles of molecular based technologies to ensure proper handling of the specimens and appropriate interpretation and significance of results *(131)*, hence, specific training must be given |

## 5.2. Appraisal of Molecular Diagnostics in Clinical Bacteriology

Adoption of molecular diagnostics into routine bacteriology has advantages and disadvantages (**Table 7**). Their use is, at present, largely confined to specialized or reference laboratories, but various technologies, including PCR, real-time PCR, and pulsed-field gel electrophoresis, may eventually be adopted in regional or even in district diagnostic laboratories. Such methodologies can provide valuable real-time information, and can aid outbreak management and identification of nonculturable or fastidious organisms. To date, the speed at which these assays have been taken up has been related to the relative skills base within the laboratory. Thus, although all hospital types (from district general hospitals to university teaching hospitals) utilizing such technology would experience advantages, the techniques have become common only where the skills base exists. This gives potential for hospital research facilities, which have often been the custodians and developers of these research techniques, to become overloaded with a molecular diagnostic workload, simply because they have the necessary skills. In addition, molecular techniques are perceived to be more expensive than "traditional" techniques. While this is true if assessed head to head (**Table 8**), the overall value of the test result should take into account other parameters, including time-to-detection and ability to detect a causal agent. When assessed by these criteria, the greater initial cost of the molecular test is, arguably, offset because it may yield results that could potentially reduce significantly the costs of patient management further downstream.

**Table 8**
**Comparison of Financial Cost of Identification of a Bacterial Culture Employing Phenotypic and Genotypic Identification Schemes**

| Routine/conventional (API) identification | | Molecular (16S rDNA PCR and sequencing) identification | |
|---|---|---|---|
| Consumable item | Approx cost (GBP £) (ex VAT) | Consumable item | Approx cost (GBP £) (ex VAT) |
| 1X API20NE strip | 3.95 | DNA extraction kit | 2.50 |
| API diagnostic reagents[a] | 0.80 | PCR reagents (not including primers) | 0.12 (*Taq*) + 0.14 (dNTPs) |
| | | Gel electrophoresis | 1.09 |
| | | Plasticware consumables | 1.50 |
| | | Specialist reagents (TAE, Tris, EtBr, etc.) | 0.10 |
| | | PCR primers (forward and reverse) | 0.10 |
| | | *PCR subtotal costs* | $5.55 \times 2 = 11.10$ |
| | | Sequencing kit | 1.96 |
| | | Polyacrylamide gel[b] | 0.93 |
| | | Plasticware consumables | 1.50 |
| | | Sequencing primers (Cy-5' labeled) | 0.20 |
| | | *Sequencing subtotal costs* | $4.59 \times 2 = 9.18$ |
| *Total* | 4.75 | | 20.28 |

[a]Assuming one set of reagents are adequate for two kits
[b]Assuming the amplicon is sequenced as part of a ten-set batch.

In conclusion, molecular diagnostic techniques have a significant role to play in clinical bacteriology. However, they will never replace conventional methodologies, which continue to be the cornerstone of modern bacteriology. Indeed, molecular diagnostic assays will be implemented primarily in specialized laboratories to enhance laboratory diagnostic efficiency, and will be confined mainly to diagnosis, identification, and genotyping, where current conventional approaches are grossly inadequate. Adoption of molecular technologies in bacteriology has occurred at a much slower rate than in clinical virology; the inadequacies of conventional virology accelerated adoption of molecular methods. Integration of molecular approaches in clinical bacteriology will be enhanced through the production of a greater range of diagnostic kits, and the existence of more accredited laboratories.

## References

1. Farkas, D. H., Drevon, A. M., Kiechle, F. L., DiCarlo, R. G., Heath, E. M., and Crisan, D. (1996) Specimen stability for DNA-based diagnostic testing. *Diagn. Mol. Pathol.* **5,** 227–235.
2. Carpousis, A. J. (2002) The *Escherichia coli* RNA degradosome: structure, function and relationship in other ribonucleolytic multienzyme complexes. *Biochem. Soc. Trans.* **30,** 150–155.
3. Steege, D. A. (2000) Emerging features of mRNA decay in bacteria. *RNA* **6,** 1079–1090.
4. Patel, J. B. (2001) 16S rRNA gene sequencing for bacterial pathogen identification in the clinical laboratory. *Mol. Diagn.* **6,** 313–321.
5. Kolbert, C. P. and Persing, D. H. (1999) Ribosomal DNA sequencing as a tool for identification of bacterial pathogens. *Curr. Opin. Microbiol.* **2,** 299–305.
6. Therese, K. L., Anand, A. R., and Madhavan, H. N. (1998) Polymerase chain reaction in the diagnosis of bacterial endophthalmitis. *Br. J. Ophthalmol.* **82,** 1078–1082.
7. Xu, J., Moore, J. E., Millar, B. C., Crowe, M., McClurg, R., and Heaney, L. (2003) Identification of a novel α-*Proteobacterium* causing bacteraemia in an immunocompetent patient. *J. Infect.* **47,** 167–169.
8. Tunney, M. M., Patrick, S., Curran, M. D., Ramage, G., Hanna, D., Nixon, J. R., et al. (1999) Detection of prosthetic hip infection at revision arthroplasty by immunofluorescence microscopy and PCR amplification of the bacterial 16S rRNA gene. *J. Clin. Microbiol.* **37,** 3281–3290.
9. Schabereiter-Gurtner, C., Lubitz, W., and Rolleke, S. (2003) Application of broad-range 16S rRNA PCR amplification and DGGE fingerprinting for detection of tick-infecting bacteria. *J. Microbiol. Methods* **52,** 251–260.
10. Woo, P. C., Fung, A. M., Lau, S. K., Chan, B. Y., Chiu, S. K., Teng, J. L., et al. (2003) *Granulicatella adiacens* and *Abiotrophia defectiva* bacteraemia characterized by 16S rRNA gene sequencing. *J. Med. Microbiol.* **52,** 137–140.
11. Moore, J. E., Millar, B. C., Yongmin, X., Woodford, N., Vincent, S., Goldsmith, C. E., et al. (2001) A rapid molecular assay for the detection of antibiotic resistance determinants in causal agents of infective endocarditis. *J. Appl. Microbiol.* **90,** 719–726.
12. Goldenberger, D., Kunzli, A., Vogt, P., Zbinden, R., and Altwegg, M. (1997) Molecular diagnosis of bacterial endocarditis by broad-range PCR amplification and direct sequencing. *J. Clin. Microbiol.* **35,** 2733–2739.
13. Millar, B., Moore, J., Mallon, P., Xu, J., Crowe, M., McClurg, R., et al. (2001) Molecular diagnosis of infective endocarditis—a new Duke's criterion. *Scand. J. Infect. Dis.* **33,** 673-680.
14. Hryniewiecki, T., Gzyl, A., Augustynowicz, E., and Rawczynska-Englert, I. (2002) Development of broad-range polymerase chain reaction (PCR) bacterial identification in diagnosis of infective endocarditis. *J. Heart Valve Dis.* **11,** 870–874.

15. Mueller, N. J., Kaplan, V., Zbinden, R., and Altwegg, M. (1999) Diagnosis of *Cardiobacterium hominis* endocarditis by broad-range PCR from arterio-embolic tissue. *Infect.* **27**, 278–279.

16. Siqueira, J. F. Jr., Rjcas, I. N., Oliveira, J. C., and Santos, K. R. (2001) Detection of putative oral pathogens in acute periradicular abscesses by 16S rDNA–directed polymerase chain reaction. *J. Endod.* **27**, 164–167.

17. Ley, B. E., Linton, C. J., Bennett, D. M., Jalal, H., Foot, A. B., and Millar, M. R. (1998) Detection of bacteraemia in patients with fever and neutropenia using 16S rRNA gene amplification by polymerase chain reaction. *Eur. J. Clin. Microbiol. Infect. Dis.* **17**, 247–253.

18. Harris, K. A., Fidler, K. J., Hartley, J. C., Vogt, J., Klein, N. J., Monsell, F., et al. (2002) Unique case of *Helicobacter* sp. osteomyelitis in an immunocompetent child diagnosed by broad-range 16S PCR. *J. Clin. Microbiol.* **40**, 3100–3103.

19. Jalava, J., Mantymaa, M. L., Ekblad, U., Toivanen, P., Skurnik, M., Lassila, O., et al. (1996) Bacterial 16S rDNA polymerase chain reaction in the detection of intra-amniotic infection. *Br. J. Obstet. Gynaecol.* **103**, 664–649.

20. Carroll, N. M., Jaeger, E. E. M., Choudhury, S., Dunlop, A. A. S., Matheson, M. M., Adamson, P., et al. (2000) Detection of and discrimination between Gram-positive and Gram-negative bacteria in intraocular samples by using nested PCR. *J. Clin. Microbiol.* **38**, 1753–1757.

21. Westergren, V., Bassiri, M., and Engstrand, L. (2003) Bacteria detected by culture and 16S rRNA sequencing in maxillary sinus samples from intensive care unit patients. *Laryngoscope* **113**, 270–275.

22. Saravolatz, L. D., Manzor, O., VanderVelde, N., Pawlak, J., and Belian, B. (2003) Broad-range bacterial polymerase chain reaction for early detection of bacterial meningitis. *Clin. Infect. Dis.* **36**, 40–45.

23. Xu, J., Millar, B. C., Moore, J. E., Murphy, K., Webb, H., Fox, A. J., et al. (2003) Employment of broad-range 16S rRNA PCR to detect aetiological agents of infection from clinical specimens in patients with acute meningitis—rapid separation of 16S rRNA PCR amplicons without the need for cloning. *J. Appl. Microbiol.* **94**, 197–206.

24. Bucholtz, G. A., Salzman, S. A., Bersalona, F. B., Boyle, T. R., Ejercito, V. S., Penno, L., et al. (2002) PCR analysis of nasal polyps, chronic sinusitis, and hypertrophied turbinates for DNA encoding bacterial 16S rRNA. *Am. J. Rhinol.* **16**, 169–173.

25. Bailey, E. A., Solomon, L. R., Berry, N., Cheesbrough, J. S., Moore, J. E., Jiru, X., et al. (2002) *Ureaplasma urealyticum* CAPD peritonitis following insertion of an intrauterine device: diagnosis by eubacterial polymerase chain reaction. *Perit. Dial.* **22**, 422–424.

26. Berger, C., Altwegg, M., Meyer, A., and Nadal D. (2001) Broad range polymerase chain reaction for diagnosis of rat-bite fever caused by *Streptobacillus moniliformis*. *Pediatr. Infect. Dis. J.* **20**, 1181–1182.

27. Cuchacovich, R., Japa, S., Huang, W. Q., Calvo, A., Vega, L., Vargas, R.B., et al. (2002) Detection of bacterial DNA in Latin American patients with reactive

arthritis by polymerase chain reaction and sequencing analysis. *J. Rheumatol.* **29,** 1426–1429.

28. van der Heijden, I. M., Wilbrink, B., Vije, A. E., Schouls, L. M., Breedveld, F. C., and Tak, P. P. (1999) Detection of bacterial DNA in serial synovial samples obtained during antibiotic treatment from patients with septic arthritis. *Arthritis Rheum.* **42,** 2198–2203.

29. Hill, K. E., Davies, C. E., Wilson, M. J., Stephens, P., Harding, K. G., and Thomas, D. W. (2003) Molecular analysis of the microflora in chronic venous leg ulceration. *J. Med. Microbiol.* **52,** 365–369.

30. Woese, C. R. (1987) Bacterial evolution. *Microbiol. Rev.* **51,** 221–271.

31. Gurtler, V. and Stanisich, V. A. (1996) New approaches to typing and identification of bacteria using the 16S–23S rDNA spacer region. *Microbiol.* **142,** 13–16.

32. Shang, S., Fu, J., Dong, G., Hong, W., Du, L., and Yu, X. (2003) Establishment and analysis of specific DNA patterns in 16S–23S rRNA gene spacer regions for differentiating different bacteria. *Chin. Med. J. (Engl.)* **116,** 129–133.

33. Houpikian P. and Raoult, D. (2001) 16S/23S rRNA intergenic spacer regions for phylogenetic analysis, identification, and subtyping of *Bartonella* species. *J. Clin. Microbiol.* **39,** 2768–2778.

34. Madico, G., Quinn ,T. C., Boman, J., and Gaydos, C. A. (2000) Touchdown enzyme time release–PCR for detection and identification of *Chlamydia trachomatis, C. pneumoniae,* and *C. psittaci* using the 16S and 16S–23S spacer rRNA genes. *J. Clin. Microbiol.* **38,** 1085–1093.

35. Geissdorfer, W., Wittmann, I., Rollinghoff, M., Schoerner, C., and Bogdan, C. (2001) Detection of a new 16S–23S rRNA spacer sequence variant (type 7) of *Tropheryma whippelii* in a patient with prosthetic aortic valve endocarditis. *Eur. J. Clin. Microbiol. Infect. Dis.* **20,** 762–763.

36. Roth, A., Reischl, U., Streubel, A., Naumann, L., Kroppenstedt, R. M., Habicht, M., et al. (2000) Novel diagnostic algorithm for identification of mycobacteria using genus-specific amplification of the 16S–23S rRNA gene spacer and restriction endonucleases. *J. Clin. Microbiol.* **38,** 1094–1104.

37. Bakshi, C. S., Singh, V. P, Malik, M., Sharma, B., and Singh, R. K. (2002) Polymerase chain reaction amplification of 16S–23S spacer region for rapid identification of *Salmonella* serovars. *Acta. Vet. Hung.* **50,** 161–166.

38. Anthony, R. M., Brown, T. J, and French, G. L. (2000) Rapid diagnosis of bacteremia by universal amplification of 23S ribosomal DNA followed by hybridization to an oligonucleotide array. *J. Clin. Microbiol.* **38,** 781–788.

39. Whitby, P. W., Carter, K. B., Burns, J. L., Royall, J. A., LiPuma, J. J., and Stull, T. L. (2000) Identification and detection of *Stenotrophomonas maltophilia* by rRNA-directed PCR. *J. Clin. Microbiol.* **38,** 4305–4309.

40. Patel, J. B., Leonard, D. G. B., Pan, X., Musser, J. M., Berman, R. E., and Nachamkin, I. (2000) Sequence-based identification of *Mycobacterium species* using the MicroSeq 500 16S rDNA bacterial identification system. *J. Clin. Microbiol.* **38,** 246–251.

41. Goh, S. H., Potter, S., Wood, J. O., Hemmingsen, S. M., Reynolds, R. P., and Chow, A. W. (1996) HSP60 gene sequences as universal targets for microbial

species identification: studies with coagulase-negative staphylococci. *J. Clin. Microbiol.* **34,** 818–823.

42. Woo, P. C., Woo, G. K., Lau, S. K., Wong, S. S., and Yuen, K. (2002) Single gene target bacterial identification. groEL gene sequencing for discriminating clinical isolates of *Burkholderia pseudomallei* and *Burkholderia thailandensis. Diagn. Microbiol. Infect. Dis.* **44,** 143–149.

43. Feltham, J. L. and Gierasch, L. M. (2000) groEL-substrate interactions: molding the fold, or folding the mold? *Cell* **100,** 193–196.

44. Matsui, T., Matsuda, M., Murayama, O., Millar, B. C., and Moore, J. E. (2001) recA genotyping of *Salmonella enteritidis* phage type 4 isolates by restriction fragment length polymorphism analysis. *Lett. Appl. Microbiol.* **32,** 424–427.

45. Francis, K. P. and Stewart, G. S. (1997) Detection and speciation of bacteria through PCR using universal major cold-shock protein primer oligomers. *J. Ind. Microbiol. Biotechnol.* **19,** 286–293

46. Moore, J. E., Millar, B. C., Jiru, X., McCappin, J., Crowe, M., and Elborn, J. S. (2001) Rapid characterization of the genomovars of the *Burkholderia cepacia* complex by PCR-single-stranded conformational polymorphism (PCR-SSCP) analysis. *J. Hosp. Infect.* **48,** 129–134.

47. Slater, E. R. and Owen, R. J. (1997) Restriction fragment length polymorphism analysis shows that the hippuricase gene of *Campylobacter jejuni* is highly conserved. *Lett. Appl. Microbiol.* **25,** 274–278.

48. Guiver, M., Borrow, R., Marsh, J., Gray, S. J., Kaczmarski, E. B., Howells, D., et al. (2000) Evaluation of the Applied Biosystems automated Taqman polymerase chain reaction system for the detection of meningococcal DNA. *FEMS Immunol. Med. Microbiol.* **28,** 173–179.

49. Such, J., Frances, R., Munoz, C., Zapater, P., Casellas, J. A., Cifuentes, A., et al. (2002) Detection and identification of bacterial DNA in patients with cirrhosis and culture-negative, nonneutrocytic ascites. *Hepatology* **36,** 135–141.

50. Apfalter, P., Assadian, O., Blasi, F., Boman, J., Gaydos, C. A., Kundi, M., et al. (2002) Reliability of nested PCR for detection of *Chlamydia pneumoniae* DNA in atheromas: results from a multicenter study applying standardized protocols. *J. Clin. Microbiol.* **40,** 4428–4434.

51. Gamboa, F., Manterola, J. M., Lonca, J., Matas, L., Vinado, B., Gimenez, M., et al. (1997) Detection and identification of mycobacteria by amplification of RNA and DNA in pretreated blood and bone marrow aspirates by a simple lysis method. *J. Clin. Microbiol.* **35,** 2124–2128.

52. Schmidt, B. L., Aberer, E., Stockenhuber, C., Klade, H., Breier, F., and Luger, A. (1995) Detection of *Borrelia burgdorferi* DNA by polymerase chain reaction in the urine and breast milk of patients with lyme borreliosis. *Diag. Microbiol Infect. Dis.* **21,** 121–128.

53. Ersch, J., Speich, R., Weber, R., Altwegg, M., and Hauser, M. (2000) Value of bronchoalveolar lavage in the diagnostic work-up of HIV-associated lung disease. *Dtsch. Med. Wochenschr.* **125,** 789–793.

54. Lehmann, M., Groh, A., Rodel, J., Nindl, I., and Straube, E.. (1999) Detection of *Chlamydia trachomatis* DNA in cervical samples with regard to infection by human papillomavirus. *J. Infect.* **38**, 12–17.

55. Kabir, S. (2001) Detection of *Helicobacter pylori* in faeces by culture, PCR and enzyme immunoassay. *J. Med. Microbiol.* **50**, 1021–1029.

56. Wu, L., Patten, N., Yamashiro, C. T., and Chui B. (2002) Extraction and amplification of DNA from formalin-fixed, paraffin-embedded tissues. *Appl. Immunohistochem. Mol. Morphol.* **10**, 269–274.

57. Gauduchon, V., Chalabreysse, L., Etienne, J., Celard, M., Benito, Y., Lepidi, H., et al. (2003) Molecular diagnosis of infective endocarditis by PCR amplification and direct sequencing of DNA from valve tissue. *J. Clin. Microbiol.* **41**, 763–766.

58. Yamada, T., Eishi, Y., Ikeda, S., Ishige, I., Suzuki, T., Takemura, T., et al. (2002) In situ localization of *Propionibacterium acnes* DNA in lymph nodes from sarcoidosis patients by signal amplification with catalysed reporter deposition. *J. Pathol.* **198**, 541–547.

59. Jero, J., Alakarppa, H., Virolainen, A., Saikku, P., and Karma, P. (1999) Polymerase chain reaction assay for detecting *Chlamydia pneumoniae* in middle ear fluid of children with otitis media with effusion. *Pediatr. Infect. Dis. J.* **18**, 939–940.

60. Kox, L. F., van Leeuwen, J., Knijper, S., Jansen, H. M., and Kolk, A. H. (1995) PCR assay based on DNA coding for 16S rRNA for detection and identification of mycobacteria in clinical samples. *J. Clin. Microbiol.* **33**, 3225–3233.

61. Klaschik, S., Lehmann, L. E., Raadts, A., Book, M., Hoeft, A., and Stuber, F. (2002) Real-time PCR for detection and differentiation of Gram-positive and Gram-negative bacteria. *J. Clin. Microbiol.* **40**, 4304–4307.

62. Yanagihara, K., Tomono, K., Sawai, T., Miyasaki,Y., Hirakata, Y., Kadota, J., et al. (2002) *Mycobacterium avium* complex pleuritis. *Respiration* **69**, 549–551.

63. Sakamoto, M., Huang, Y., Umeda, M., Ishikawa, I., and Benno Y. (2002) Detection of novel oral phylotypes associated with periodontitis. *FEMS Microbiol. Lett.* **217**, 65–69.

64. Gdoura, R., Keskes-Ammar, L., Bouzid, F., Eb, F., Hammami, A., and Orfila, J. (2001) *Chlamydia trachomatis* and male infertility in Tunisia. Eur. *J. Contracept. Reprod. Health Care* **6**, 102–107.

65. Torres, P., Camarena, J. J., Gomez, J. R., Nogueira, J. M., Gimeno, V., Navarro, J. C., et al. (2003) Comparison of PCR mediated amplification of DNA and the classical methods for detection of *Mycobacterium leprae* in different types of clinical samples in leprosy patients and contacts. *Lepr. Rev.* **74**, 18–30.

66. Enzensberger, R., Hunfeld, K. P., Elshorst-Schmidt, T., Boer, A., and Brade, V. (2002) Disseminated cutaneous *Mycobacterium marinum* infection in a patient with non-Hodgkin's lymphoma. *Infect.* **30**, 393–395.

67. McDowell, A., Mahenthiralingam, E., Moore, J. E., Dunbar, K. E., Webb, A. K., Dodd, M. E., et al. (2001) PCR-based detection and identification of *Burkholderia*

*cepacia* complex pathogens in sputum from cystic fibrosis patients. *J. Clin. Microbiol.* **39,** 4247–4255.

68. Mahony, J. B., Jang, D., Chong, S., Luinstra, K., Sellors, J., Tyndall, M., et al. (1997) Detection of *Chlamydia trachomatis, Neisseria gonorrhoeae, Ureaplasma urealyticum,* and *Mycoplasma genitalium* in first-void urine specimens by multiplex polymerase chain reaction. *Mol. Diagn.* **2,** 161–168.

69. Obata-Yasuoka, M., Ba-Thein, W., Hamada, H., and Hayashi, H. (2002) A multiplex polymerase chain reaction-based diagnostic method for bacterial vaginosis. *Obstet. Gynecol.* **100,** 759–764.

70. Sharma, S., Das, D., Anand, R., Das, T., and Kannabiran, C. (2002) Reliability of nested polymerase chain reaction in the diagnosis of bacterial endophthalmitis. *Am. J. Ophthalmol.* **133,** 142–144.

71. Millar, B. C., Jiru, X., Moore, J. E., and Earle, J. A. (2000) A simple and sensitive method to extract bacterial, yeast and fungal DNA from blood culture material. *J. Microbiol. Methods* **42,** 139–147.

72. Raggam, R. B., Leitner, E., Muhlbauer, G., Berg, J., Stocher, M., Grisold, A. J., et al. (2002) Qualitative detection of *Legionella* species in bronchoalveolar lavages and induced sputa by automated DNA extraction and real-time polymerase chain reaction. *Med. Microbiol. Immunol. (Berl.)* **191,** 119–125.

73. Clarke, L., Millar, B. C., and Moore, J. E. (2003) Extraction of genomic DNA from *Pseudomonas aeruginosa*: a comparison of three methods. *Br. J. Biomed. Sci.* **60,** 34–35.

74. Moore, J. E., Lanser, J., Heuzenroeder, M., Ratcliff, R. M., Millar, B. C., and Madden, R. H. (2002) Molecular diversity of *Campylobacter coli* and *C. jejuni* isolated from pigs at slaughter by flaA-RFLP analysis and ribotyping. *J. Vet. Med. B Infect. Dis. Vet. Public Health* **49,** 388–393.

75. Tang, Y. W., Procop, G. W., and Persing, D. H. (1997) Molecular diagnostics of infectious diseases. *Clin. Chem.* **43,** 2021–2038.

76. Moore, J. E., Xu, J., Millar, B. C., Crowe, M., and Elborn, J. S. (2002) Improved molecular detection of *Burkholderia cepacia* genomovar III and *Burkholderia multivorans* directly from sputum of patients with cystic fibrosis. *J. Microbiol. Methods* **49,** 183–191.

77. Hinrikson, H. P., Dutly, F., and Altwegg, M. (2000) Evaluation of a specific nested PCR targeting domain III of the 23S rRNA gene of *"Tropheryma whippelii"* and proposal of a classification system for its molecular variants. *J. Clin. Microbiol.* **38,** 595–599.

78. Weaver, J. W. and Rowe, M. T. (1997) Effect of non-target cells on the sensitivity of the PCR for *Escherichia coli* O157:H7. *Lett. Appl. Microbiol.* **25,** 109–112.

79. Zheng, X., Kolbert, C. P., Varga-Delmore, P., Arruda, J., Lewis, M., Kolberg, J., et al. (1999) Direct mecA detection from blood culture bottles by branched-DNA signal amplification. *J. Clin. Microbiol.* **37,** 4192–4193.

80. Jin, L. and Lloyd, R. V. (1997) In situ hybridization: methods and applications. *J. Clin. Lab. Anal.* **11,** 2–9.

81. Blocker, M. E., Krysiak, R. G., Behets, F., Cohen, M. S., and Hobbs, M. M. (2002) Quantification of *Chlamydia trachomatis* elementary bodies in urine by ligase chain reaction. *J. Clin. Microbiol.* **40**, 3631–3634.

82. Bachmann, L. H., Desmond, R. A., Stephens, J., Hughes, A., and Hook, E. W. 3rd. (2002) Duration of persistence of gonococcal DNA detected by ligase chain reaction in men and women following recommended therapy for uncomplicated gonorrhea. *J. Clin. Microbiol.* **40**, 3596–3601.

83. Wang, S. X. and Tay, L. (2002) Early identification of *Mycobacterium tuberculosis* complex in BACTEC cultures by ligase chain reaction. *J. Med. Microbiol.* **51**, 710–712.

84. Gilpin, C. M., Dawson, D. J., O'Kane, G., Armstrong, J. G., and Coulter, C. (2002) Failure of commercial ligase chain reaction to detect *Mycobacterium tuberculosis* DNA in sputum samples from a patient with smear-positive pulmonary tuberculosis due to a deletion of the target region. *J. Clin. Microbiol.* **40**, 2305–2307.

85. Castriciano, S., Luinstra, K., Jang, D., Patel, J., Mahony, J., Kapala, J., et al. (2002) Accuracy of results obtained by performing a second ligase chain reaction assay and PCR analysis on urine samples with positive or near-cutoff results in the LCx test for *Chlamydia trachomatis*. *J. Clin. Microbiol.* **40**, 2632–2634.

86. Rajo, M. C., Perez Del Molina, M. L., Lado Lado, F. L., Lopez, M. J., Prieto, E., and Pardo, F. (2002) Rapid diagnosis of tuberculous meningitis by ligase chain reaction amplification. *Scand. J. Infect. Dis.* **34**, 14–16.

87. Freise, J., Gerard, H. C., Bunke, T., Whittum-Hudson, J. A., Zeidler, H., Kohler, L., et al. (2001) Optimised sample DNA preparation for detection of *Chlamydia trachomatis* in synovial tissue by polymerase chain reaction and ligase chain reaction. *Ann. Rheum. Dis.* **60**, 140–145.

88. Anthony, R. M., Brown, T. J., and French, G. L. (2001) DNA array technology and diagnostic microbiology. *Expert. Rev. Mol. Diagn.* **1**, 30–38.

89. Cook, N. (2003). The use of NASBA for the detection of microbial pathogens in food and environmental samples. *J Microbiol Methods* **53**, 165–174.

90. Cook, N., Ellison, J., Kurdziel, A. S., Simpkins, S., and Hays, J. P. (2002) A nucleic acid sequence–based amplification method to detect *Salmonella enterica* serotype enteritidis strain PT4 in liquid whole egg. *J. Food Prot.* **65**, 1177–1178.

91. Maibach, R. C., Dutly, F., and Altwegg, M. (2002) Detection of *Tropheryma whipplei* DNA in feces by PCR using a target capture method. *Clin. Microbiol.* **40**, 2466–2471.

92. An, Q., Liu, J., O'Brien, W., Radcliffe, G., Buxton, D., Popoff, S., et al. (1995) Comparison of characteristics of Q beta replicase-amplified assay with competitive PCR assay for *Chlamydia trachomatis*. *J. Clin. Microbiol.* **33**, 58–63.

93. O'Mahony, J. and Hill, C. (2002) A real time PCR assay for the detection and quantitation of *Mycobacterium avium* subsp. *paratuberculosis* using SYBR Green and the Light Cycler. *J. Microbiol. Methods* **51**, 283–293.

94. Ellerbrok, H., Nattermann, H., Ozel, M., Beutin, L., Appel, B., and Pauli, G. (2002) Rapid and sensitive identification of pathogenic and apathogenic *Bacillus anthracis* by real-time PCR. *FEMS Microbiol. Lett.* **214,** 51–59.

95. Ge, B., Larkin, C., Ahn, S., Jolley, M., Nasir, M., Meng, J., et al. (2002) Identification of *Escherichia coli* O157:H7 and other enterohemorrhagic serotypes by EHEC-hlyA targeting, strand displacement amplification, and fluorescence polarization. *Mol. Cell. Probes* **16,** 85–92.

96. Hill, C. S. (2001) Molecular diagnostic testing for infectious diseases using TMA technology. *Expert. Rev. Mol. Diagn.* **1,** 445–455.

97. Moreno, C., Kutzner, H., Palmedo, G., Goerttler, E., Carrasco, L., and Requena, L. (2003) Interstitial granulomatous dermatitis with histiocytic pseudorosettes: a new histopathologic pattern in cutaneous borreliosis. Detection of *Borrelia burgdorferi* DNA sequences by a highly sensitive PCR-ELISA. *J. Am. Acad. Dermatol.* **48,** 376–384.

98. Brown, P. D. and Levett, P. N. (1997) Differentiation of *Leptospira* species and serovars by PCR-restriction endonuclease analysis, arbitrarily primed PCR and low-stringency PCR *J. Med. Microbiol.* **46,** 173–181.

99. Kerr, J. R. and Curran, M. D. (1996) Applications of polymerase chain reaction single stranded conformational polymorphism to microbiology. *J. Clin. Path. Mol. Path.* **49,** M315–M320.

100. Hein, I., Mach, R. L., Farnleitner, A. H., and Wagner, M. (2003) Application of single-strand conformation polymorphism and denaturing gradient gel electrophoresis for fla sequence typing of *Campylobacter jejuni*. *J. Microbiol. Methods* **52,** 305–313.

101. Moreno, Y., Ferrus, M. A., Vanoostende, A., Hernandez, M., Montes, R. M., and Hernandez, J. (2002) Comparison of 23S polymerase chain reaction–restriction fragment length polymorphism and amplified fragment length polymorphism techniques as typing systems for thermophilic campylobacters. *FEMS Microbiol. Lett.* **211,** 97–103.

102. Dabrowski, W., Czekajlo-Kolodziej, U., Medrala, D., and Giedrys-Kalemba, S. (2003) Optimisation of AP-PCR fingerprinting discriminatory power for clinical isolates of *Pseudomonas aeruginosa*. *FEMS Microbiol. Lett.* **218,** 51–57.

103. Enright, M. C. and Spratt, B. G. (1999) Multilocus sequence typing. *Trends Microbiol.* **7,** 482–487.

104. Wu, F. and Della-Latta, P. (2002) Molecular typing strategies. *Semin. Perinatol.* **26,** 357–366.

105. Power, E. G. (1996) RAPD typing in microbiology—a technical review. *J. Hosp. Infect.* **34,** 247–265.

106. Gillespie, S. H. (1999) The role of the molecular laboratory in the investigation of *Streptococcus pneumoniae* infections. *Semin. Respir. Infect.* **14,** 269–275.

107. Marty, N. (1997) Epidemiological typing of *Stenotrophomonas maltophilia*. *Hosp. Infect.* **36,** 261–266.

108. Baldy-Chudzik, K. (2001) Rep-PCR—a variant to RAPD or an independent technique of bacteria genotyping? A comparison of the typing properties of rep-PCR

with other recognised methods of genotyping of microorganisms. *Acta. Microbiol. Pol.* **50,** 189–204.

109. Fluit, A. C., Visser, M. R., and Schmitz, F. J. (2001) Molecular detection of antimicrobial resistance. *Clin. Microbiol. Rev.* **14,** 836–871.

110. Kobayashi, N., Kojima, K., Taniguchi, K., Urasawa, S., Uehara, N., Omizu, Y., et al. (1994) Detection of *mecA, femA,* and *femB* genes in clinical strains of staphylococci using polymerase chain reaction. *Epidemiol. Infect.* **113,** 259–266.

111. Towner, K. J., Talbot, D. C. S., Curran, R., Webster, C. A., and Humphreys, H. (1998) Development and evaluation of a PCR-based immunoassay for the rapid detection of methicillin-resistant *Staphylococcus aureus. J. Med. Microbiol.* **47,** 1–7.

112. Høiby N. (1991) Cystic fibrosis: infection. *Schweiz. Med. Wochenschr.* **121,** 105–109.

113. Doring, G., Jansen, S., Noll, H., Grupp, H., Frank, F., Botzenhart, K., et al. (1996) Distribution and transmission of *Pseudomonas aeruginosa* and *Burkholderia cepacia* in a hospital ward. *Pediat. Pulmonol.* **21,** 90–100.

114. Mahenthiralingam, E., Campbell, M. E., and Speert, D. P. (1997) Identification and characterization of a novel DNA marker associated with epidemic strains of *Burkholderia cepacia* recovered from patients with cystic fibrosis. *J. Clin. Microbiol.* **35,** 808–816.

115. Sajjan, U. S., Sun, L., Goldstein, R., and Forstner, J. F. (1995) Cable (Cbl) type II pili of cystic fibrosis–associated *Burkholderia (Pseudomonas) cepacia*: nucleotide sequence of the cblA major subunit pilin gene and novel morphology of the assembled appendage fibers. *J. Bacteriol.* **177,** 1030–1038.

116. Anon. Infection Control Guidelines: *Burkholderia cepacia* (1999) UK Cystic Fibrosis Trust, Bromley, Kent, England.

117. De Soyza, A., Corris, P. A., Archer, L., McDowell, A., Moore, J., Elborn, S., et al. (2000) Pulmonary transplantation for CF: the effect of *B. cepacia genomovars* on outcomes. *Thorax* **55 (Suppl 3),** S35.

118. Segonds, C., Heulin, T., Marty, N., and Chabanon, G. (1999) Differentiation of *Burkholderia* species by PCR-restriction fragment length polymorphism analysis of the 16S rRNA gene and application to cystic fibrosis isolates. *J. Clin. Microbiol.* **37,** 2201–2208.

119. LiPuma, J. J., Dulaney, B. J., McMenamin, J. D., Whitby, P. W., Stull, T. L., Coenye, T., et al. (1999) Development of rRNA-based PCR assays for identification of *Burkholderia cepacia* complex isolates recovered from cystic fibrosis patients. *J. Clin. Microbiol.* **37,** 3167–3170.

120. Mahenenthiralingam, E., Bischof, J., Byrne, S. K., Radomski, C., Davies, J. E., Av-Gay, Y., et al. (2000) DNA-based diagnostic approaches for identification of *Burkholderia cepacia* complex, *Burkholderia vietnamiensis, Burkholderia multivorans, Burkholderia stabilis,* and *Burkholderia cepacia* genomovars I and III. *J. Clin. Microbiol.* **38,** 3165–3173.

121. Millar, B. C., Xu, J., and Moore, J. E. (2002) Risk assessment models and contamination management: implications for broad-range ribosomal DNA PCR as a diagnostic tool in medical bacteriology. *J. Clin. Microbiol.* **40,** 1575–1580.

122. Bastien, P., Chabbert, E., Lauchaud, L., Millar, B. C., Xu, J., and Moore, J. E. (2003) Contamination management of broad-range or specific PCR: Is there any difference? *J. Clin. Microbiol.* **41,** 2272.

123. Anon. (1999) Additional guidance for inspectors' use of molecular biology techniques in clinical pathology. Clinical Pathology Accreditation (UK) Ltd., London. pp. 1–9.

124. Wilson, I. G. (1997) Inhibition and facilitation of nucleic acid amplification. *Appl. Environ. Microbiol.* **63,** 3741–3751.

125. Struelens, M. J. and the Members of the European Study Group on Epidemiological Markers (ESGEM) of the European Society for Clinical Microbiology and Infectious Diseases (ESCMID) (1996) Consensus guidelines for appropriate use and evaluation of microbial epidemiologic typing systems. *Clin. Microbiol. Infect.* **2,** 2–11.

126. Deplano, A., Schuermans, A., Van Eldere, J, Witte, W., Meugnier, H., Etienne, J., et al. (2000) Multicenter evaluation of epidemiological typing of methicillin-resistant *Staphylococcus aureus* strains by repetitive-element PCR analysis. *J. Clin. Microbiol.* **38,** 3527–3533.

127. Dijkshoorn, L., Towner, K. J., and Struelens M., eds. (2001) *New Approaches for the Generation and Analysis of Microbial Typing Data.* Elsevier, Amsterdam.

128. Fry, N. K., Alexiou-Daniel, S., Bangsborg, J. M., Bernander, S., Castellani Pastoris, M., Etienne, J., et al. (1999) A multicenter evaluation of genotypic methods for epidemiologic typing of *Legionella pneumophila* serogroup 1: results of a pan-European study. *Clin. Microbiol. Infect.* **5,** 462–477.

129. Grundmann, H. J., Towner, K. J., Dijkshoorn, L., Gerner-Smidt, P., Maher, M., Seifert, H., et al. (1997) Multicenter study using standardized protocols and reagents for evaluation of reproducibility of PCR-based fingerprinting of *Acinetobacter* spp. *J. Clin. Microbiol.* **35,** 3071–3077.

130. Van Belkum, A., Van Leeuwen, W., Kaufmann, M. E., Cookson, B., Forey, F., Etienne, J., et al. (1998) Assessment of resolution and intercenter reproducibility of results of genotyping *Staphylococcus aureus* by pulsed-field gel electrophoresis of *Sma*I macrorestriction fragments: a multicenter study. *J. Clin. Microbiol.* **36,** 1653–1659.

131. Moore, J. E. and Millar, B. C. (2002) Need for improved molecular biology training for biomedical scientists in NHS microbiology laboratories. *Br. J. Biomed. Sci.* **59,** 180.

# 8

## Molecular Diagnostics

*Future Probe-Based Strategies*

### Peter Marsh and Donald L. N. Cardy

### Summary

Nucleic acid amplification technologies (NAATs) represent powerful tools in clinical microbiology, particularly in areas where traditional culture-based methods alone prove insufficient. A notable advantage is in reducing the time from taking samples to reporting results. This, and the specificity and sensitivity imparted by NAATs, can help to improve patient care. Both thermal and isothermal NAATs have been adapted to aid diagnosis in clinical laboratories. Current molecular diagnostic assays are generally high-tech, and are expensive to buy and perform. Easy-to-use NAATs are beginning to appear, not only facilitating acceptable throughput in clinical laboratories, but also allowing tests to move out of the laboratory, closer to the point of care. Demand for simpler, miniaturized equipment and assays, and the trend toward personalized medicine, is leading towards the development of fully integrated automation and home-use kits. The integration of diverse disciplines, such as genomics, molecular biology, microelectromechanical systems, microfluidics, microfabrication, and organic chemistry, is behind the emerging DNA microarray technology. Development of DNA microchips allows the simultaneous detection of potentially thousands of target sequences, not only favoring high throughput, but also the potential for genotyping patient subsets with respect to their response to particular drug types (pharmakogenomics). It is envisaged that the future of probe-based technologies will see the development of fully integrated assays and devices suitable for nonskilled users.

**Key Words:** Isothermal amplification; signal-mediated amplification; cycling probe technology; invader technology; Q-beta replicase; point-of-care testing.

From: *Methods in Molecular Biology, vol. 266: Genomics, Proteomics, and Clinical Bacteriology:*
*Methods and Reviews*
Edited by: N. Woodford and A. Johnson © Humana Press Inc., Totowa, NJ

## 1. Introduction

Technologies based on nucleic acid probes have developed over the past two decades into essential tools in molecular microbiology. In terms of speed, specificity, and sensitivity, probe-based technologies generally outperform traditional microbiology and other nonmolecular technologies such as serological tests (e.g., latex agglutination) in specific detection of organisms. Many probe-based technologies have already been developed into diagnostic kits and assays for use in clinical microbiology. However, health care decision-makers weigh the advantages of molecular biology in diagnostics against the perceived lower cost of traditional methods. Many routine methods, which depend on microbial recovery and growth, are sufficient for general patient screening tests for organisms that are easily grown and identified using laboratory culture media (e.g., coliforms). Therefore, at present and in the near future, the costs of probe-based technologies enforce a high degree of selectivity as to when and to what sample type they can be applied.

The areas of clinical diagnosis where probe-based technologies will have most impact will be those where they offer solutions to diagnostic problems that are currently insurmountable by the use of traditional methods. The higher costs must be justified by advantages unique to, or easily imparted by, probe-based technologies. Target and/or signal amplification techniques are the most attractive of probe-based technologies in terms of applicability to diagnostics, in that the sensitivity and speed required is readily facilitated by nucleic acid amplification technologies (NAATs). NAATs provide the assay with an amplified mass of nucleic acid, which can be manipulated with relative ease for end-detection. The main problem with the use of microbial recovery and growth for diagnosis is the time between obtaining the sample and reporting the identification. If a NAAT can be interpolated into a pathogen-detection procedure that significantly reduces this time, then it is feasible to employ the NAAT. The value of such a change in diagnostic strategy would be further increased if, by virtue of its nature, the NAAT imparted improved reproducibility, specificity, and sensitivity. **Table 1** illustrates the typical differences between use of culture-based techniques and the use of probe-based techniques for the detection of methicillin-resistant *Staphylococcus aureus* (MRSA) in patient swab samples from routine screening in a typical hospital clinical laboratory. Current standard tests can take 3–4 d before results are available for reporting, whereas a NAAT-based test (i.e., one which can be completed within 1 working day) can yield results on the day following the initial overnight enrichment broth. Thus medical management of patients can be improved, with earlier administration of correct therapy and a resulting reduction in morbidity and mortality, and more efficient use of hospital resources (e.g., beds). Consequently, the money saved by health services might balance favorably against

**Table 1**
**Comparison of Time and Activities Required for Traditional Culture-Based Detection of MRSA With NAAT-Based Detection of MRSA in Hospital Patient-Screening Swab Samples**

| Time taken | Culture-based MRSA detection | MRSA detection using NAATs |
|---|---|---|
| Day 0 | Patient swabs taken and enrichment broth inoculated: incubate | Patient swabs taken and enrichment broth inoculated |
| Day 1 | Selective plates inoculated with broth cultures: incubate | NAAT-based test, e.g., PCR, CytAMP-MRSA Results reported |
| Day 2 | Selective plates observed for any growth: incubation continued | |
| Day 3 | Confirmatory tests, e.g., Staph latex, Mastalex Results reported Includes possible further culture-based tests, e.g., antibiotic disc diffusion assays | |
| Day 4 | Further incubation of culture-based confirmatory tests Results reported | |

The details for processing specimens shown the middle column are based on the standard operating procedure of a large hospital in the United Kingdom.

the cost of NAATs as diagnostic aids. Even more desirable are NAATs that can detect specific target sequences directly from samples without enrichment culture. The polymerase chain reaction (PCR) and other target-amplifying technologies have the sensitivity to detect very low copy numbers of target, and are hence suitable for such an application. Sample matrices (blood, urine, and so on) can inhibit reactions and can therefore make such approaches unpredictable. However, the sensitivity of PCR lends this technology in particular to direct detection, and there are reported applications in clinical microbiology, described in **Subheading 2.**

Many clinical laboratories are still not equipped or staffed for undertaking molecular biological methods such as PCR, which requires expensive thermal cycling equipment for amplification and (usually) electrophoresis equipment for end-detection. A diagnostic version of PCR (the Roche Amplicor kit) requires a dedicated machine incorporating a thermal cycler, while in more advanced versions, dedicated equipment for real-time detection of fluorescent amplicons (e.g., the Roche LightCycler; *see* Chapter 9) is required. At the present time, the use of these and other high-tech diagnostics, such as nucleic acid sequence-based amplification (NASBA), ligase chain reaction (LCR), and strand displacement amplification (SDA), is confined to larger central or reference laboratories, which process samples that include those sent from outlying hospitals. This point rather negates one of the main advantages that molecular biology should impart to clinical diagnostics—namely, that of rapid identification of clinically important targets.

## 2. Polymerase Chain Reaction

PCR is the first-generation NAAT and can be viewed as suitable for diagnostic applications, due to its ability to amplify multiple copies of a target sequence in a short time. PCR can be of particular value for the detection of pathogens that are not easily or quickly culturable, such as *Mycobacterium tuberculosis,* where sputum samples collected on filter paper can be used for direct PCR *(1)*. Manual PCR protocols are, however, somewhat laborious and unsuitable for general use in clinical diagnostics. Hence, the few commercial diagnostic versions of PCR are generally automated, as is the case with assays marketed by Roche and Johnson & Johnson *(2)*.

End-product detection is of critical importance when using a NAAT as a diagnostic test. Electrophoresis of PCR products and their visualization in stained gels is impractical in busy clinical laboratories. The use of real-time PCR offers many benefits, but the expense and technical requirements render this option relevant only in larger central laboratories. The skill of developing a NAAT into a feasible diagnostic test lies in its ability to simplify the hands-on aspects of the assay, particularly end-product detection. One example of adapting PCR into a user-friendly assay is the PCR-immunoassay method for detection of MRSA *(3)*.

Although it is highly desirable to use NAATs directly on patient material without prior microbiological recovery and growth, the possible need to detect more than one target sequence can present problems. For example, in the case of MRSA detection, it is desirable both to detect the *mecA* gene, which confers methicillin resistance and is considered the gold standard for differentiating MRSA from susceptible *S. aureus,* and to differentiate *S. aureus* from other *mecA*-positive staphylococci (e.g., *S. epidermidis*). Target genes with unique regions suitable for differentiating *S. aureus* from coagulase-negative staphylococci include *coa* and *femB (3,4)*. However, while a multiplex PCR could demonstrate *mecA* and *femB* in a patient sample, the possibility that the positive results reflect the presence of both a methicillin-susceptible *S. aureus* (giving a positive *femB* result) and a *mecA*-positive, coagulase-negative strain (giving a positive *mecA* result) cannot be excluded. In such a case, it would be advisable to use a selective enrichment broth containing oxacillin prior to multiplex PCR or other multiplex NAATs, to ensure preferential growth of MRSA, if present. Although this seems somewhat self-defeating in the light of the stated advantages of NAATs over culture-based techniques, reference to **Table 1** reveals that even if patient samples undergo an overnight selective culture step prior to NAAT analysis, reporting of results would still be more rapid than that achievable using current methods of MRSA detection.

## 2.1. Real-Time PCR

PCR assays in the research laboratory presently require dedicated equipment such as thermal cyclers, and electrophoresis apparatus for gel analysis. The recent development of the LightCycler® with the concurrent development of Taqman® *(6)*, molecular beacons *(7)*, and other fluorescence-based probes for fluorescence resonance energy transfer (FRET) *(8)* has facilitated the ability to detect positive reaction signals as they develop during NAAT reactions. Real-time PCR has been assessed for use in many aspects of clinical diagnosis, such as detection of *Neisseria gonorrhoeae* in urine samples *(9)*. In addition to the real-time nature of signal detection, this method also dispenses with the need to open reaction tubes once the reaction is complete (as required for gel electrophoresis), thus minimizing the risk of interbatch contamination. The latter point makes real-time PCR especially attractive as a diagnostic tool.

The speed of LightCycler reactions also means that real-time PCR is a very rapid method. When results are required with particular urgency, the rapid generation of results allows remedial strategies to be put in place in a timely manner. This would be especially useful in settings such as a bioterrorism attack, where urgent testing for potential pathogens such as *Bacillus anthracis* would be of crucial importance *(10)*. Real-time PCR has been used for detection of *B. anthracis* directly from air-filter samples *(11)*, and multiplex PCR for simultaneous detection of four different biological agents has been performed in military field testing *(12)*. This setting illustrates an important feature of future molecular diagnostics application—that of point of care (PoC), where quick detection of pathogens is desirable; their use may not be confined to the setting of the clinical laboratory.

## 3. Beyond PCR: Examples of Novel Isothermal Amplification Technologies

Although the recent trend in the use of molecular diagnostics in clinical microbiology has usually seen the use of PCR in central laboratories, the demand for less complicated isothermal technologies has resulted in the development of second-generation amplification technologies. The primary advantage of isothermal amplification technologies is, as the name suggests, that amplification reactions take place at one temperature without the need for the thermal cyclers necessary for PCR. Hence, isothermal amplification techniques have the potential to be performed using standard laboratory equipment such as incubators, water baths, or hot plates, and are also more amenable to automation. The lack of need for expensive incubation equipment makes diagnostic kits based on isothermal amplification technology more attractive to routine laboratories, especially when such laboratories possess standard incubation equipment already. They are also more amenable to PoC applications than are PCR-based methods.

### 3.1. Nucleic Acid Sequence-Based Amplification

Nucleic acid sequence-based amplification (NASBA; **Fig. 1**) *(13)* has been incorporated into a diagnostic kit (the Nuclisens® kit) by Organon Teknika (now bioMerieux), which is largely applicable to RNA targets, such as RNA viruses, although bacterial targets such as *Escherichia coli* in drinking water have also been successfully detected *(14)*. Some clinical studies have concluded that NASBA is more user-friendly and sensitive than PCR for the detection of enteroviruses and respiratory viruses *(15,16)*. The technology is based on the transcription of multiple RNA copies of a DNA product from an RNA target sequence using reverse transcription. Although NASBA is a complex technology, this disadvantage is offset by its greater suitability for automation than PCR. For detection of DNA target sequences, an extra step is required, complicating the assay further. Therefore its strength and main market niche is in detection of RNA targets, notably those found in RNA viruses, and it has been used to detect hepatitis C virus *(17)*.

### 3.2. Q-Beta Replicase

Q-beta replicase (QBR, originally developed by Gene-Trak, now owned by Neogen) is based on the RNA-dependent RNA polymerase from the bacteriophage Q beta. The binding of RNA probes to RNA target initiates replication of the probe sequence (**Fig. 2**) *(18)*. Although billion-fold amplification levels within 30 min are possible, compared with the ten million-fold achievable with PCR, QBR suffers from the amplification of nonspecifically bound probe. Despite this, an automated version has been successfully used in clinical detection of *M. tuberculosis* *(19)*.

### 3.3. Third-Generation Isothermal NAATs

One of the major drawbacks of PCR, NASBA, QBR, and their associated applications is the fact that they involve the amplification of target sequences. In areas of molecular biology that require target sequences for purposes of

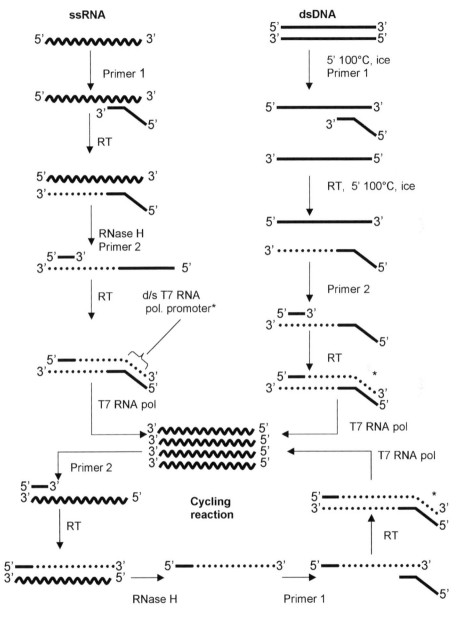

Fig. 1. NASBA isothermal amplification scheme, showing amplification of single-stranded RNA (ssRNA) and double-stranded DNA (dsDNA). Note that in the case of the latter, two 100°C steps are required to produce transcription templates suitable for the isothermal cycling reaction. Primer 1 contains a T7 RNA polymerase promoter sequence. RT, AMV reverse transcriptase; dotted lines, new DNA polymerized by RT. From **ref. 13** with permission from Elsevier Science.

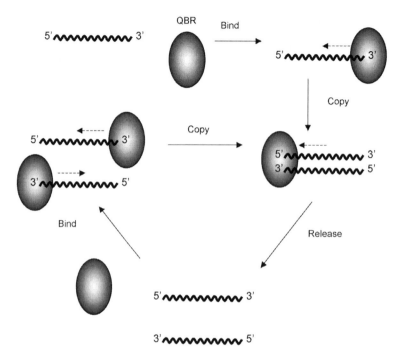

Fig. 2. Q-beta replicase amplification. The replicase enzyme (shaded rhomboid, labeled QBR) binds to the 3' end of the RNA probe molecule, which is subsequently copied. Both the template and nascent RNA molecules serve as substrates for further copying in an isothermal amplification cycle. A portion of the RNA probe is specific to the target. On hybridization, a cleavage site for ribonuclease III is disrupted, allowing the full probe to be copied. Nonhybridized probe, with the intact cleavage site, is digested on addition of ribonuclease III, and hence unbound probe is not amplified. From **ref. *18*** with permission from Elsevier Science.

gene research, forensic identification, and so on, such target amplification technologies are ideal. However, in diagnosis, a simple positive or negative signal indicating presence or absence of the target is often all that is required. Furthermore, the generation of high levels of DNA (or RNA) amplicons in *target*-amplifying technologies is a considerable problem, because these amplicons may persist in laboratories (e.g., on equipment or in reagents), and can contaminate other test samples, resulting in false-positive results. There is therefore a need for an amplification technology that can deliver the considerable advantages of PCR and NASBA (relatively short reaction time, specificity, and sensitivity), but without the disadvantage of high target-sequence amplification. Ideally, such a technology would detect target sequences with the specificity of PCR, and with an amplified signal that can be easily detected, but which does not pose the risk of cross-contamination of future assays associated with high target amplification. For PCR to function in the diagnostic arena, extreme care and cleanliness are required to avoid target contamination. Due to practical and economic realities, clinical laboratories (with the diversity of work carried out therein) tend to be more prone to aerosols than dedicated molecular biology suites, and use many different types of assay. Tests involving microbial recovery and growth of pathogens from clinical material may also be performed in the same laboratory areas in which a NAAT-based diagnostic kit may be used. This requires that the NAAT be robust enough to work under such conditions, and not be sensitive to target contamination. The demand for a third generation of NAATs has resulted in several, currently under development, that amplify signal nucleic acid molecules rather than target sequences.

### 3.3.1. Cycling Probe Technology

Cycling probe technology (CPT: ID Biomedical) is a truly nontarget-amplifying NAAT (**Fig. 3**) *(20)*. The chimeric target detection probe consists of two single-stranded DNA oligonucleotides, linked by a single strand of RNA. Hybridization to the target results in RNase H (an enzyme that digests RNA in RNA-DNA hybrids) releasing the two DNA portions and freeing the target site for further hybridization. This cycle continues until there is a sufficient population of nontarget-sequence DNA probe to allow differentiation between positive and negative reactions. Advantages of CPT are its lack of interbatch cross-contamination and the fact that it is isothermal and simple, lending itself to automation. However, CPT is a nonexponential, linear amplification system, so close attention must be paid to controls. End-detection relies on the capture (in streptavidin-coated microtiter plate wells) of a biotin molecule linked to one of the DNA portions of the probe. A detection molecule or enzyme, such as horseradish peroxidase (HRP), is linked at the end of the other DNA portion. Therefore, end-detection, if using colorimetry, relies on a "negative" color reaction, while appearance of a color indicates intact probe (i.e., no target-dependent RNase H digestion and loss of the end-detection molecule during plate washes). Assays in which a "negative color reaction indicates a positive result" are not generally desired by laboratory staff, who would rather equate a positive color reaction with a positive result. Although cosmetic, this point becomes important when marketing such kits. CPT has been successfully used as a culture confirmation assay in detection of *mecA* in preisolated MRSA colonies *(21)*.

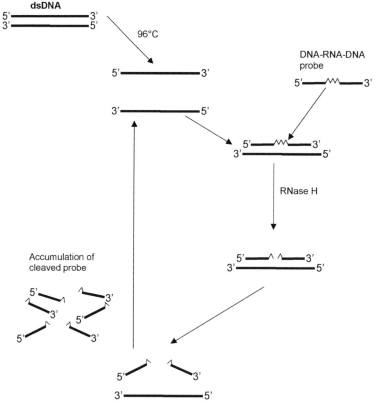

Fig. 3. Cycling probe technology (CPT). The double-stranded (ds) DNA target is made single-stranded by a 96°C step. This is followed by addition of a probe/enzyme mix containing RNase H and scissile-link probe (chimera of DNA–RNA–DNA). Hybridization of the probe to the target results in cleavage by RNase H (this enzyme is specific to RNA–DNA duplexes); the reduction in Tm allows the remaining DNA portions to dissociate from the target, allowing hybridization of intact scissile probe, thus perpetuating the isothermal amplification cycle. Detection molecules, such as biotin and HRP, are attached to 5' and 3' ends of the probes. From **ref. 20** with permission.

### 3.3.2. Invader Technology

Invader® technology (Third Wave) is another linear signal amplification technology (**Fig. 4**) *(22)*. A cleavase releases a nontarget portion of one of two probes after hybridization of the two overlapping oligonucleotide probes to a target nucleic acid strand. The released oligonucleotide fragment is further amplified in overlapping/cleaving reactions using synthetic nontarget-sequence oligonucleotides. The accumulated cleaved nontarget fragments are then end-detected by various appropriate means. The strength of Invader has been shown

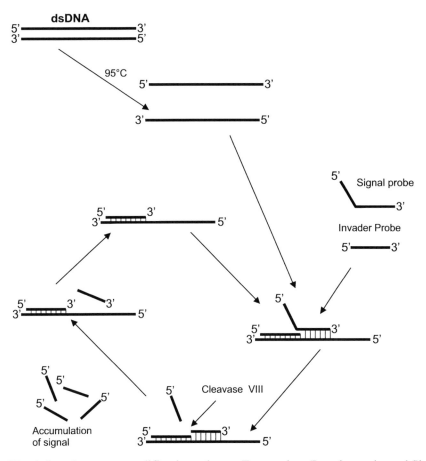

Fig. 4. Invader assay amplification scheme. Two probes (Invader probe and Signal probe) hybridize to the DNA target following heat denaturation. They hybridize in such a way as to form a single base overlap. This overlap is recognized by cleavase VIII, which cleaves the signal probe at the point of the overlapping base. With the reaction temperature kept close to the melting temperature of the signal probe, the latter is destabilized by the cleaving and the portion hybridized to the target dissociates. This allows an intact signal probe to hybridize in its place, facilitating a continued cycle. The cleaved nontarget portion of the signal probe accumulates, and end-detection can be via a suitable molecule linked to the 5' end. From **ref. 20** with permission.

to be in detection of single nucleotide polymorphisms (SNPs), such as Factor V Leiden (a hereditary blood-coagulation disorder) *(23)*, and thus the technology has found application mainly in human genomics. The target-dependent structure formed by the two probes (**Fig. 4**) on hybridization to a target containing a SNP, such as Factor V Leiden, is designed so as to be specific to the

SNP. Thus, in the presence of the wild-type, no target-dependent structure is formed and therefore no cleavage and signal accumulation occurs. It also has potential in bacteriology, such as for detection of the SNP mutations responsible for resistances to rifampin and isoniazid in *M. tuberculosis (22)*.

### 3.3.3. Signal-Mediated Amplification of RNA Technology

Signal Mediated Amplification of RNA Technology (SMART: British Biocell International Ltd.) is an isothermal amplification technology currently under development that amplifies a signal RNA sequence rather than the target sequence *(24)*. The amplification system utilizes DNA polymerase and T7 RNA polymerase for the isothermal generation of a target-dependent RNA signal. A unique target-dependent structure called a three-way junction (3WJ) is formed by the hybridization of DNA oligonucleotide probes that are partially complementary to each other and partially complementary to the target sequence, which can be DNA or RNA. The T7 RNA polymerase promoter encoded within this structure is functionalized by being made double stranded in the presence of target by DNA polymerization of the "extension probe," following which transcription of a unique, short, nontarget RNA sequence occurs ("RNA-1"; **Fig. 5**). Unique to this technology is the fact that signal sequences of RNA are amplified: at no time is any target sequence amplified by the SMART reaction. The lability of RNA, in addition to the smaller amounts amplified by SMART as compared with PCR, minimize target contamination problems in the working clinical laboratory. Further amplification is required following the 3WJ stage. This is made possible by inclusion of a third DNA oligonucleotide ("linear oligo") in the reaction. RNA-1 anneals to the linear oligonucleotide as shown in **Fig. 6A.** DNA polymerization of this hybrid, followed by transcription, leads to the amplification of a second RNA signal molecule (RNA-2). As the linear oligonucleotide is in excess relative to the 3WJ oligos, a stepped amplification of signal occurs. For detection of small amounts of target, further amplification can be attained by use of alternative amplification modules with the same front end. An example is shown in **Fig. 6B**, whereby RNA-2 anneals to a second linear oligonucleotide, initiating transcription of RNA-3. Here, a three-level amplification system ensures detection of at least $10^4$ copies. **Figure 6C** shows an isothermal cycling version of SMART. The RNA-1 transcript from the target-detection three-way junction acts as a target for a second three-way junction, which transcribes RNA-2 when formed. RNA-2 in turn acts as a target for a third three-way junction, which transcribes RNA-1 when formed. The latter, of course, acts as a target for further second three-way junction formation, and the cycle continues. As the oligonucleotides for the second and third three-way junctions are in excess, a cycling reaction is possible. The level of signal amplification facilitates detec-

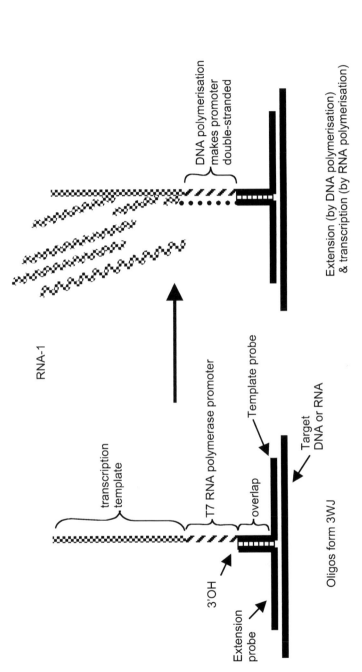

Fig. 5. Formation of SMART three-way junction on target nucleic acid, and transcription of target-dependent RNA-1 signal molecules.

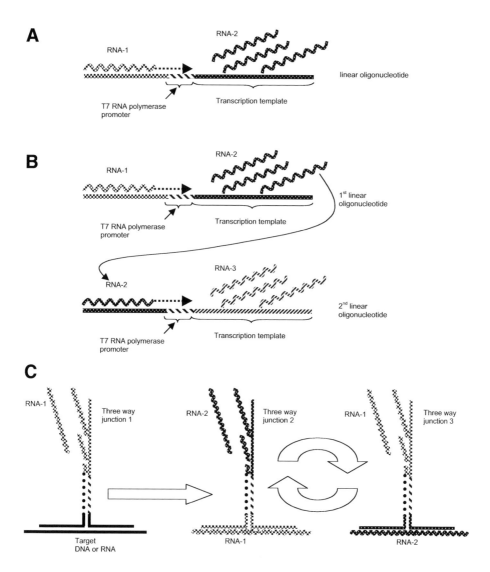

Fig. 6. Alternative modules, for further amplification of RNA-1 signal from target-dependent three-way junction formation. **(A)** Hybridization of RNA-1 to the linear oligonucleotide, followed by DNA and RNA polymerization to transcribe RNA-2 (different sequence from RNA-1). This will detect about $10^4$ single copies of target sequence. **(B)** Hybridization of RNA-2 to a second linear oligonucleotide, and transcription of RNA-3. This will detect about $10^3$ copies of target sequence, **(C)** Cycling SMART reaction. Transcripts RNA-1 and RNA-2 act as targets for three-way junctions 2 and 3. This has the potential to detect about $10^1$ copies of target sequence.

tion of target molecules at a sensitivity level comparable to that of PCR in clinical applications. The use of the RNA signal as a three-way junction target also illustrates the capability of SMART to be used for detection of RNA targets, without recourse to reverse transcriptase, because the DNA and RNA polymerization steps occur using DNA oligonucleotide templates of unique generic sequences.

Signal RNA (be it RNA-1, -2, and/or -3) is transcribed in quantities detectable by a number of alternative end-detection techniques, any of which can be applied to kits depending on the intended use. Some examples are given in **Fig. 7,** which illustrates a method suitable for a manual ELISA-like assay (**Fig. 7A**) and use of molecular beacons for real-time detection of signal (**Fig. 7B**) in applications of SMART amenable to automation. **Fig. 7C** illustrates adaptation to a lateral flow format, which could either facilitate application to automation or, at the other end of the end-user spectrum, to PoC.

SMART has been used for the specific discrimination of closely related marine cyanophages in studies of algal blooms *(25)*, for the detection of MRSA in clinical samples *(26)*, for salmonellae in a variety of food matrices, and for *Chlamydia trachomatis* 23S rRNA (authors' unpublished data). A SMART kit (CytAMP™-MRSA, British Biocell International) with suitable sample preparation and end-detection has been developed as a viable alternative for screening of overnight enrichment broths for the presence or absence of MRSA *(27)*.

## 4. Emerging Applications and Adaptations

When technologies are invented, they generally evolve, becoming adapted and improved in response to the demand for increasingly broad applications. The demand for rapid and reliable methods for diagnosing infectious diseases is ever increasing; hence, most or all of the technologies mentioned in this chapter are likely to become established, and will continue to evolve in response to future needs. Moreover, looking to the future, one can envisage that techniques that are currently expensive and complicated, such as real-time PCR, are likely to become easier for nonspecialist users, and more affordable, with smaller, more convenient equipment than that available now. A crude analogy is the television, which, although it is a highly complex electronic appliance, is widely used by individuals with no specialist knowledge. Therefore, we are likely to see many current laboratory-based technologies being developed into more and more PoC applications with time. For example, the future may see the development of hand-held PCR-, NASBA-, or SMART-based devices which, following manual addition of a drop of blood or urine, would produce simple read-outs of the results, thereby facilitating rapid and accurate diagnosis.

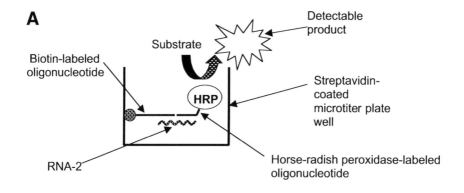

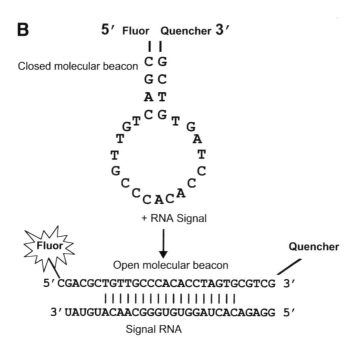

Fig. 7. Examples of methods for end-detection of RNA signal transcripts (RNA-1, RNA-2, and/or RNA-3). **(A)** Colorimetric or luminescent capture/signal/probe sandwich. Signal RNA is hybridized to a biotin-labeled oligonucleotide, which in turn is immobilized on the wall of a streptavidin-coated microtiter plate well. The unannealed sequence of the signal RNA is then hybridized to a horseradish peroxidase (HRP)-

**C**

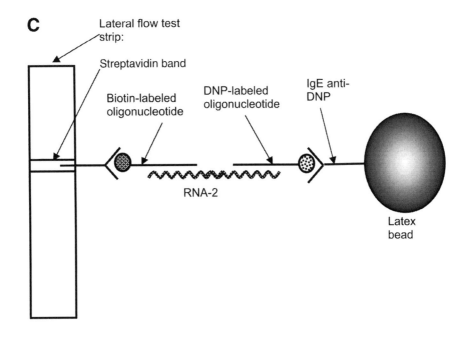

Lateral flow test strip:

Streptavidin band

Biotin-labeled oligonucleotide

DNP-labeled oligonucleotide

IgE anti-DNP

RNA-2

Latex bead

labeled detection probe. After plate washes, addition of a color substrate (e.g., BioFX's TM Blue) or a luminescent substrate can be used for detection of positive SMART reactions. **(B)** Real-time detection of the RNA signal has been achieved by addition of a molecular beacon to the reaction mix, which anneals to signal RNA from the three-way junction as soon as transcription occurs. Under optimal conditions, a positive reaction can be detected within 15 min. **(C)** A lateral flow dipstick has been developed. During the SMART reaction signal transcript (RNA-2) hybridizes to a DNP-labeled oligonucleotide and a biotin-labeled oligonucleotide. During diffusion of this hybrid up the test strip after the SMART reaction, the DNP binds to an anti-DNP-labeled blue latex bead. The whole sandwich is bound to a streptavidin band, resulting in a blue stripe appearing, indicating a positive reaction.

## 4.1. Point of Care and Home Testing

The optimal evolution of biotechnological assays will see them move from central reference laboratories (where specialized techniques currently tend to be concentrated) to what are currently less-specialized regional or local laboratories, where they will become established as routine clinical tests. In addition to this, the ultimate goal for many technologies is their use by physicians in the community for home testing of patients ("personalized medicine"). Of approximately 500 home-based test kits on the FDA over-the-counter devices list, 65% are blood-glucose, ovulation-predicting, or pregnancy-test kits *(28)*. There is an obvious demand for home-testing kits, which give people the opportunity for absolute privacy. Moreover, a major advantage of such tests is release of resources for health services, as the detection of negative results by patients negates the need to seek medical consultation (e.g., women generally do not seek medical advice following a negative result using a pregnancy-test kit). As with clinical laboratories, it is generally the generation of positive results that leads to further actions in health care management. Therefore, the adaptability of NAATs to simple, fool-proof devices that could be used by nonspecialists, in the home, in trauma and emergency rooms, GP/nurses surgeries, and so on, is a major goal for molecular biology companies and researchers. The thermal-cycling nature of PCR is an enormous drawback in this instance, although demand for convenient versions is driving development of more rapid and more user-friendly formats (*see* **Subheadings 4.2.** and **5.**).

Increasing longevity has resulted in an expanding percentage of the elderly in many Western populations. Because elderly people are generally more susceptible to infectious diseases, there will be an increasing demand on health-service resources, especially a huge increase in nursing-home occupancy, where epidemic strains of pathogens such as MRSA are already becoming prevalent, much as in the case of hospital super-bugs. However, for home testing to have an impact on clinical bacteriology, a suitable platform is required. The types of sample that are used in clinical diagnostic laboratories for bacterial examination are typically blood, urine, feces, skin swabs, and orifice swabs. A primary hurdle facing the application of many NAATs, often overlooked in the overall evaluation of particular technologies, is the front-end sample preparation and delivery to the NAAT itself. Many NAATs can work very efficiently if the sample material is sufficiently purified; and although there are certain sample types which can be directly applied to NAATs, it is often the case that environmental matrices such as blood, urine, sputum, feces, and so on contain inhibitory compounds (e.g., hemoglobin). Hence, rapid sample preparation is an ancillary area of research in the field of NAAT diagnostics development. There are several lysis buffers currently available that can deliver amplifiable nucleic acids to NAATs. For example, microLYSIS™ (a detergent-based lysis buffer manufactured by microZONE) has been used to release total RNA from *C. trachomatis* for detection by SMART (*see* **Subheading 3.3.3.**). Therefore a challenge exists for either the development of sample preparation methods that ensure delivery of NAAT-friendly target material, or the adaptation and optimization of NAATs to cope with environmental matrices.

### 4.2. DNA Microelectronic Array Technologies

DNA microelectronic arrays (usually shortened to *microarrays,* often referred to as DNA chips) are based on the integration of nucleic-acid hybridization technology (often NAAT) and computer microelectronics, i.e., microchips. This area of technological development encompasses a synergy of several disciplines, including microelectromechanical systems (MEMS), microfluidics, microfabrication, organic chemistry, genomics, and molecular biology. Many thousands of DNA probes can be immobilized on DNA chips, enabling the potential detection of thousands of different sequences simultaneously. The scope for rapid typing of target organisms in detection of bacterial pathogens is obvious. Current drawbacks of DNA chips are the high concentrations of target required for hybridization, and the necessity for uniform stringency across the array, which presents a considerable challenge in designing oligonucleotide probes for high-density arrays. NAAT reactions within DNA chips can provide sufficiently high concentrations of nucleic acid for hybridization to the array probes. Anchoring the primers required for the NAAT reaction can physically separate potentially cross-reacting multiplex probes in an array targeting thousands of sequences. PCR can be adapted in this way, but isothermal amplification technologies are more easily adapted to simple (and therefore miniaturized) instrumentation. Nanogen, Inc., has developed miniaturized assays in their Lab-on-a-Chip, and have used anchored strand displacement amplification (SDA) in a DNA chip for detection of *Pseudomonas, Salmonella, E. coli,* and *C. trachomatis (29)*. The rate-limiting nature of passive hybridization in DNA chips can be overcome by exploitation of the charged nature of nucleic acid. Development of an electronically addressable electrode array facilitated movement of charged nucleic acid fragments to microlocations, resulting in concentration over the immobilized probes and subsequent hybridization. For differentiation of SNPs, reversal of the electric field is used to repel selectively mismatched hybridizations *(30)*.

The end-detection of positive signals from hybridization events at microlocations in arrays can be achieved by several technologies. Integration of hybrid duplex-chelating fluorescent dyes can be detected by microfluorimetric detectors in the array-containing device. Hybridization itself can be exploited to complete electric circuits in arrays. Currently in its early stages of development, liquid-phase DNA probes labeled with gold nanoparticles hybridize to the single-stranded portion of a target nucleic acid, which is itself hybridized to another immobilized probe between two titanium electrodes at the micro-location in the DNA chip. Liquid-phase silver stain aggregates around the gold nanoparticle, closing the electric circuit between the electrodes resulting in a signal (Park et al., 2002, cited in **ref. *31***). It can clearly be seen that the linkage of hybridization events to generation of electric signals has potential for the adaptation of DNA chips to simple, hand-held devices that may receive a sample and give a positive/negative read-out.

## 5. The Future for Probe-Based Technologies

Nucleic acid diagnostics, PoC testing, automation in the laboratory, and microminiaturization are among the current technological developments that are going to have major impact on the in vitro diagnostics (IVD) market in the next few years. Molecular technologies are likely to displace at least 15% of the IVD tests currently on the market, and will occupy a large and fast-growing segment of the IVD market by 2010 *(28)*. The novelty of NAATs becomes increasingly diminished by their increasing commercial availability. Hence the pressure is on increasing reliability, ease of use, reduction in complexity (including reduction of reaction times in relation to non-NAAT clinical tests), and reduction in cost. Semiautomated modules, such as Roche COBAS, enable improvements to be approached, but fully integrated automation is likely to be one of the major ways in which simple, reliable NAATs make an impact on diagnostics. Much automation is likely to be focused on the use of microarrays and advances in microfluidics, with integrated heaters and optical sensors facilitating internal thermal cycling or other heating-stage technologies and real-time or instant end-detection. Evidence of movement towards fully automated PoC NAAT-based diagnostics can be seen in the development of fast thermal cycling machines; e.g., a new machine by Megabase uses pressurized helium and carbon dioxide for rapid heating and cooling, enabling amplification of target within 78 s. A portable light-cycling PCR machine successfully detected foot-and-mouth disease virus in veterinary trials *(32)*, illustrating how the demand for use of a currently complicated technology is leading to simplification and convenience for application in the field. Demonstration of scalability to multiples of target assays, and transferability from laboratory bench to clinical and other PoC settings, is going to be a fundamental factor in the survival and success of NAATs in the future.

The term *pharmakogenomics* refers to the potential use of NAATs in genotyping subsets of patients who will respond to certain subsets of drug types. In addition to their application to this field of human genomics, theranostic use of NAATs in clinical microbiology represents a large part of the future of probe-based technologies; NAATs can be employed rapidly to genotype pathogens and epidemic strains, leading to rapid application of appropriate drugs, such as antibiotics. Pharmaceutical companies are likely to codevelop pharmakogenomics with new drugs subsets.

In conclusion, it appears that NAATs are set to become significant players in the future of clinical diagnostics, not least in the field of theranostics, where genome projects are providing previously untapped data that make it possible to identify genotypes with concurrent development of patient/target subset-specific drugs. An essential ingredient in the success of a NAAT will be its adaptability to microminiaturization and automation, leading to utility by unskilled end-users. Ironically, a measure of the success of NAATs in diagnostics will be the extent to which they become taken for granted as central technologies in clinical science, much as "traditional" culture-based techniques are now.

## References

1. Tansuphasiri, U. (2001) Detection of *Mycobacterium tuberculosis* from sputum collected on filter paper and stored at room temperature for 5 days by PCR assay and culture. *J. Med. Assoc. Thai.* **84,** 1183–1189.
2. Diagnostics Intelligence. (1996) *Profiting from Gene-Based Diagnostics.* CTB International Publishing, Mapelwood, USA.
3. Towner, K. J., Talbot, D. C. S., Curran, R., Webster, C. A., and Humphreys, H. (1998) Development and evaluation of a PCR-based immunoassay for the rapid detection of methicillin resistant *Staphylococcus aureus. J. Med. Microbiol.* **47,** 607–613.
4. Hookey, J. V., Richardson, J. F., and Cookson, B. D. (1998) Molecular typing of *Staphylococcus aureus* based on PCR restriction fragment length polymorphism and DNA sequence analysis of the coagulase gene. *J. Clin. Microbiol.* **36,** 1083–1089.
5. Jonas D., Grundmann H., Hartung D., Daschner F. D., and Towner K. J. (1999) Evaluation of the *mecA femB* duplex polymerase chain reaction for detection of methicillin resistant *Staphylococcus aureus. Eur. J. Clin. Microbiol. Infect. Dis.* **18,** 643–647.
6. Heid, C. A., Stevens, J., Livak, K. J., and Williams, P. M. (1996) Real time quantitative PCR. *Genome Res.* **6,** 986–994.
7. Tyagi, S. and Kramer, F. R. (1996) Molecular beacons: probes that fluoresce upon hybridization. *Nature Biotech.* **14,** 303–308.
8. Sixou, S., Szoka, F. C. Jr., Green, G. A., Giusti, B., Zon, G., and Chin, D. J. (1994) Intracellular oligonucleotide hybridization detected by fluorescence resonance energy transfer (FRET). *Nucleic Acids Res.* **22,** 662–668.
9. Whiley, D. M., LeCornec, G. M., Mackay, I. M., Siebert, D. J., and Sloots, T. P. (2002) A real-time PCR assay for the detection of *Neisseria gonorrhoeae* by LightCycler. *Diagn. Microbiol. Infect. Dis.* **42,** 85–89.

10. Qi, Y., Patra, K., Liang, X., Williams, L. E., Rose, S., Redkar, R. J., et al. (2001) Utilization of the *rpo*B gene as a specific chromosomal marker for real-time PCR detection of *Bacillus anthracis*. *Appl. Environ. Microbiol.* **67,** 3720–3727.

11. Makino, S. I, Cheun, H. I., Watari, M., Uchida, I., and Takeshi, K. (2001) Detection of anthrax spores from the air by real-time PCR. *Lett. Appl. Microbiol.* **33,** 237–240.

12. McDonald, R., Cao, T., and Borschel, R. (2001) Multiplexing for the detection of multiple biowarfare agents shows promise in the field. *Mil. Med.* **166,** 237–239.

13. Kievits, T., van Gemen, R., van Strijp, D., Schukkink, R., Dircks, M., Adriaanse, H., et al. (1991) NASBA isothermal enzymatic in vitro nucleic acid amplification optimized for the diagnosis of HIV-1 infection. *J. Virol. Methods* **35,** 273–286.

14. Min, J. and Baeumner, A. J. (2002) Highly sensitive and specific detection of viable *Escherichia coli* in drinking water. *Anal. Biochem.* **303,** 186–193.

15. Foolen, H., Sillikens, P., van de Wiel-van de Meer, M., and Fox, J. D. (2001) Nuclisens Basic Kit NASBA including an internal control and "real-time" detection for diagnosis of enterovirus infections. *J. Microbiol. Methods* **47,** 96.

16. Fox, J. D., Hibbitts, S., Rahman, A., and Westmorland, D. (2001) Nuclisens Basic Kit NASBA with "end point" and "real-time" detection for diagnosis of respiratory virus infections. *J. Microbiol. Methods* **47,** 96.

17. Damen, M., Sillekens, P., Cuypers, H. T., Frantzen, I., and Melsert, R. (1999) Characterization of the quantitative HCV NASBA assay. *J. Virol. Methods* **82,** 45–54.

18. Lizardi, P. M. and Kramer, F. R. (1991) Exponential amplification of nucleic acids: new diagnostics using DNA polymerases and RNA replicases. *Trends Biotech.* **9,** 53–58.

19. Smith, J. H., Radcliffe, G, Rigby, S., Mahan, D., Lane, D. J., and Klinger, J. D. (1997) Performance of an automated Q-beta replicase amplification assay for *Mycobacterium tuberculosis* in a clinical trial. *J. Clin. Microbiol.* **35,** 1484–1491.

20. Beggs, M. L., Cave, M. D., Marlowe, C., Cloney, L., Duck, P., and Eisenach, K. D. (1996) Characterization of *Mycobacterium tuberculosis* complex direct repeat sequence for use in cycling probe reaction. *J. Clin. Microbiol.* **34,** 2985–2989.

21. Cloney, L., Marlowe, C., Wong, A., Chow, R., and Bryan R. (1999) Rapid detection of *mecA* in methicillin resistant *Staphylococcus aureus* using Cycling Probe Technology. *Mol. Cell Probes* **13,** 191–197.

22. Cooksey, R. C., Holloway, B. P., Oldenburg, M. C., Listenbee, S., and Miller, C. W. (2000) Evaluation of the invader assay, a linear signal amplification method, for identification of mutations associated with resistance to rifampin and isoniazid in *Mycobacterium tuberculosis*. *Antimicrob. Agents Chemother.* **44,** 1296–1301.

23. Hessner, M. J., Budish, M. A., and Friedman, K. D. (2000) Genotyping of factor V G1691A (Leiden) without the use of PCR by invasive cleavage of oligonucleotide probes. *Clin. Chem.* **46,** 1051–1056.

24. Wharam, S. D., Marsh, P., Lloyd, J. S., Ray, T. D., Mock, G. A., Assenberg, R., et al. (2001) Specific detection of DNA and RNA targets using a novel isothermal

nucleic acid amplification assay based on the formation of a three-way junction structure. *Nucleic Acids Res. Methods Online* **29,** E54.

25. Hall, M. J., Wharam, S. D., Weston, A., Cardy, D. L. N., and Wilson, W. H. (2002) Use of signal-mediated amplification of RNA technology (SMART) to detect marine cyanophage DNA. *BioTechniques* **32,** 604–611.

26. Levi, K., Bailey, C., Marsh, P., Cardy, D. L. N., and Towner, K. J. (2001) Evaluation of a novel isothermal amplification assay (CytAMP™) for direct detection of methicillin resistant *Staphylococcus aureus* (MRSA). *J. Microbiol. Methods* [Abstract] **47,** 100.

27. Levi, K., Bailey, C., Bennett, A., Marsh, P., Candy, D. L., and Towner, K. J. (2003) Evaluation of an isothermal signal amplification method for rapid detection of methicillin resistant Staphylococcus aureus from patient-screening swabs. *J. Clin. Microbiol.* **41,** 3187–3191.

28. Crosby, L. (2002) IVDs for home-use testing: issues in diagnosis and monitoring. *IVD Technology* **8,** 18–21.

29. Westin, L., Xu, X., Miler, C., Wang, L., Edman, C. F., and Nerenberg, M. (2000) Anchored multiplex amplification on a microelectronic chip array. *Nature Biotech.* **18,** 199–204.

30. Sosnowski, R. G., Tu, E., Butler, W. F., O'Connell, J. P., and Heller, M. J. (1997) Rapid determination of single base mismatch mutations in DNA hybrids by direct electric field control. *Proc. Natl. Acad. Sci. USA* **94,** 1119–1123.

31. Syvänen, A. C. and Söderlund, H. (2002) DNA sandwiches with silver and gold. *Nature Biotech.* **20,** 349–350.

32. Donaldson, A. L., Hearps, A., and Alexandersen, S. (2001) Evaluation of a portable, "real-time" PCR machine for FMD diagnosis. *Veterinary Record* **149,** 430.

# 9

## Real-Time PCR

### Nicholas A. Saunders

**Summary**

The development of instruments that allow real-time monitoring of fluorescence within PCR reaction vessels is a significant advance in clinical bacteriology. The technology is very flexible, and many alternative instruments and fluorescent probe systems are currently available. Real-time PCR assays can be completed very rapidly, because no manipulations are required after amplification. Identification of amplification products by probe detection in real time is highly accurate compared with size analysis on gels. Analysis of the progress of the reaction allows accurate quantification of the target sequence over a very wide dynamic range, provided suitable standards are available. Finally, probe melting analysis can detect sequence variants including single base mutations.

**Key Words:** Real-time detection; polymerase chain reaction (PCR); fluorescence; quantification; fluorescent resonant energy transfer; mutation detection; multiplex.

## 1. Introduction

PCR using thermostable polymerase was first described in 1988 *(1)* and since then has become an essential tool in molecular microbiology. The power of PCR lies in its ability to provide nanogram quantities of essentially identical DNA molecules (approx $10^{13}$ copies of a 100- to 200-bp product) starting from a single copy of a known sequence. The amplified material (the PCR amplicon) is available in sufficient quantity to be identified by size analysis, sequencing, or by probe hybridization. It can also be cloned readily or used as a reagent.

Standard PCR protocols are now very simple and user-friendly. Most of the technical effort involved is directed toward sample extraction and positive recognition of the amplicons. For clinical applications it may not be sufficient or

From: *Methods in Molecular Biology, vol. 266: Genomics, Proteomics, and Clinical Bacteriology: Methods and Reviews*
Edited by: N. Woodford and A. Johnson © Humana Press Inc., Totowa, NJ

convenient to use gel electrophoresis for verification of product identity on the basis of size alone. Characterization of the product by its sequence is far more reliable and precise. Probe hybridization assays for this purpose are available, but many are multistep procedures. Such methods are time-consuming, and care must be taken to ensure that amplicons accidentally released into the laboratory environment do not contaminate other DNA preparations. A further disadvantage of PCR in the clinical laboratory is that, in its standard formats, it is not quantitative.

All current instruments designed for real-time PCR measure the progress of amplification by monitoring changes in fluorescence within the PCR tube. Changes in fluorescence can be linked to product accumulation by a variety of methods. Real-time PCR overcomes the problem of the nonquantitative nature of standard PCR that relies on end-point analysis. It also allows rapid and convenient product verification by probe hybridization. Real-time PCR instruments measure the kinetics of product accumulation in each PCR reaction tube. Generally, products are not detected during the first few temperature cycles, since the fluorescent signal is below the detection threshold of the instrument. However, most combinations of machine and fluorescence reporter are capable of detecting the accumulation of amplicons before the end of the exponential amplification phase. During this time the efficiency of PCR is often (but not always) close to 100%, giving a doubling of the quantity of product at each cycle. As product concentrations approach the nanogram per microliter level, the efficiency of amplification is reduced. This occurs primarily because of the tendency of amplicons to re-associate during the annealing step, rather than to bind primer. Efficiency also declines at this stage due to limited availability of *Taq* polymerase and other reaction components. This leads to a phase during which the accumulation of product is approximately linear (i.e., a constant level of net synthesis at each cycle) because the increase in available template is balanced by decreasing amplification efficiency. Finally, a plateau is reached when net synthesis approximates zero. Quantification in real-time PCR is achieved by measurement of the number of cycles required for the fluorescent signal to cross a threshold level.

There are two general approaches used to obtain a fluorescent signal following the synthesis of product in PCR. The first depends on the property of fluorescent dyes, such as SYBR green I, to bind to double-stranded DNA and undergo a conformational change that results in an increase in their fluorescence. The second approach is to use fluorescence resonance energy transfer (FRET). This technique uses a variety of means to alter the relative spatial arrangement of photon donor and acceptor molecules. These molecules are attached to probes, primers, or the PCR product, and are usually selected so that amplification of a specific DNA sequence brings about an increase in fluorescence at a particular wavelength.

A major advantage of the real-time PCR instruments and signal transduction systems currently available is that it is possible to characterize the PCR amplicon by analysis of its melting temperature and/or probe hybridization. In the intercalating dye system, the melting temperature of the amplicon can be estimated by measuring the level of fluorescence emitted by the dye as the temperature is increased from below to above the expected melting temperature. Methods that rely on probe hybridization to produce a fluorescent signal are generally less liable to produce false-positive results than those that use intercalating dyes to detect net synthesis of dsDNA. Hybridization, ResonSense, and hydrolysis probe systems (*see* below) produce fluorescent signals only when the target sequence is amplified, and are unlikely to give false-positive results. An additional feature of these methods is that it is also possible to measure the temperature at which the probes dissociate from their complementary sequences, giving further verification of the specificity of the amplification reaction.

A significant advantage of the real-time format is that analysis can be performed without opening the reaction tubes, which can therefore be disposed of without the risk of dissemination of PCR amplicons or other target molecules into the laboratory environment. Although alternative methods for avoiding PCR contamination are available, containment within the PCR vessel is likely to be the most efficient and cost-effective.

In recent years, real-time PCR has found many applications in clinical bacteriology. These include the detection of pathogens in a range of clinical specimens, bacterial quantification (*2–5*), schemes for identification and typing of strains using markers including those that confer drug resistance (*6–8*), and immuno-PCR (*9*).

## 2. Materials

### 2.1. Oligonucleotides (Probes and Primers)

Oligonucleotide stocks should be dissolved in nuclease-free water and stored frozen ($-20°C$) at a concentration of 100 $\mu M$. To avoid repeated freeze/thaw cycles, which have a detrimental effect on oligonucleotide stability, small aliquots should be prepared. Design and synthesis of oligonucleotides is discussed in **Note 1.**

### 2.2. Intercalating Dyes

Stock solutions of SYBR Green I or SYBR Gold in dimethyl sulfoxide (Molecular Probes Inc., Eugene, OR) should be stored frozen ($-20°C$).

Dilutions of these dyes in aqueous buffers have limited stability; therefore only small quantities should be prepared. The usual working solution of SYBR dye is a 1/1000 dilution, since the final concentration in PCR is most frequently 1/10,000.

### 2.3. Buffers and PCR Reaction Mixtures

Buffers used for PCR and reverse transcriptase (RT) PCR in standard thermal cyclers are generally also suitable for use in real-time instruments. The exception to this is that reactions in the LightCycler need bovine serum albumin (BSA) to stabilize the polymerase owing to interactions with the glass surface of the reaction capillaries. In addition, the optimum magnesium concentration for PCR in capillaries is often higher (usually approx 1 m$M$) than in plastic tubes, because of magnesium adsorption at the glass surface. A typical 10X buffer for use in capillaries contains 500 m$M$ Tris-HCl pH8.3, and 5 µg/µL BSA. This buffer is not suitable for reactions that include an RT step. Kits for homogenous RT-PCR are available commercially.

### 2.4. Enzymes

Many different thermostable polymerases are suitable for use in real-time PCR. The choice depends on the particular application. There may be good reasons to choose an enzyme that has proofreading (3' to 5' exonuclease) activity—e.g., if the product is to be sequenced. However, generally the choice reduces to whether to use a "hot-start" or standard form of *Taq* polymerase (*see* **Note 2**). Because many enzyme formulations perform similarly, price may be a key factor in the choice of reagent.

### 2.5. Template DNA

Extraction methods for real-time PCR are identical to those used in standard nucleic-acid amplification reactions. However, real-time PCR may provide information concerning the purity and yield of target molecules in the extract that may be difficult to obtain by other means. This means that for real-time PCR it is straightforward to assess minimal extraction conditions that are sufficient to remove all significant PCR inhibitors and give a reliable quantitative yield.

The minimum conditions necessary for extraction/purification clearly depend on the specimen type. Methods such as boiling in the presence of Chelex *(10)* are rapid and convenient, but a full work-up using a method such as those that depend upon adsorption to silica particles *(11)* in high-salt buffer may be required.

## 2.6. Standards and Controls

The most helpful controls in PCR are included within the reaction mixture containing the unknown material. Such internal controls indicate whether the PCR reaction is functional under the precise conditions applied, and should therefore reveal the presence of inhibitors or defective reagents. In real-time PCR, the control product can be distinguished from the target by its different melting temperature, or more usually by using probes that emit fluorescence at a distinct wavelength. Unfortunately, internal control target competes for resources within the reaction mixture and may therefore limit PCR sensitivity. The most appropriate internal control material is single- or double-stranded DNA or RNA that corresponds with the target. Such molecules carry the same primer-binding sites as the authentic target, and are therefore amplified by the same primers. The central portion of the control template is modified so that the product is easily distinguishable from the target. The amplified control sequence should be as similar to the target sequence as possible, since length, GC content, and primary structure each influence the efficiency of the PCR.

If an internal control system is unavailable, external control reactions should be used. These provide assurance that the reagents used in the PCR are in good condition and that the instrument is functioning correctly. Generally the template sequence used as an external control should be identical to the target sequence, but with a modified central sequence that can be distinguished in case it becomes necessary to investigate a contamination problem. Quantitative standards are external controls containing a known quantity of the target sequence. A convenient method of preparing such standards is to clone the appropriately modified or unmodified PCR amplicon into a plasmid vector such as pCRII-TOPO (Invitrogen Ltd., Paisley, UK). The DNA is suitable for use as a standard following linearization of the purified plasmid at a restriction site within the polylinker. Alternatively, RNA transcripts may be prepared using either the T7 or SP6 promotor on the vector. Quantitative standards can be calibrated by measuring absorbance or by comparison of the observed band intensities with a range of commercial molecular size standards in DNA-binding dye-treated gels.

Low concentrations of quantitative standards are liable to change on storage due to adsorption of material to the walls of the storage vessel. Consequently a DNA (e.g., salmon-sperm DNA) or protein (e.g., BSA) carrier should be added to the diluent.

## 3. Fluorescence Chemistries

### 3.1. DNA-Binding Dye Reporter Methods

The addition of intercalating dyes to reaction mixtures is a simple but powerful and effective method of monitoring net DNA synthesis in real-time PCR

*(12,13)*. This approach has several advantages over competing methods that depend upon FRET: intercalating dyes are inexpensive and readily available; it is not necessary to optimize the concentration of dye for different reactions; and the dye itself has no significant effects on PCR efficiency or on kinetics of annealing, extension, and denaturation at the low levels used for detection in real-time machines.

SYBR dyes are stable under PCR conditions and give relatively high levels of fluorescent signal, so that it is possible to detect amplification during the exponential amplification phase. SYBR green I or SYBR gold (Molecular Probes Inc.) are the DNA-binding dyes of choice for real-time PCR owing to their relatively high sensitivities and low toxicities compared with ethidium bromide. SYBR Gold is apparently more sensitive than SYBR green I, and may well become the first-choice DNA-binding dye for real-time PCR. Higher concentrations of the SYBR dyes (>1 in 5000 final concentration) inhibit PCR and should be avoided. Both SYBR dyes have similar excitation and emission spectra (497 nm excitation and 520 nm emission maxima for SYBR green I, and 495 nm excitation and 537 nm emission maxima for SYBR gold). When free in solution, both dyes exhibit very low levels of fluorescence, which increase by up to 1000-fold when the dye is bound to DNA. Unfortunately, both SYBR dyes bind to oligonucleotides; consequently, the background level of fluorescence in PCR mixtures prior to amplification is contributed to by the primers.

During PCR cycling, SYBR dye fluorescence levels change depending on the temperature and the quantity of DNA present. It is therefore best to measure the level of fluorescence once per cycle at a fixed point, which is usually the end of the extension phase, when the quantity of double-stranded DNA reaches a maximum level.

### 3.2. Melting-Curve Analysis With DNA-Binding Dyes

Melting-curve analysis is very useful when SYBR dyes are used as reporters in real-time PCR *(14)*. The temperature is gradually raised at a constant rate from a point at which the PCR product is known to be double-stranded to one at which all DNA strands are fully dissociated. The temperature at which this equilibrium is reached depends on the stability of the base pairing between the two strands in the PCR buffer and also on the amplicon concentration.

During melting analysis, fluorescence is measured as frequently as possible, and the temperature at which the maximum decrease in fluorescence (or "melting peak") occurs is measured. Melting peaks are visualized on plots of $-dF/dT$ against $T$ ($F$ = fluorescence and $T$ = temperature). If there is more than one PCR product, more than one melting peak may be observed, depending on their physical properties. The temperature of the melting peak for any product

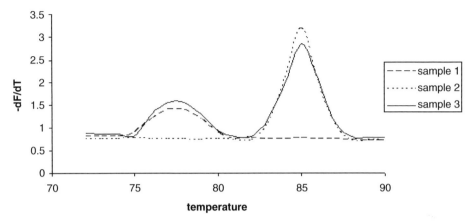

Fig. 1. Melting analysis. Melting curves of the reaction products of three HBV surface gene PCRs are shown. The reporter was SYBR green I. Sample 1 was a control containing no copies of the template; sample 2 contained 2000 copies; and sample 3 contained 20 copies. The well-resolved peaks at 85°C and at 77°C correspond to the surface gene product and the primer dimers, respectively.

corresponds to the point at which 50% of the DNA is associated. At this point, the population of molecules will consist of fully dissociated, partially associated, and fully associated DNA strands. One consequence of this is that the temperature of the melting peak for a particular DNA molecule is not fixed, but is concentration dependent. As melting peaks occur within temperature ranges, it is difficult to distinguish between PCR products when their ranges overlap. However, it is usually possible to identify primer artifacts, such as dimers, since these usually have melting-peak temperature ranges quite different from those of the products. For example, **Fig. 1** shows a typical melting analysis used to distinguish between a product and primer artefact.

### 3.3. Probe-Reporter Methods

The probe-reporter methods rely upon the detection of specific sequences within the amplicon. A nonspecific signal from the reporter is unlikely since both primers and the probe must hybridize to the target sequence in the correct orientation. Consequently, probe methods are highly specific. Several different approaches for translating probe hybridization into signal are available; these are discussed in **Notes 3–7**. In each case the donor and acceptor fluors are either brought into closer average proximity or are separated in the presence of a specific PCR product. When specific amplification reduces the average separation, the best approach is to select a pair of fluors so that the donor becomes excited by the light source and subsequently emits a photon that is able to

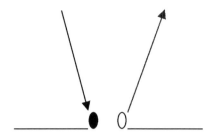

Fig. 2. Hybridization probes. The incident light (dark arrow) excites the donor fluor (dark ellipse) on the first probe (short line on left) hybridized to the target sequence (long line). A proportion of the photons emitted by the donor fluor are capable of exciting the acceptor fluor (open ellipse), and this occurs efficiently when both probes are hybridized. Photons emitted by the acceptor fluor (gray arrow) can be distinguished from more energetic photons emitted by the donor.

excite the acceptor, which fluoresces at a wavelength that is picked up by the detector. The hybridization probe method (*see* **Fig. 2** and **Note 3**) is a good example of this approach. When the effect of amplification is to separate the donor/acceptor pair, then the acceptor acts as a quencher (it may also be a fluor) that absorbs photons emitted by the donor. As amplification proceeds, the donor/quencher pair are separated and the fluorescent signal increases. For example, in the TaqMan system (*see* **Fig. 3** and **Note 4**), the donor and acceptor molecules are attached to different ends of a single probe that is cleaved by the 5'–3' exonuclease activity of *Taq* polymerase. Notes on the most frequently encountered probe reporter methods, hybridization probes (**Note 3**), ResonSense probes (*see* **Fig. 4** and **Note 5**), molecular beacons (*see* **Fig. 5** and **Note 6**), and hydrolysis probes (*see* **Fig. 3** and **Note 4**), are included below. Variations of these methods have been developed in which the probe sequence is covalently bound to one of the primers. The advantage of these intramolecular probe systems is that the rate of probe hybridization is greatly increased (*see* **Note 7**).

## 4. Thermal Cycling for Real-Time PCR

### 4.1. Fluorescence Measurement

Thermal profiles for real-time PCR are modified to allow measurement of the level of fluorescence within the reaction tube at one or more points in the cycle. Generally, it is best to make the measurement when the signal-to-noise ratio is at its maximum level. However, care must be taken to avoid compromising the PCR by introducing temperature holds at or below the optimum

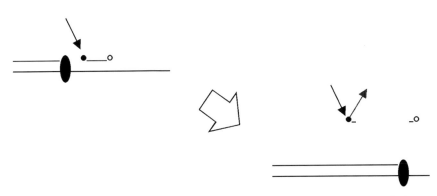

Fig. 3. Hydrolysis probes. The left-hand figure shows incident light (dark arrow) exciting the donor fluor (dark spot) attached to the probe. Light emitted by the donor is absorbed efficiently by the acceptor fluor (open spot) when the probe is intact. In the presence of the target sequence, the probe hybridizes and becomes a substrate for the 5'–3' exonuclease activity of the *Taq* polymerase (large dark ellipse). Cleavage of the probe (right of figure) results in the separation of donor and acceptor fluors so that photons emitted by the donor fluor (gray arrow) are not transferred to the acceptor fluor efficiently and can be detected by the fluorimeter.

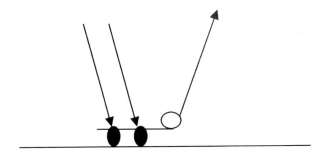

Fig. 4. ResonSense probes. The incident light (dark arrows) excites the SYBR dye (dark ellipses) bound to the double-stranded DNA formed by hybridization of the probe (short line) to its target sequence (long line). A proportion of the photons emitted by the SYBR dye are capable of exciting the acceptor fluor (open ellipse). Photons emitted by the acceptor fluor (gray arrow) can be distinguished from the shorter-wavelength light emitted by the SYBR dye.

annealing temperature. For hybridization probes, ResonSense probes, molecular beacons, and similar systems, fluorescence is often measured during a hold step following annealing and at a temperature approx 5°C higher than that used for annealing. DNA-binding dye and hydrolysis probe reporters are usually assayed immediately following extension.

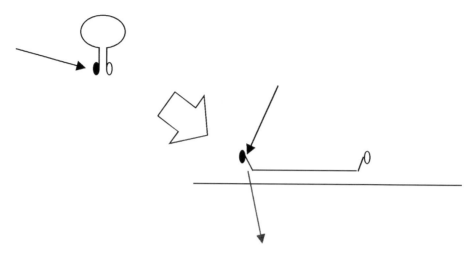

Fig. 5. Molecular beacons. The left hand figure shows incident light (dark arrow) exciting the donor fluor (dark ellipse) attached to the probe. Light emitted by the donor is absorbed efficiently by the acceptor (open ellipse) when the probe is in the "hairpin" conformation shown. When the probe hybridizes to target sequence (right of figure) photons emitted by the donor fluor (gray arrow) are not transferred to the acceptor efficiently and can be detected by the fluorimeter.

### 4.2. Two-Step PCR

Two-step PCR, in which annealing and extension steps are performed at the same temperature, may shorten the time required to complete the cycle, and is often used with the hydrolysis probe system. A factor contributing to the popularity of the two-step method for hydrolysis probes is that 5' nuclease cleavage may be favored over displacement at lower extension temperatures. Because the optimum temperatures for annealing and extension are therefore likely to be very similar, there is a tendency to select the primers so that a single value can be used.

### 4.3. Rapid PCR

Some real-time PCR instruments (e.g., LightCycler® and SmartCycler®) allow very fast and accurate temperature transitions, and have therefore become associated with rapid PCR (*see* **Note 8**). For example, it is possible to run a complete PCR cycling protocol in approx 15 min using the LightCycler. Rapid cycling is achieved in these machines because the designs that allow fluorescent measurements to be made also avoid the use of heating blocks and standard PCR tubes. Rapid temperature transition is an additional advantage of these cyclers, but this is not a requirement for real-time PCR.

## 4.4. Multiplex PCR

There are two possible approaches to the design of real-time multiplex PCRs. One method is to use DNA-binding dyes to label all products and then to distinguish among them by melting curve analysis. However, this is only likely to give clear results if the amplicons differ significantly in %GC content, and separation of more than two amplicons would be very difficult to achieve. A more successful approach is to use a reporter-probe method to produce a different fluorescent signal for each component PCR. In standard multiplex PCR, each product is designed to be identified by its size as determined by gel electrophoresis. This means that not all amplicons can be of the size optimal for PCR. An advantage of the reporter-probe multiplex PCR method is that the size of the products generated does not affect detection, so primers that produce amplicons of similar size may be selected.

All of the reporter-probe methods and multidetection channel PCR instruments described in **Note 9** have some potential for use in real-time multiplex PCR. The major constraints on the degree of multiplexing are the spectral separation of the available fluors and the optics of the machine. For example, the LightCycler can be used for duplex PCRs with hybridization probes in which fluorescein is the universal donor for LC Red 640 and LC Red 705 *(15)*. The inclusion of additional PCR reactions in this case would require the provision of further detection channels on the instrument, and suitable fluors. Here the major difficulty is finding fluors that are excited by fluorescence emitted from fluorescein and that have emissions that may be distinguished. Systems that use a universal acceptor system are more amenable to multiplexing, since the detectors may cover well-separated portions of the light spectrum. A hydrolysis probe multiplex comprised of six PCRs has been described by Lee et al. *(16)*. In this case, a range of donor fluors were combined with a universal dark acceptor (nitrothiazole blue).

## 5. Quantitative Real-Time PCR

The lack of correlation between the number of copies of the target sequence and the final yield of amplicons is due to several factors. The main reason is that the yield generally reaches a plateau when synthesis of new amplicons is sufficient only to replace DNA lost by chemical and enzymatic hydrolysis. PCR efficiency declines as product accumulates primarily owing to reassociation of the amplicon strands, which blocks primer annealing. Minor variations between the samples and reaction components of different PCR tubes affect the efficiency of amplification and the plateau amplicon concentration. Post-PCR analyses do not distinguish between reactions that have been at the product plateau for different numbers of cycles. Equally, efficient reactions

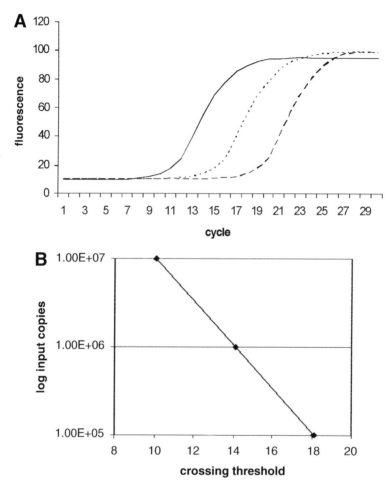

Fig. 6. Quantitative PCR. (**A**) Amplification curves obtained for 107, 106, and 105 copies of the HBV surface gene in a gene-specific PCR using SYBR Green I as reporter. An appropriate noise band or significance threshold would be parallel to the *x* axis at a fluorescence value of 15–20 units. (**B**) A plot for input copies vs the crossing threshold (i.e., the number of cycles required for the amplification curve to cross the noise band).

that have not reached the plateau may contain the same level of amplicons as inefficient reactions that reached their maxima after relatively few cycles.

Real-time PCR overcomes the problem of the nonquantitative nature of standard PCR that relies on end-point analysis. Quantification in real-time PCR is achieved by measurement of the number of cycles required for the fluorescent signal to cross a threshold level. The threshold is usually set at a value at least three standard deviations above the baseline noise level. Crossing threshold values are inversely proportional to $\log_{10}$ of the initial target copy number (**Fig. 6**). Quantification values for unknown samples are calculated from standard curves constructed using the data obtained for standards containing dilutions of the target sequence.

## 5.1. Robustness in Quantitative PCR

To achieve accurate quantification the real-time PCR should be robust. The following facets should be considered:

- Primer artifact formation should occur only after many cycles. If PCR artifacts form early in the reaction, the dynamic range of the quantification reaction will be limited to higher copy numbers. At the lowest template input values, quantification will be impossible since no product will be made. In addition, accuracy will be reduced in the presence of artifacts even when the specific product is made.
- Sequence-specific detection is preferable for quantitative real-time PCR since reasonable accuracy is then possible even in the presence of modest levels of artifact formation. The ResonSense system is ideal since the formation of primer artifacts can be detected and unreliable results identified.
- The efficiency of the reaction should be predictable. Samples may contain inhibitors of the PCR that impair the efficiency of reactions to differing degrees. Such samples should be diluted or purified prior to analysis. When samples might contain small quantities of magnesium or magnesium-binding compounds, it is important to ensure that the magnesium concentration in the reaction has been carefully selected.
- The primer and probe sequences should match the target perfectly. Where this is impossible it might be possible to choose initial annealing conditions that allow reasonable quantification (i.e., 2–3 cycles at a lower annealing temperature or with the annealing temperature held for longer). A better alternative is to use multiple quantification reactions based on different portions of the target genome.

## 5.2. Quantification of DNA and RNA Templates

Real-time instrument quantification software constructs standard curves of cycle number at the noise band crossing threshold (i.e., the number of cycles required before a statistically significant increase in fluorescent signal) for a dilution series of PCR template (*see* **Subheading 5**). Unknown copy-number values are assigned to test samples according to the cycle number at the crossing threshold estimated from the curve. Accurate quantification therefore relies on the quality of the standard curve. For the highest accuracy, the reactions used to provide the data for construction of the standard curve should be run concurrently with the unknown samples.

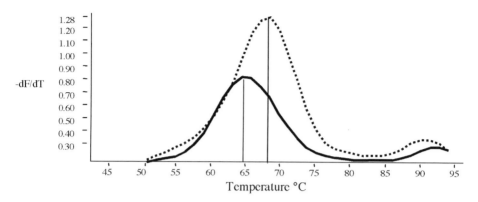

Fig. 7. Mutation detection by melting-curve analysis. In this example, the ResonSense probe melting curve for a wild-type sequence and the matching probe is shown as a dotted line. The solid line shows the curve obtained when a sequence with a single-base mutation is analyzed using the same probe. A 3 to 4°C difference in the position of the -dF/dT maximum is seen for the matched and mismatched sequences.

When the target of the quantitative real-time PCR is an RNA molecule, it is necessary to choose either DNA or RNA standards. The latter can be prepared conveniently in quantity by T7 RNA polymerase transcription of the template sequence cloned downstream of a T7 promotor, but must be handled carefully to avoid loss through nuclease damage. The accuracy of RNA quantification using DNA standards depends on the efficiency of the RT reaction, but if this is close to 100%, then DNA standards may be appropriate. If RT efficiency is low, RNA standards must be used to control for the RT reaction. We have found that RT reactions that rely on random hexamer priming are very efficient, and for this reason RNA and DNA standards give indistinguishable results. Reverse transcription using specific primers appears to be less efficient. Real-time PCR has been used successfully for quantification of bacterial mRNA (e.g., *17*).

## 6. Detection of Sequence Variants by Real-Time PCR

Sequence variations (including single-base polymorphisms) at the site of probe hybridization result in significant differences in melting temperature, so these values reliably indicate the presence of single-point or multiple mutations within the PCR amplicon (**Fig. 7**).

## 7. Notes

1. Primers and probes for real-time PCR. Oligonucleotides suitable for use in real-time PCR can be made in-house or by any of the companies offering custom synthesis. In the author's experience, oligonucleotides with the minimal level of purification offered by the custom service (i.e., desalting for standard unmodified oligonucleotides or HPLC for modified oligonucleotides) are likely to perform as well as reagents purified further. However, opinions differ on the level of purity required for PCR primers and probes. Use of a low synthesis scale does not seem to present a barrier to good performance.

2. Hot-start enzymes. Most PCR reactions give higher sensitivity and specificity when a hot start is used since the formation of artifacts is avoided or delayed. Formulations that use an antipolymerase antibody to inactivate the enzyme are effective, and full activity is restored after a short period of denaturation *(18)*. *Taq* polymerase structural variants that require incubation at elevated temperatures for extended periods for enzyme activity to reach a maximum level are slightly less convenient. However, the slow activation of these variants may be exploited to raise specificity of the reaction, by limiting enzyme activity during the early PCR cycles before full activation of the enzyme for later cycles.

3. Hybridization probes. Hybridization probes were first described by Wittwer and colleagues *(19)* and have been used most widely on the LightCycler platform. Hybridization probes rely on hybridization of a fluorescent-labeled oligonucleotide probe to a target sequence within the PCR product. The second fluor is brought into proximity with the probe in one of two ways. In the first, a second probe is hybridized adjacent to the first so that the two fluors are brought close together for efficient FRET. For the best fluorescent signal the two probes hybridize next to each other with a gap of 0 or 1 base between the adjacent 5' and 3' termini, which carry the fluors. The 3' end of the 5'-derivitized probe must be blocked, usually either with a phosphate or biotin group, to prevent extension. The alternative method is to attach either fluor to one of the primers, internally, close to its 3' terminus. The second fluor is attached to the 3' terminus of a probe that hybridizes to the strand synthesized from the modified primer. For the best levels of FRET the fluors are positioned 6 bp apart, so that they are in proximity on the double-helical structure. The intensity of fluorescent signal is dependent on the number of copies of the target sequence. For this reason short-target sequences (approx 100 bp), which are amplified more efficiently, should be chosen wherever possible.

Convenient fluors that allow FRET between hybridization probes using the LightCycler are fluorescein, LC Red 640, and LC Red 705. Fluorescein acts as the donor excited by the blue LED, and LC Red 640 and 705 are acceptors that emit photons that can be monitored in channels 2 or 3 of the LightCycler. Hybridization probes should be designed to have temperatures of disassociation ($T_d$) from their target sequences that are a few degrees higher than the $T_d$s of the primers. The probe $T_d$s must be significantly lower than the extension temperature to avoid inhibition of the PCR reaction.

Melting analysis of adjacent hybridization probes primarily gives a value for the probe with the lower $T_d$, as the fluorescent signal is dependent on both probes being bound simultaneously. However, the apparent $T_d$ is the temperature at which 50% of the probe with the lowest $T_d$ is hybridized adjacent to the alternate probe. Thus, the $T_d$ of the more stable probe may affect the value significantly. In addition, because $T_d$ measurements depend on the equilibrium between probe binding and melting, values may vary according to the quantities of PCR product and probe available. Matters are simpler when one of the fluors is attached to a primer. In this system, only the $T_d$ of the probe is measured.

Wherever possible, hybridization probes are designed so that they do not interfere significantly with PCR amplification. It is particularly important to avoid proximity between the 3' end of the primer and the 5' end of the probe that hybridizes closest to that primer. This can be achieved by allowing a gap of approx 10 bases between annealing sites, which allows the primers to be stabilized by extension before the temperature is raised for the extension step. At the extension temperature both probes should be dissociated from the template, allowing completion of the complementary strand. The hybridization signal is concentration dependent at lower levels of probe. However, very high concentrations of probe should be avoided, as the effective melting temperature of the probe is raised under such conditions and the chance of interference with primer extension is increased. As a starting point for optimization, probes should be used at a concentration of 0.25 pmol/μL.

4. Hydrolysis probes. The hydrolysis or 5' nuclease system relies upon the 5' exonuclease activity of *Taq* polymerase *(20)*. Hydrolysis probes have structures similar to molecular beacons (**Note 6**) except that the self-complementary ends are absent. When the probe is free in solution, the donor and acceptor are in relative proximity due to secondary structure, and FRET occurs. Hydrolysis of the probe occurs during temperature cycling. The probe binds to the target sequence at the annealing phase and remains bound during the extension phase until the *Taq* polymerase encounters the free 5' end, which is then cleaved from the probe. The remaining part of the probe is then displaced or cleaved further by the polymerase. When the donor and acceptor are separated during hydrolysis, the efficiency of FRET between them is greatly reduced, and net fluorescence at the emission wavelength of the donor fluor increases. The intensity of the fluorescent signal is dependent upon the accumulation of the target sequence during amplification. As for the other probe-reporter methods, the target sequence should be as short as possible (approx 100-bp) because such molecules are amplified to higher final concentrations.

For the hydrolysis probe system, PCR conditions and probe sequences are chosen so that probe hydrolysis rather than displacement is favored. Generally, high magnesium ion concentrations and low extension temperatures favor hydrolysis. Guidelines for the selection of good hydrolysis probes have been developed by Applied Biosystems *(21,22)*. Briefly, the GC content of the probe should be close to 50% and in the range of 40–60%. The probe should be selected to hybridize to the strand with fewer C than G residues (i.e., the probe should be relatively rich in Cs) and G should be absent from the bases adjacent to the acceptor. The melting temperature of the probe should be at least 5°C higher than that of either primer. The probe should neither hybridize nor overlap with the forward and reverse primers.

Convenient fluors for use in hydrolysis probes are the dyes used in automated sequencing. TAMRA or a dark quencher such as DABCYL can be used as universal FRET acceptors in combination with a large number of donors, including FAM, JOE, TET, and HEX. As a starting point for optimization, hydrolysis probes are added to the reaction mixture at a concentration of 0.20 pmol/µL.

5. ResonSense probes. The ResonSense probe system (also known as bi-probe) uses a DNA-binding dye such as SYBR Gold as the donor together with a single oligonucleotide probe labeled with one of many possible acceptor fluors, such as Cy3, Cy5, and Cy5.5 *(6–8,23)*. When the probe binds to the single-stranded template DNA at the annealing step, a short stretch of double-stranded DNA is created, which binds to the DNA-binding dye. In this way, the DNA-binding dye and the probe carrying the acceptor fluor are brought into proximity, resulting in an increase in FRET. It is essential to choose a short-target sequence (approx 100 bp) in order to increase the signal from the probe and at the same time reduce the background fluorescence from the DNA-binding dye. The design criteria for ResonSense probes are as described in **Note 3** for hybridization probes. Melting analysis of ResonSense probes gives the temperature at which the probe is 50% bound to the target and also gives a melting peak when amplicon that has reannealed finally melts, releasing the DNA-binding dye. Thus, melting curves produced for the ResonSense system may be particularly informative. As a starting point for optimization, probes should be used at a concentration of 0.25 pmol/ µL and SYBR dyes at a dilution of 1/10,000.

6. Molecular beacons. Molecular beacons are oligonucleotide probes with complementary ends that adopt a hairpin conformation. Each probe molecule carries either two fluors, or one fluor and a quencher attached at its termini so that they are held together by the hairpin. Efficient FRET occurs between the two residues when the oligonucleotide is in its hairpin form. One fluor acts as the donor and the other fluor (or quencher) as the acceptor. A common approach *(24)* is to use a dark quencher (e.g., DABCYL) as the universal acceptor, and a range of different donors that have different emission spectra, so that multiplexing is possible.

During temperature cycling in the presence of the PCR product, the central portion of the molecular beacon binds to its target sequence at the annealing temperature. This prevents the formation of the less stable hairpin conformation of the probe and results in an increase in observed fluorescence from the acceptor fluor. The probe should dissociate from the template at below the extension temperature used in order to prevent inhibition of the reaction. The fluorescent signal is dependent upon the number of copies of the target sequence; consequently, short target sequences (approx 100 bp), which are amplified more efficiently, are preferred. As a starting point for optimization, molecular beacons are added to the reaction mixture at a concentration of 0.25 pmol/μL.

7. Intramolecular probe systems. Scorpion probes *(25)* are modified molecular beacons with a donor fluor at the 5' end and an acceptor at the 3' end. However, the 3' end of the molecular beacon is attached to the 5' end of one of the primers via a blocker residue (typically hexethylene glycol) that cannot be read through during synthesis of the other strand during PCR cycling. The target sequence for the probe portion of the Scorpion is part of the strand synthesized by the primer portion of the molecule. As for molecular beacons, there is an increase in fluorescence when the probe is hybridized to its target sequence, because the fluor is removed from proximity to the quencher. Hybridization is rapid and efficient because the probe and target sequences are part of a single molecule. The fluorescent signal is dependent on the number of copies of the target sequence. For this reason, short-target sequences (approx 100 bp), which are amplified more efficiently, are preferred. As a starting point for optimization, the Scorpion's probe/primer is used at a concentration of 0.5 pmol/μL.

   Other reporter probe systems have been adapted to be intramolecular using similar principles. Angler *(23)* is a modified version of the ResonSense system; Solinas et al. *(26)* have reported a duplex Scorpion primer method; and IntraTaq *(27)* is a version of the hydrolysis probe method.

8. Rapid PCR. The annealing and denaturation steps of the PCR process occur very rapidly in comparison with the times required for the sample to reach the desired temperature. Unfortunately, this does not mean that annealing or denaturation holds can be reduced to near to zero seconds, or that the extension step can be reduced to the length of time required for DNA elongation, since thermal cyclers with heating blocks require time to equilibrate, and there is also a delay while heat is conducted into the sample. However, these problems are greatly reduced in the LightCycler and SmartCycler instruments, where very short holds become possible. In the LightCycler, turbulent air is used as a heating/cooling medium, and glass capillaries as the reaction vessels. In the SmartCycler, individually controlled elements are used to heat and cool special plastic vessels that have a high surface area-to-volume ratio. Rapid temperature transitions and short holds have advantages beyond the obvious reduction in assay times. The major advantage is that annealing accuracy is increased, as the sample spends less time at the lower temperatures, where mispriming may occur. Rapid transition to the annealing temperature reduces the amount of target reannealing prior to primer binding. Rapid denaturation steps reduce the loss of product due to hydrolysis and result in less cumulative destruction of the polymerase. All in all, these factors increase the sensitivity and yield of PCR reactions.

9. Real-time machines. A wide range of real-time PCR instruments are now available. They differ in two main areas. First, in the method used to heat and cool the samples, and second, in the optical system employed.

Most real-time machines (i.e., Applied Biosystems 7700 and 7000, BioRad iCycler, MJ DNA Engine Opticon, Stratagene Mx4000) use advanced heating block technology to pump heat in and out of standard thin-walled plastic reaction vessels. Consequently, sample uniformity is poor and temperature transitions are relatively slow compared with the machines that employ alternative heat-exchange technologies (i.e., Roche and Idaho Technologies Light-Cyclers™, Cepheid SmartCycler® and Corbett Rotor-Gene™). However, the use of blocks has the advantage that standard PCR plates can be employed.

Light sources for excitation of the fluors in real-time PCR can be broad or narrow spectrum. If a broad-spectrum light source is used (e.g., Bio-Rad iCycler, Stratagene Mx4000) filters can be used to provide light tuned to the excitation spectrum of individual fluors. This approach widens the choice of fluors that are available, so that those with well-separated emission maxima can be selected. The disadvantage of this approach is that the intensity of light passing through the filters is limited and, in theory, the sensitivity of detection may therefore be limited. The narrow-spectum machines use LED (e.g., LightCyclers, Cepheid SmartCycler and Corbett Rotor-Gene) or laser (e.g., Applied Biosystems 7700) light sources. Both the SmartCycler and Corbett Rotor-Gene have multiple LEDs of different colors that, in theory, allow a wide choice of fluors to be selected (the SmartCycler has limited flexibility), as for the machines with broad-spectrum sources. The Applied Biosystems 7700 and LightCycler have single light sources and a limited choice of fluors.

Real-time fluorescence detectors are generally factory set to respond to narrow bands of the spectrum, although filter sets that can be changed by the user are available for the iCycler. The number of detection channels that can be useful depends on whether a range of excitation wavelengths are available. When only a single narrow-range light source is used, one approach is to employ similar fluors that are all excited to some extent in the same range, and then to use software correction to deconvolute the emitted light. This system is used successfully in the Applied Biosystems 7700. The LightCycler narrow-band light source excites fluors such as the SYBR dyes and fluorescein, and has three spectrally well-separated detectors. Two of these detect fluors emitting long wavelength light that are only minimally excited by the blue LED light source. Instead the LightCycler relies on FRET to excite the long-wavelength fluors (i.e., Cy5, LC Red 640 and 705).

## References

1  Saiki, R. K., Gelfand, D. H., Stoffel, S., Scharf, S. J., Higuchi, R., Horn, G. T., et al. (1988) Primer-directed enzymatic amplification of DNA with a thermostable DNA polymerase. *Science* **239,** 487–491.

2.  Bach, H. J., Tomanova, J., Schloter, M., and Munch, J. C. (2002) Enumeration of total bacteria and bacteria with genes for proteolytic activity in pure cultures and in environmental samples by quantitative PCR mediated amplification. *J. Microbiol. Meth.* **49,** 235–245.

3. Eishi, Y., Suga, M., Ishige, I., Kobayashi, D., Yamada, T., Takemura, T., et al. (2002) Quantitative analysis of mycobacterial and propionibacterial DNA in lymph nodes of Japanese and European patients with sarcoidosis. *J. Clin. Microbiol.* **40,** 198–204.

4. Piesman, J., Schneider, B. S., and Zeidner, N. S. (2001) Use of quantitative PCR to measure density of *Borrelia burgdorferi* in the midgut and salivary glands of feeding tick vectors. *J. Clin. Microbiol.* **39,** 4145–4148.

5. Brechtbuehl, K., Whalley, S. A., Dusheiko, G. M., and Saunders, N. A. (2001) A rapid real-time quantitative polymerase chain reaction for hepatitis B virus. *J. Virol. Meth.* **93,** 105–113.

6. Edwards, K. J., Metherell, L. A., Yates, M., and Saunders, N. A. (2001) Detection of *rpo*B mutations in *Mycobacterium tuberculosis* by biprobe analysis. *J. Clin. Microbiol.* **39,** 3350–3352.

7. Edwards, K. J., Kaufmann, M. E., and Saunders, N. A. (2001) Rapid and accurate identification of coagulase-negative staphylococci by real-time PCR. *J. Clin. Microbiol.* **39,** 3047–3051.

8. Whalley, S. A., Brown, D., Teo, C. G., Dusheiko, G. M., and Saunders, N. A. (2001) Monitoring the emergence of hepatitis B virus polymerase gene variants during lamivudine therapy using the LightCycler. *J. Clin. Microbiol.* **39,** 1456–1459.

9. McKie, A., Samuel, D., Cohen, B., and Saunders, N. A. (2002) A quantitative immuno-PCR assay for the detection of mumps-specific IgG. *J. Immunol. Meth.* **270,** 135–141.

10. Mygind, T., Birkelund, S., Birkebaek, N., Oestergaard, L., Jensen, J., and Christiansen, G. (2002) Determination of PCR efficiency in chelex-100 purified clinical samples and comparison of real-time quantitative PCR and conventional PCR for detection of *Chlamydia pneumoniae*. *BMC Microbiol.* **2,** 17.

11. Boom, R., Sol, C. J., Salimans, M. M., Jansen, C. L., Wertheim-van Dillen, P. M., and van der Noordaa, J. (1990) Rapid and simple method for purification of nucleic acids. *J. Clin. Microbiol.* **28,** 495–503.

12. Higuchi, R., Fockler, C., Dollinger, G., and Watson, R. (1993) Kinetic PCR analysis: real-time monitoring of DNA amplification reactions. *Biotechnology* **11,** 1026–1030.

13. Higuchi, R., Dollinger, G., Walsh, P. S., and Griffith, R. (1992) Simultaneous amplification and detection of specific DNA sequences. *Biotechnology* **10,** 413–417.

14. Ririe, K. M., Rasmussen, R. P., and Wittwer, C. T. (1997) Product differentiation by analysis of DNA melting curves during the polymerase chain reaction. *Anal. Biochem.* **245,** 154–160.

15. Reischl, U., Linde, H. J., Metz, M., Leppmeier, B., and Lehn, N. (2000) Rapid identification of methicillin-resistant *Staphylococcus aureus* and simultaneous species confirmation using real-time fluorescence PCR. *J. Clin. Microbiol.* **38,** 2429–2433.

16. Lee, L. G., Livak, K. J., Mullah, B., Graham, R. J., Vinayak, R. S., and Woudenberg, T. M. (1999) Seven-color, homogeneous detection of six PCR products. *Biotechniques* **27**, 342–349.
17. Edwards, K. J. and Saunders, N. A. (2001) Real-time PCR used to measure stress-induced changes in the expression of the genes of the alginate pathway of *Pseudomonas aeruginosa. J. Appl. Microbiol.* **91**, 29–37.
18. Kellogg, D. E., Rybalkin, I., Chen, S., Mukhamedova, N., Vlasik, T., Siebert, P. D., et al. (1994) TaqStart Antibody: "hot start" PCR facilitated by a neutralizing monoclonal antibody directed against *Taq* DNA polymerase. *Biotechniques* **16**, 1134–1137.
19. Wittwer, C. T., Ririe, K. M., Andrew, R. V., David, D. A., Gundry, R. A., and Balis, U. J. (1997) The LightCycler: a microvolume multisample fluorimeter with rapid temperature control. *Biotechniques* **22**, 176–181.
20. Holland, P. M., Abramson, R. D., Watson, R., and Gelfand, D. H. (1991) Detection of specific polymerase chain reaction product by utilizing the 5' to 3' exonuclease activity of *Thermus aquaticus* DNA polymerase. *Proc. Natl. Acad. Sci. USA* **88**, 7276–7280.
21. Perkin-Elmer. (1995) *TaqMan Probe Design, Synthesis and Purification.* Applied Biosystems, Foster City, CA.
22. Gelmini, S., Orlando, C., Sestini, R., Vona, G., Pinzani, P., Ruocco, L., et al. (1997) Quantitative polymerase chain reaction-based homogeneous assay with fluorogenic probes to measure *c-erbB-2* oncogene amplification. *Clin. Chem.* **43**, 752–758.
23. Lee, M. A., Siddle, S. L., and Hunter, R. P. (2002) ResonSense: simple linear probes for quantitative homogenous rapid polymerase chain reaction. *Anal. Chimic. Acta.* **457**, 61–70.
24. El-Hajj, H. H., Marras, S. A., Tyagi, S., Kramer, F. R., and Alland, D. (2001) Detection of rifampin resistance in *Mycobacterium tuberculosis* in a single tube with molecular beacons. *J. Clin. Microbiol.* **39**, 4131–4137.
25. Whitcombe, D., Theaker, J., Guy, S. P., Brown, T., and Little, S. (1999) Detection of PCR products using self-probing amplicons and fluorescence. *Nat. Biotechnol.* **17**, 804–807.
26. Solinas, A., Brown, L. J., McKeen, C., Mellor, J. M., Nicol, J., Thelwell, N., et al. (2001) Duplex Scorpion primers in SNP analysis and FRET applications. *Nucl. Acids Res.* **29**, E96.
27. Solinas, A., Thelwell, N., and Brown, T. (2002) Intramolecular TaqMan probes for genetic analysis. *Chem. Commun.* **2002**, 2272–2273.

# 10

## Microarrays for Bacterial Typing

*Realistic Hope or Holy Grail?*

### Carola Van Ijperen and Nicholas A. Saunders

### Summary

Microbiology has entered the postgenomic era and it is clear that bacterial typing should aim to be based on analysis of complete genomes. Although complete genome sequencing for epidemiological typing remains unrealistic for the present, microarrays that provide information on gene content are now becoming available. Microarrays comprised of several thousand probes on glass slides can now be manufactured in the laboratory using robotic arrayers. The gene probes are either PCR products or synthetic oligonucleotides that can be irreversibly attached to a reactive glass surface. The target nucleic acids to be hybridized to the probe array are tagged with fluorescent dyes. Relative probe hybridization signals can be measured when two or more different preparations are labeled with distinguishable fluorophores. Microarrays that include probes for every gene within a genome provide excellent comparative data, although a focus on variable genes may be more useful for typing purposes. Composite arrays of variable genes are under development.

**Key Words:** Bacterial typing; microarray; probe hybridization; comparative genomics; genetic diversity.

## 1. Introduction
### 1.1. Bacterial Typing

The traditional methods of typing bacteria are based on phenotypic characteristics such as serotype and phage susceptibility. Although these methods are widely used and remain important, they are gradually being replaced by methods that differentiate strains on the basis of genotypic differences, such as

From: *Methods in Molecular Biology, vol. 266: Genomics, Proteomics, and Clinical Bacteriology: Methods and Reviews*
Edited by: N. Woodford and A. Johnson © Humana Press Inc., Totowa, NJ

pulsed-field gel electrophoresis (PFGE), amplified fragment-length polymorphism (AFLP), and mulilocus sequence typing (MLST; *see* Chapter 15). Both genotypic and phenotypic methods can be tailored to give information about strains that may be useful for disease management (e.g., drug resistance markers), but current genotypic methods generally rely on the analysis of anonymous markers that have no known phenotypic correlates. To illustrate this point, one of the features of MLST is that the products of the genes analyzed have essential "housekeeping" functions, and therefore it is highly unlikely that polymorphisms in the open reading frames would have gross phenotypic consequences. A disadvantage of many of the early genotyping methods such as PFGE and restriction fragment-length polymorphisms (RFLP) was that laboratories found it difficult to standardize methods and compare data. Despite this, the genotypic methods are finding favor because they offer greater flexibility and discriminatory ability than the phenotypic methods.

### 1.2. Arrays

The term *array* is applied to a range of different technological platforms. A common feature of all arrays is that they comprise a set of defined nucleic acid probes, each placed at specific *X*, *Y* coordinates on a surface. The probes on the array can be exposed to labeled target nucleic acids, and hybridization occurs if complementary sequences are present. The term *microarray* is usually applied when probes are placed a very small distance apart (approx 200 μm). Robotic laboratory arrayers that exploit split-pin or ring-and-pin technology (**Fig. 1**; *see* **Note 1**) can produce microarrays comprised of > 20,000 precisely positioned probes on standard low-fluorescence microscope slides. The construction, handling, and potential applications of this type of microarray are described and discussed in this chapter. High-density microarrays are available commercially and comprise up to 100 to 200 times more probes per unit area than can be achieved using a laboratory arrayer. These arrays are produced using photolithography to build the DNA oligonucleotide probes *in situ* (*see,* for example, http://www.affymetrix.com).

Microarrays represent a powerful new method for analyzing bacterial genotypes. The main advantage of arrays is their flexibility in both the format of the array and the versatility of the probes to test sample material. The great promise of microarray technology lies in the fact that the probes can be used to detect DNA sequences linked to specific phenotypic characteristics. Future arrays might be used to generate typing data for epidemiological purposes linking disease presentation, transmissibility, and changes in the bacterial and host populations.

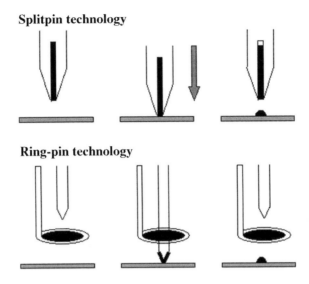

Fig. 1. Schematic diagram of probe deposition onto the slides using split-pin technology (top) and ring-pin technology (bottom).

### *1.3. Comparative Genomics*

The number of complete bacterial genome sequences available is expanding rapidly. The technical and logistical problems presented by sequencing a few megabases of DNA appear to have been solved (*see* Chapter 3), principally owing to advances in robotics and the software for assembly of the data. Currently, the major bottleneck appears to be the full annotation of the available sequence data (**Fig. 2**). Many of the major bacterial pathogens that infect humans now have several fully sequenced strains, and precise comparison of these genomes is now possible (for example, *see* Baba et al. *(1)*; *see* Chapter 4). This work illustrates the significance of lateral gene transfer and gene deletions as mechanisms in the evolution of bacterial species. Although genera vary in the degree of genomic plasticity evident, a common feature is the presence of a core of genes that are always present, together with a category of "divergent" genes that are present in only some strains. These genes are usually clustered, indicating that they are lost or acquired as a unit, and are often found in the vicinity of chromosomal elements that are associated with gene translocation (e.g., transposons and bacteriophage components) (*see* Chapter 5). Comparison of complete genome sequences is likely to remain the most informative and sensitive method for typing bacterial strains. Unfortunately, for the present, genome sequencing remains far from being a practical typing

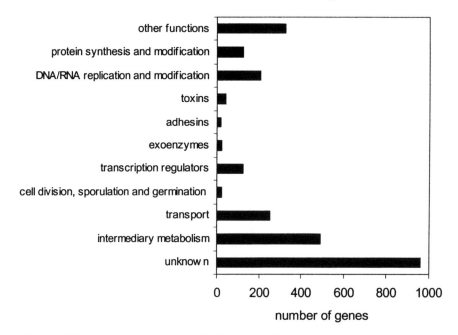

Fig. 2. The functions of the 2595 ORFs of *Staphylococcus aureus* strain
N315 are shown. Approximately two-thirds of the ORFs encode proteins with
function that are either known or suspected from homology. The data were
from Kuroda and colleagues *(18)*.

method for most purposes, because of its high cost. However, the rapid decrease
in the cost per base of sequencing shows no current sign of coming to an end.
Furthermore, technologies for resequencing (i.e., finding variants in a known
sequence) in particular, are very promising in terms of low cost and high
capacity. For this reason it seems entirely plausible that in the not-too-distant
future, comparison of complete genome sequences will be the primary method
used for epidemiological typing. Bacterial gene microarrays use the newly
acquired wealth of sequence information to provide a practical shortcut to this
ultimate goal of whole-genome comparison. Using arrays, genomes are compared gene by gene instead of base by base. This approach provides data comparable with that of sequencing in terms of the presence and absence of genes,
but at a fraction of the current cost.

## 1.4. Arrays for Genomotyping

*Genomotyping* is a new term *(2)* applied to describe the analysis of bacteria
by comparison of their genomes using microarrays. For example, Dorrell and

colleagues *(3)* have described the use of arrays comprised of probes derived by amplification of clones produced for sequencing the genome of *Campylobacter jejuni*. Each probe was thus specific for a part of the genome of the strain used to generate the clones. When DNA from different strains was hybridized to the array, it was possible to identify sequences that were present only in the strain used to construct the array, from the pattern of probe hybridization. The advantage of this approach is that it is relatively inexpensive, because a single primer pair (with annealing sites in the cloning vector) can be used to amplify all of the necessary probes. Unfortunately this method has the significant disadvantage that a single probe may include sequences derived from more than one gene and, consequently, the data may be misleading. An alternative method of array construction, which avoids this drawback, is to design probes specific for each gene in a bacterial genome sequence *(4,5)*. Probes for individual genes may be produced by amplification using specific PCR primers, or may be oligonucleotides *(6)*. Computer software for the design of large numbers of PCR primer pairs *(7)* and oligonucleotides *(8)* is available. An array of probes for every open reading frame (ORF) in a bacterial genome will include only a strain-specific subset of the "divergent" genes maintained by the species. Since these are, in some ways, the most interesting genes, most array designs now include additional probes for the "divergent" genes present in other key strains (e.g., *see* Smoot et al.; *5*). These composite arrays give a more sensitive and complete genomotype than arrays based on the genes of a single strain.

### 1.5. Applications in Clinical Bacteriology

The analysis of bacterial genomes using genome-scale microarrays has already contributed to our understanding of the molecular evolution and diversity of human pathogens including *C. jejuni (3)*, *Vibrio cholerae (4)*, *Streptococcus pyogenes (5)*, *Helicobacter pylori (9)*, *Mycobacterium bovis (10)*, and *Staphylococcus aureus (11)*. These data are a valuable contribution to our effort to understand the pathogenic mechanisms of these bacteria. It is already clear that human pathogens rely on the appropriately timed expression of a large number of different genes, including those for survival in the host and for transmission between hosts. Diversity in these genes between different strains of a species appears to have developed in response to host defense mechanisms. Typing of certain human pathogens *(11,12)* by analysis of their virulence-associated gene content is likely to be highly discriminatory, but the most significant advantage of this approach may be that it should be possible to correlate gene content with disease transmission or presentation. From a typing viewpoint, there is likely to be a great deal of redundancy in the data provided by genome-scale microarrays; therefore, it should be advantageous to select subsets of informative genes for the arrays. This approach, which has been demonstrated by van Ijperen et al. *(12)*, reduces costs and the time required for data analysis.

### *1.6. Future Developments*

The current microarray format, in which each glass slide may carry several thousand probes, is well suited to hybridizations between target material and genome-scale probe arrays. However, although several arrays of less than 1000 probes can easily be accommodated on a single slide, it is not then a simple matter to hybridize each array to a different target. Systems that allow hybridization of multiple targets to a single slide would be very convenient for typing applications. This equipment is likely to become available as probe sets for typing are developed further.

Genomotyping has been the main focus of bacterial typing work using microarrays. However, an alternative would be to develop arrays of probes each complementary to a different amplified fragment made by the AFLP method. Hybridization of the array to targets composed of AFLP fragments from an individual strain would result in a strain-specific pattern of reactions *(13)*. The discriminatory ability of arrays of AFLPs would depend upon the choice of probes used, and could be very high. In contrast to the genomotyping approach, the probes used would be selected on the basis of their contribution to the discriminatory ability of the typing method. In AFLP, the absence of a fragment in the amplified material shows only that one of the restriction sites is absent. This may indicate that a single base mutation has occurred in one of the sites or that the sequence is entirely absent.

## 2. Materials

### *2.1. Slides*

Microarrays are usually printed onto glass slides coated with reactive groups. Slides with different coatings are available and are used in variations of the microarray method. There are two classes of slides in common use, and these are most suitable for spotting either PCR products or oligonucleotides. Slides for spotting PCR products are usually coated with polylysine or aminosilane; these surfaces bind unmodified DNA covalently via negatively charged phosphate groups (**Fig. 3A**). The second class of slides, which are modified with aldehyde (**Fig. 3B**) and epoxy (**Fig. 3C**) groups, are usually recommended for printing oligonucleotides that have a free aminolinker. The amino group of the modified oligonucleotide binds to the active groups on the slide. This has the advantage that all of the molecules are bound in the same place and have the same orientation on the slide. This is important for relatively short oligonucleotide sequences, since steric hindrance can significantly affect the specificity

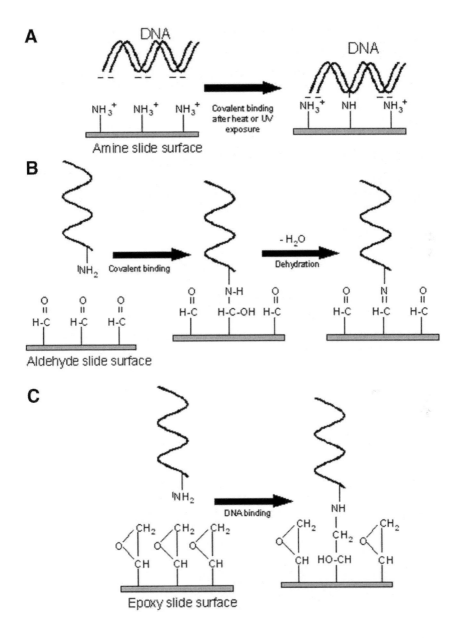

Fig. 3. Binding of DNA onto different slide surfaces. **(A)** Binding of DNA to amine slide in which positive amine groups covalently bind negatively charged DNA phosphate backbone. Binding of amino-linked oligonucleotides to **(B)** aldehyde and **(C)** epoxy slide surfaces by an active binding of the free amino group.

and stability of probe/target hybridization. Although nonmodified oligonucle-otides can be covalently linked to epoxy slides, longer probes (>50-mer) are recommended *(14)*, and care should be taken to obtain good crosslinking by UV treatment or baking *(15)*.

## 2.2. Probes for Printing

In microarray technology, the PCR products or oligonucleotides spotted onto the slide are designated *probes*. Amplified probes are usually produced by PCR using gene-specific primers, with the genomic DNA of the organism of interest as substrate. Each PCR product represents a specified ORF and may either be an ORF-specific fragment or the complete ORF. Following amplification, the probes are checked for purity and identity. The most common approach is to analyze the products of each PCR by agarose-gel electrophoresis to check for the presence of a single amplicon of the expected molecular size. A proportion of products that varies depending on the resources available, are then sequenced to provide a positive identity check prior to arraying. When oligonucleotides are used as probes, most of these steps are unnecessary. Oligonucleotides ordered from commercial companies are purified and ready to print as soon as they arrive in the laboratory. This saves the time and resources required to run PCR reactions, analyze the products, and redesign any PCRs that do not give the expected product. However, short oligonucleotides are best bound to the glass surface via an aminolink group, and this increases the cost of synthe-sis. Depending on the length, oligonucleotides have the advantage of giving more specific hybridization signals. The longer PCR-amplified products, how-ever, might detect similar genes with partial homology to the gene of interest, which may be useful for the detection of gene homologs.

The probes are dissolved in solutions that interact physically with the slide surface via the depositing device to give aliquots of consistent volume that dry as round spots of equal size and shape. The spotting solution must also be of suitable volatility to ensure rapid drying of the deposited spots while prevent-ing excessive evaporation of the sample during the printing process. This is also influenced by the local temperature and humidity. A common spotting solution is 50% dimethyl sulfoxide (DMSO), which is normally used on nonactivated surfaces. However, DMSO can interfere with some active groups on the slide surface, and it is less suitable for the printing oligonucleotides than alternatives such as saline–sodium citrate (SSC) solution. A wide range of com-mercial spotting buffers are also available. Changes in the volumes of the probe solutions dispensed into microtiter plate wells for spotting inevitably occur over time, even when care is taken to avoid the evaporation of solvent. To preserve valuable probes, plates affected by evaporation may be dried down fully and the probes redissolved in the original volume of solvent. It is usually unnecessary to adjust the volume for probe used in a previous print run.

### 2.3. Target

In the usual nomenclature, the "target" is the nucleic acid that is hybridized to the array. A wide range of target formulations are used. The well-known application of microarrays to gene-expression studies uses labeled cDNA made from mRNA extracts by reverse transcription *(16)*. Multiplex PCR products may be used *(17)*, and genomic DNA may be most appropriate for typing or screening for the presence or absence of genes.

Target nucleic acids are labeled with fluorescent dyes by either direct or indirect incorporation methods. The most common dyes used are cyanine three (Cy3, green) and cyanine five (Cy5, red). These two dyes have different absorption and emission spectra and can therefore be used in the same hybridization experiment. The control and a test sample are each labeled with a different dye, then mixed and hybridized to the array. The hybridization signal is detected with a high resolution scanner. The result at each probe is either a green, red, or yellow (mixed) signal. Green or red signals indicate that only one of the targets has hybridized to the probe, and yellow shows that both control and test sample have hybridized.

Direct incorporation of the Cy dyes by some polymerases is not always efficient due to steric hindrance by the large Cy groups. The degree of inhibition depends upon the structure of the Cy dye, and this may cause inaccuracies in experiments in which the ratio of two dyes is measured. Alternative indirect labeling methods have been developed using amino-allyl-modified dNTPs. The small amino-allyl dNTPs are incorporated into the target nucleic acid at a rate similar to that of unmodified nucleotides, and Cy dye is then bound to the target DNA by ester bonding. This approach gives good yields of highly labeled target, and thus minimizes experimental variation. Furthermore, as all Cy dyes can be incorporated equally, the normalization of signal intensities is less problematic.

### 2.4. Buffers

During the three stages of a microarray experiment, various buffers are used. The first step is to treat the slide with a blocking or prehybridization buffer before addition of the target, the purpose being to prevent nonspecific binding by blocking remaining sites on the glass surface that could interact with nucleic acids. Blocking buffers usually contain bovine serum albumin (BSA), sodium dodecyl sulphate (SDS), and SSC, but, depending on the application, competitor DNA may also be added. For example, addition of $C_0t$-1 DNA to the buffer reduces the hybridization of repetitive elements by binding to the labeled repetitive target sequences, while tRNA acts as a blocker of nonspecific hybridization. The next stage is the hybridization of target to the array. Various commercial hybridization buffers are available, but in-house buffers are also widely used. The use of formamide in the hybridization buffer has the advantage that the DNA remains denatured at lower temperatures. Hybridization at a lower temperature results in reduced evaporation from the target solution during incubation; evaporation leads to high background signals in locations where target has dried. Following hybridization, several washes are carried out to remove nonspecifically bound target. The stringency of the posthybridization washes affects the strength of the hydrogen bonding between the probe and target. Washing stringency can be adjusted by changing the temperature and SSC concentration (stringency is raised at higher temperatures and lower salt concentrations). SDS is autofluorescent and should be omitted from the buffer used for the final post-hybridization wash.

### 2.5. Hybridization

Water-tight chambers (Corning, Schiphol-Rijk, The Netherlands; and Genetix Limited, New Milton, UK) that fit the microarray slide exactly are used for hybridization. The chambers prevent the slide from drying out when used within a hybridization oven, but they can also be used in a water bath. Hybridization takes place under a coverslip so that a small amount of target is evenly spread over the array. Automatic hybridization stations are also commercially available. In these systems, hybridization takes place in a low-volume chamber with access ports for the addition of buffers. An advantage of this equipment is that the target solution can mix more freely than under a coverslip so the target has a greater opportunity to interact with the probe array; this increases signal intensities and ensures that all positions on the array are equally exposed.

### 2.6. Scanners and Software

A variety of fluorescent scanners are now available for microarray work. The machines generally include a high-resolution confocal microscope scanner (10 to 4 µm), and have lasers with the appropriate wavelengths as light sources. The standard lasers excite Cy3- (532 nm) and Cy5- (635 nm) labeled targets, but up to six different lasers are now included with recent designs. Scanners generally come with user-friendly operation software and are therefore simple to use. The Affymetrix 428™, the Axon Genepix 4000®, and the PerkinElmer ScanArray™ are most commonly used.

Similarly, a variety of software packages are available for microarray image and data analysis. Although there are differences, most programs have basic functions to allow measurement of signal intensities and to present the data in convenient formats (i.e., histograms, scatterplots, and so on). The complexity of microarray data presents great challenges to the analysis software. The available programs are under continuous development, and new upgrades contain more and better tools for normalization, analysis, and statistical analysis. Any microarray software listing will be rapidly outdated due to the frequent introduction of new programs. Software should be selected to fulfil the needs of the individual user. Bacterial comparative genomics is a relatively simple application, and consequently the major software consideration is likely to be ease of use.

## 3. Methods

### 3.1. Probe Preparation and Microarray Printing

Probes often comprise specific parts of ORFs that are amplified by PCR and then purified before printing. In the case of oligonucleotide microarrays, the probes are obtained from custom synthesis houses and may be aminolinker-modified depending on the application and slide coating employed. The probes spotted onto the slides must be very specific for the detection of the target gene or genes, since cross-hybridizations lead to array results that are very difficult to interpret. The steps required to prepare oligonucleotide or amplified probes are summarized in **Note 2**.

Microarrays are usually spotted using automated robotic arrayers, although manual instruments are available that may be suitable for producing arrays comprised of relatively few probes. Optimal environmental conditions for printing are between 40% and 50% humidity at around 20°C. These conditions prevent spots from drying too quickly. Handling of the slides after printing depends on the slide coating. Crosslinking by UV illumination is recommended for some slides, but quoted energy values vary and convincing optimization data are not available. Baking at 80°C for up to 2 h with occasional reintroduction of water by "fogging" the slide over warm water is recommended for some slides. Some protocols call for both UV and heat treatments, while others recommend curing at room temperature. Currently, most authors and slide manufacturers take an empirical approach to probe attachment, and a wide variety of conditions are satisfactory.

### 3.2. Hybridization and Posthybridization Processing of Slides

For genomotyping experiments where the aim is to show the presence or absence of genes within the genome of a particular bacterial strain, the target is comprised of labeled total DNA. Cy dyes can be incorporated directly into representative DNA by primer extension using a random priming kit (Invitrogen, Paisley, UK). Purification of the target prior to hybridization is very important. For each hybridization on a 22 mm × 22 mm surface, 15 µL of probe is needed. The procedure is summarized in **Note 3**. Labeling of nucleic acid using direct methods is inefficient for some combinations of enzyme and modified nucleotide. Cy dyes in particular have a large 3D structure and are therefore relatively inefficiently incorporated into DNA. Indirect labeling methods have been developed that use relatively small amino-allyl dNTPs as substrate for the enzymatic reaction (*see* **Subheading 2.3.**). These reactive primary amines react with NHS ester-modified Cy dyes in a nonenzymatic second step.

### 3.3. Analysis

Currently, the most time-consuming part of microarray experiments is data analysis. The reason for this is that many factors must be considered before any conclusions can be drawn from the raw data. The first steps in data analysis involve normalization of the images. Standard methods of normalization calculate either the mean or median intensity for a microarray image. This value is then used to correct the intensity of each spot. The mean or median of all corrected spot intensities is finally adjusted to center around one, and then converted to $\log_{10}$ form so that the points follow a normal distribution centered around zero. Using logarithmic transformed data helps to normalize the distribution of the data and improves the applicability of statistical tests. Most statistical tests work with differences, although scientists are more interested in a fold difference. This problem is solved by working with a log ratio, as $\log a - \log b = \log a/b$. The next step in analysis is to detect outliers. If the intensity of a spot is influenced by the strong fluorescence of a dust particle on the slide, this spot should be excluded from further analysis. In this way the data are reduced to a series of values that are in the form of ratios when different targets, each labeled with a distinct fluor, have been hybridized to the array. At this point the question to be considered is whether any differences observed are reproducible or significant. To answer this question, duplicate experiments are required, which in turn generate yet more data to analyze.

### 4. Bacterial Typing by Microarrays: Realistic Goal or Holy Grail?

Microarrays for typing is clearly not a holy grail, i.e., the objective of a quest that is unlikely to be realized without divine intervention. However, current microarray technology is not well-suited to high-throughput typing of many different strains. Considerable hands-on time is required for each sample, and the analysis can be time-consuming, especially while setting up a new application. At present, genomotyping using microarrays should be viewed as a complementary method that can provide large quantities of gene-content data and is appropriate in that context. These comparative genomic data allow the construction of detailed overviews of the relationships between strains and are extremely valuable. We anticipate rapid progress in the development of protocols and automation for microarray hybridization. These developments are likely to lead to the use of genomotyping as a routine tool for bacterial typing. This is certainly a realistic goal.

## 5. Notes

1. Array spotting technology. Various techniques for spotting small volumes of probe solutions have been described. The most common method of array printing currently is the use of capillary reservoir pins (split pins), which deposit picoliter volumes of probe onto the glass surface. The ring-pin system is similar in that pins are used to deposit the probe solution, but a loop filled with probe solution acts as the reservoir with the pin passing through the probe meniscus each time a spot is printed. Another method uses an inkjet-type spotter that fires the probe onto the surface without touching the surface.

2. Probe preparation. Probe preparation by PCR amplification includes the following steps: First, specific oligonucleotide primers have to be designed using the appropriate software, followed by the PCR amplification of the product. After removing the enzyme and unincorporated nucleotides, the identity of the products can be verified by size and/or sequencing. Finally, the concentrations are adjusted where necessary and probes are diluted in the appropriate spotting solution.

   If probes are prepared by oligonucleotide synthesis, fewer steps are involved. Oligonucleotides are designed using specialized software and bioinformatics knowledge, and users decide on the basis of the length of the oligonucleotide whether an aminolink is necessary. Oligonucleotides are produced synthetically by commercial companies. After adjustment of the concentration, probes are ready for printing.

3. Genomic DNA hybridization on microarrays. When genomic DNA is used for the preparation of target material, up to 2 µg can be labeled directly using Cy-labeled nucleotides and Klenow enzyme. Removing enzyme and unincorporated nucleotides is crucial for a low background signal after hybridization. The target is dried under vacuum and dissolved in 15 µL of hybridization buffer. After denaturing, the target can be applied to the slide, covered with a disposable coverslip, and incubated in a chamber at the appropriate temperature. After hybridization, slides are washed several times in 2X SSC buffer containing 0.1% SDS, and finally in 2X SSC buffer. Slides are dried by centrifugation in 50-mL tubes.

## References

1. Baba, T., Takeuchi, F., Kuroda, M., Yuzawa, H., Aoki, K., Oguchi, A., et al. (2002) Genome and virulence determinants of high virulence community-acquired MRSA. *Lancet* **359**, 1819–1827.

2. Lucchini, S., Thompson, A., and Hinton, J. C. D. (2001) Microarrays for microbiologists. *Microbiol.* **147,** 1403–1414.

3. Dorrell, N., Mangan, J. A., Laing, K. G., Hinds, J., Linton, D., Al-Ghusein, H., et al. (2001) Whole genome comparison of *Campylobacter jejuni* human isolates using a low-cost microarray reveals extensive genetic diversity. *Genome Res.* **11,** 1706–1715.

4. Dziejman, M., Balon, E., Boyd, D., Fraser, C. M., Heidelberg, J. F., and Mekalanos J. J. (2002) Comparative genomic analysis of *Vibrio cholerae*: genes that correlate with cholera endemic and pandemic disease. *Proc. Natl. Acad. Sci. USA* **99,** 1556–1561.

5. Smoot, J. C., Barbian, K. D., Van Gompel, J. J., Smoot, L. M., Chaussee, M. S., Sylva, G. L., et al. (2002) Genome sequence and comparative microarray analysis of serotype M18 group A *Streptococcus* strains associated with acute rheumatic fever outbreaks. *Proc. Natl. Acad. Sci. USA* **99,** 4668–4673.

6. Relógio, A., Schwager, C., Richter, A., Ansorge, W., and Valcárcel, J. (2002) Optimization of oligonucleotide-based DNA microarrays. *Nucl. Acids Res.* **30,** e51.

7. Rozen, S. and Skaletsky, H. J. (1996–1998) Primer3. Code available at http://www-genome.wi.mit.edu/genome_software/other/primer3.html

8. Rouillard, J-M., Herbert, C. J., and Zuker, M. (2002) OligoArray: genome-scale oligonucleotide design for microarrays. *Bioinformatics* **18,** 486–487.

9. Salama, N., Guillemin, K., McDaniel, T. K., Sherlock, G., Tompkins, L., and Falkow, S. (2000) A whole-genome microarray reveals genetic diversity among *Helicobacter pylori* strains. *Proc. Natl. Acad. Sci. USA* **97,** 14,668–14,673.

10. Behr, M. A., Wilson, M. A., Gill, W. P., Salamon, H., Schoolnik, G. K., Rane, S., et al. (1999) Comparative genomics of BCG vaccines by whole-genome DNA microarray. *Science* **284,** 1520–1523.

11. Fitzgerald, J. R., Sturdevant, D. E., Mackie, S. M., Gill, S. R., and Musser, J. M. (2001) Evolutionary genomics of *Staphylococcus aureus:* insights into the origin of methicillin-resistant strains and the toxic shock syndrome epidemic. *Proc. Natl. Acad. Sci. USA* **98,** 8821–8826.

12. Van Ijperen, C., Kuhnert, P., Frey, J., and Clewley, J. P. (2002). Virulence typing of *Escherichia coli* using microarrays. *Mol. Cell Probes* **16,** 371–378.

13. Hu, H., Lan, R., and Reeves, P. R. (2002) Fluorescent amplified fragment length polymorphism analysis of *Salmonella enterica* serovar Typhimurium reveals phage-type-specific markers and potential for microarray typing. *J. Clin. Microbiol.* **40,** 3406–3415.

14. Kane, M. D., Jatkoe, T. A., Stumpf, C. R., Lu, J., Thomas, J. D., and Madore, S. J. (2000). Assessment of the sensitivity and specificity of oligonucleotide (50mer) microarrays. *Nucl. Acids Res.* **28,** 4552–4557.

15. Massimi, A., Harris, T., Childs, G., and Somerville, S. (2003) *DNA Microarray: A Molecular Cloning Manual.* (Bowtell, D. and Sambrook, J., eds.), Cold Spring Harbor Laboratory, Cold Spring Harbor, New York. p. 78.

16. Watson, A., Mazumder, A., Stewart, M., and Balasubramanian, S. (1998) Technology for microarray analysis of gene expression. *Curr. Opin. Biotechnol.* **9,** 609–614.
17. Chizhikov, V., Rasooly, A., Chumakov, K., and Levy, D. D. (2001) Microarray analysis of microbial virulence factors. *Appl. Environ. Microbiol.* **67,** 3258–3263.
18. Kuroda, M., Ohta, T., Uchiyama, I., Baba, T., Yuzawa, H., Kobayashi, I., et al. (2001) Whole genome sequencing of meticillin-resistant *Staphylococcus aureus.* *Lancet* **357,** 1225–1240.

# III

## INTERROGATING BACTERIAL GENOMES

# 11

# Genomic Approaches to Antibacterial Discovery

David J. Payne, Michael N. Gwynn, David J. Holmes,
and Marty Rosenberg

## Summary

This chapter describes two key strategies for the discovery of new antibacterial agents and illustrates the critical role played by genomics in each. The first approach is genomic target-based screening. Comparative genomics and bioinformatics are used to identify novel, selective antibacterial targets of the appropriate antibacterial spectrum. Genetic technologies integral for the success of this approach, such as essentiality testing, are also described. An unprecedented number of novel targets have been discovered via this approach, and a plethora of examples are discussed. This section concludes with the case history of a target successfully progressed from identification by genomics, to high-throughput screening, and onto proof of concept in curing experimental infections. The second approach is based on screening for compounds with antibacterial activity and then employing a broad variety of newer technologies to identify the molecular target of the antibacterial agent. The advantage of this approach is that compounds already possess antibacterial activity, which is a property often challenging to engineer into molecules obtained from enzyme-based screening approaches. The recent development of novel biochemical and genomic technologies that facilitate identification and characterization of the mode of action of these agents has made this approach as attractive as the genomic target-based screening strategy.

**Key Words:** Genomics; bioinformatics; essentiality; chorismate biosynthesis; isoprenoid biosynthesis; TCSTs; FtsZ; FtsA; MurA; fatty acid biosynthesis; FabI; enzyme based screening; whole-cell screening.

From: *Methods in Molecular Biology, vol. 266: Genomics, Proteomics, and Clinical Bacteriology:
Methods and Reviews*
Edited by: N. Woodford and A. Johnson © Humana Press Inc., Totowa, NJ

## 1. Introduction: Antibiotic Research

Traditionally, antibiotics have been identified by exposing bacteria to novel compounds and assaying their bacteriostatic or bactericidal activity. Once antibacterial activity was confirmed, a lengthy series of studies would determine their mode of action (MOA) with regard to which essential cellular function was being inhibited. While this approach was highly successful for several decades, only one new class of antibiotic has been commercialized in the last 20 yr. It is a reflection of the methods and the relatively limited chemical diversity employed, that currently available antibiotics target a small number of cellular processes involved in DNA, RNA, protein, and cell-wall biosynthesis *(1)*.

Although antibiotic discovery declined during the 1980s and early 1990s, the same cannot be said for the ability of pathogenic bacteria to develop resistance. The appearance of high-level vancomycin-resistant *Staphylococcus aureus* (VRSA) *(2)* and vancomycin-resistant *Enterococcus faecium* (VREF), along with multiple drug resistance in many other clinically important bacteria, has made it essential that new antibiotics be found *(3)*.

We discuss here two key strategies utilizing interchangeable technologies that have been applied to the discovery of novel antibiotics. The first strategy is target-based screening, where comparative genomics and bioinformatic analysis are used to select targets with the appropriate spectrum and selectivity. Genetic technologies integral for the success of this approach, such as essentiality testing, are also described. An unprecedented number of novel targets have been discovered via this approach, and a number of examples are discussed. This section concludes with the case history of a target successfully progressed through identification by genomics, high-throughput screening (HTS), and proof of concept in curing experimental infections. The second approach is to screen for compounds that possess antibacterial activity, and then employ genomic and bioinformatic tools to identify the molecular target. The advantage of this approach is that the compounds already possess antibacterial activity, which is a property often difficult to engineer into molecules identified by enzyme-based screening approaches. The recent development of novel biochemical and genomic technologies that facilitate identification and characterization of the MOA of these agents have made this approach as attractive as the genomic target-based screening strategy. Both of these approaches have advantages and unique challenges associated with them, but genomics and genomic technologies are integral to the success of either approach.

## 2. Strategies in Antibacterial Discovery

### 2.1. Strategy I: Target-Based Screening

#### 2.1.1. Target Selection

Comparison of the many bacterial genome sequences available (**Table 1**) allows the identification of targets that are present in most, if not all, clinically relevant bacteria. Further comparison with the genome sequences of eucaryotes allows the selection of targets that are sufficiently different or unique that they would be targets to inhibit bacteria *(4,5)*. Indeed, this is particularly relevant given the recent publication of the human genome sequence *(6,7)*. Moreover, subset comparisons among different bacterial species suggest the possibility of more targeted chemotherapy approaches that would allow the development of antibacterial agents selectively active against only Gram-positive or Gram-negative bacteria, or even against individual pathogens (e.g., *Helicobacter pylori, Chlamydia*). There is an enormous benefit in knowing the phylogenetic distribution of potential target systems early in the drug-discovery process. It provides some predictive value of the potential spectrum of activity that is achievable and, perhaps more importantly, allows targets of defined spectrum to be sought. Targets specific to only one or a few bacterial species would allow narrow-spectrum agents to be created, thereby limiting the side-effects, often inherent in the use of broad-spectrum agents, that are often caused by altering normal bacterial flora.

#### 2.1.2. Target Validation: Essentiality Testing

Valid antimicrobial targets are those that are essential for cell viability during growth and infection. To determine the role of various gene products in cell viability and pathogenesis, a number of techniques have been developed for testing essentiality both in vitro and in infection models. For example, random mutants can be generated in which growth is conditional. In one case, a signature-tagged transposon-based system has been developed to identify genes that are essential for infection, but which are dispensable for growth in vitro and thus encode virulence factors (*see* Chapter 13) *(8,9)*. Temperature-sensitive mutants can also be generated in which in vitro growth is conditional, but these require rather tedious procedures to identify the responsible allele. In the past, due to the paucity of genetic information in pathogenic bacteria, these random mutagenic methodologies were the only way to identify genes vital for cell viability. Such approaches are limited by the fact that they do not generate truly random libraries of mutants. Transposons have preferred sites of integration, and there are regions of chromosomes into which they apparently do not insert. Equally, conditional lethal mutations cannot be generated in all essen-

**Table 1**
**Publicly Available Bacterial Genomes**

| Organism | Domain | Contigs | Genes | Total bases | Source |
|---|---|---|---|---|---|
| *Actinobacillus actino-mycetemcomitans* HK1651 | Bacteria | 9 | 1988 | 2,095,439 | University of Oklahoma |
| *Aeropyrum pernix* K1 | Archaea | 1 | 1840 | 1,669,695 | NCBI |
| *Agrobacterium tumefaciens* C58 | Bacteria | 4 | 5299 | 5,673,563 | NCBI |
| *Agrobacterium tumefaciens* C58 (Dupont) | Bacteria | 4 | 5402 | 5,674,064 | NCBI |
| *Aquifex aeolicus* VF5 | Bacteria | 2 | 1553 | 1,590,791 | NCBI |
| *Archaeoglobus fulgidus* DSM4304 | Archaea | 1 | 2420 | 2,178,400 | NCBI |
| *Bacillus anthracis* | Bacteria | 2 | 226 | 277,885 | NCBI |
| *Bacillus anthracis* Ames | Bacteria | 1 | 5287 | 5,227,297 | TIGR |
| *Bacillus halodurans* C-125 | Bacteria | 1 | 4066 | 4,202,353 | NCBI |
| *Bacillus stearothermophilus* 10 | Bacteria | 380 | 3342 | 3,269,999 | University of Oklahoma |
| *Bacillus subtilis* 168 | Bacteria | 1 | 4112 | 4,214,814 | NCBI |
| *Bordetella pertussis* Tohama I | Bacteria | 1 | 3892 | 4,086,186 | Sanger Center |
| *Borrelia burgdorferi* B31 | Bacteria | 22 | 1637 | 1,519,856 | NCBI |
| *Brucella melitensis* 16M | Bacteria | 2 | 3198 | 3,294,931 | NCBI |
| *Buchnera sp.* APS | Bacteria | 3 | 574 | 655,725 | NCBI |
| *Caenorhabditis elegans* | Eukaryota | 0 | 22,407 | 0 | NCBI |
| *Campylobacter jejuni* NCTC 11168 | Bacteria | 1 | 1633 | 1,641,481 | Sanger Center |
| *Caulobacter crescentus* | Bacteria | 1 | 3737 | 4,016,947 | TIGR |
| *Chlamydia muridarum* | Bacteria | 2 | 916 | 1,076,912 | NCBI |
| *Chlamydia trachomatis* D/UW-3/Cx | Bacteria | 1 | 921 | 1,042,519 | NCBI |
| *Chlamydophila pneumoniae* AR39 | Bacteria | 1 | 1109 | 1,229,853 | NCBI |
| *Chlamydophila pneumoniae* CWL-029 | Bacteria | 1 | 1052 | 1,230,230 | Incyte |
| *Chlamydophila pneumoniae* J138 | Bacteria | 1 | 1069 | 1,226,565 | NCBI |
| *Chlamydophila pneumoniae* L2 | Bacteria | 2 | 1042 | 1,234,390 | TIGR |
| *Chlorobium tepidum* TLS | Bacteria | 24 | 3178 | 2,196,918 | TIGR |

*(continued)*

**Table 1** *(Continued)*

| Organism | Domain | Contigs | Genes | Total bases | Source |
|----------|--------|---------|-------|-------------|--------|
| *Clostridium acetobutylicum* ATCC824 | Bacteria | 1 | 3672 | 3,940,880 | TIGR |
| *Clostridium perfringens* 13 | Bacteria | 2 | 2723 | 3,085,740 | NCBI |
| *Corynebacterium diphtheriae* NCTC13129 | Bacteria | 1 | 2128 | 2,488,600 | Sanger Center |
| *Corynebacterium glutamicum* | Bacteria | 1 | 2989 | 3,309,400 | EBI |
| *Deinococcus radiodurans* R1 | Bacteria | 4 | 3103 | 3,284,156 | TIGR |
| *Desulfovibrio vulgaris* | Bacteria | 1 | | 3,571,425 | TIGR |
| *Enterococcus faecalis* V583 | Bacteria | 4 | 3148 | 3,359,973 | TIGR |
| *Enterococcus faecium* | Bacteria | 365 | 3309 | 2,928,706 | DOE/JGI |
| *Escherichia coli* K-12 | Bacteria | 1 | 4279 | 4,639,221 | NCBI |
| *Escherichia coli* O157:H7 | Bacteria | 1 | 5361 | 5,498,450 | NCBI |
| *Escherichia coli* O157:H7 EDL933 | Bacteria | 1 | 5324 | 5,528,445 | NCBI |
| *Fusobacterium nucleatum* ATCC 25586 | Bacteria | 1 | 2067 | 2,174,500 | NCBI |
| *Haemophilus influenzae* KW20 | Bacteria | 1 | 1714 | 1,830,138 | NCBI |
| *Halobacterium sp.* NRC-1 | Archaea | 3 | 2605 | 2,571,010 | NCBI |
| *Helicobacter pylori* 26695 | Bacteria | 1 | 1576 | 1,667,867 | NCBI |
| *Helicobacter pylori* J99 | Bacteria | 1 | 1491 | 1,643,831 | NCBI |
| *Lactococcus lactis subsp.* | Bacteria | 1 | 2266 | 2,365,589 | NCBI |
| *Listeria innocua* Clip11262 | Bacteria | 1 | 2968 | 3,011,208 | NCBI |
| *Listeria monocytogenes* EGD | Bacteria | 1 | 2846 | 2,944,528 | NCBI |
| *Mesorhizobium loti* | Bacteria | 3 | 7281 | 7,596,300 | NCBI |
| *Methanobacterium thermo- autotrophicum delta* H | Archaea | 1 | 1869 | 1,751,379 | NCBI |
| *Methanococcus jannaschii* DSM 2661 | Archaea | 3 | 1770 | 1,739,927 | NCBI |
| *Methanopyrus kandleri* AV19 | Archaea | 1 | 1687 | 1,694,969 | NCBI |
| *Methanosarcina acetivorans* | Archaea | 1 | 4540 | 5,751,492 | NCBI |
| *Mycobacterium leprae* | Bacteria | 1 | 2157 | 3,268,203 | Sanger Center |
| *Mycobacterium tuberculosis* CDC1551 | Bacteria | 1 | 4187 | 4,403,836 | NCBI |
| *Mycobacterium tuberculosis* H37Rv (lab strain) | Bacteria | 1 | 3927 | 4,411,529 | Sanger Center |
| *Mycoplasma genitalium* G-37 | Bacteria | 1 | 484 | 580,074 | NCBI |
| *Mycoplasma pneumoniae* M129 | Bacteria | 1 | 688 | 816,394 | NCBI |

*(continued)*

**Table 1** *(Continued)*

| Organism | Domain | Contigs | Genes | Total bases | Source |
|---|---|---|---|---|---|
| *Mycoplasma pulmonis* UAB CTIP | Bacteria | 1 | 782 | 963,879 | NCBI |
| *Neisseria gonorrhoeae* FA 1090 | Bacteria | 1 | 2129 | 2,146,879 | University of Oklahoma |
| *Neisseria meningitidis* MC58 | Bacteria | 1 | 2025 | 2,272,351 | TIGR |
| *Neisseria meningitidis* Z2491 | Bacteria | 1 | 2065 | 2,184,406 | Sanger Center |
| *Nostoc sp.* PCC 7120 | Bacteria | 7 | 6129 | 7,211,789 | NCBI |
| *Pasteurella multocida* PM70 | Bacteria | 1 | 2014 | 2,257,487 | NCBI |
| *Porphyromonas gingivalis* W83 | Bacteria | 1 | 1777 | 2,343,478 | TIGR |
| *Prochlorococcus marinus* MED4 | Bacteria | 26 | 1716 | 1,674,813 | DOE Joint Genome |
| *Pseudomonas aeruginosa* PAO1 | Bacteria | 1 | 5565 | 6,264,403 | University of W |
| *Pyrobaculum aerophilum* | Archaea | 1 | 2275 | 2,222,890 | UCLA Dept. Micr |
| *Pyrococcus abyssi* | Archaea | 1 | 1765 | 1,765,118 | NCBI |
| *Pyrococcus furiosus* | Archaea | 1 | 2208 | 1,908,253 | Utah |
| *Pyrococcus horikoshii* OT3 | Archaea | 1 | 2058 | 1,738,505 | NCBI |
| *Ralstonia solanacearum* | Bacteria | 2 | 5116 | 5,810,922 | NCBI |
| *Rhizobium sp.* NGR234 | Bacteria | 1 | 417 | 536,165 | NCBI |
| *Rickettsia conorii* | Bacteria | 1 | 1374 | 1,268,755 | NCBI |
| *Rickettsia prowazekii* Madrid E | Bacteria | 1 | 834 | 1,111,523 | NCBI |
| *Saccharomyces cerevisiae* S288C | Eukaryota | 16 | 6261 | 12,057,849 | NCBI |
| *Salmonella typhi* CT18 | Bacteria | 3 | 4633 | 5,133,712 | Sanger Center |
| *Salmonella typhimurium* LT2 (strain AZ1516) | Bacteria | 2 | 4553 | 4,951,371 | Washington Univ |
| *Shewanella putrefaciens* | Bacteria | 2 | 4221 | 5,131,063 | TIGR |
| *Sinorhizobium meliloti* 1021 | Bacteria | 3 | 6205 | 6,691,694 | NCBI |
| *Staphylococcus aureus* EMRSA-16 | Bacteria | 1 | 2679 | 2,902,619 | Sanger Center |
| *Staphylococcus aureus* MW2 | Bacteria | 1 | 2632 | 2,820,462 | NCBI |
| *Staphylococcus aureus* Mu50 | Bacteria | 2 | 2748 | 2,903,147 | NCBI |

*(continued)*

**Table 1** *(Continued)*

| Organism | Domain | Contigs | Genes | Total bases | Source |
|---|---|---|---|---|---|
| *Staphylococcus aureus* N315 | Bacteria | 2 | 2625 | 2,838,294 | NCBI |
| *Staphylococcus aureus* NCTC 8325 | Bacteria | 2 | 2631 | 2,836,373 | University of Oklahoma |
| *Staphylococcus epidermidis* | Bacteria | 367 | 2920 | 2,701,014 | GSK |
| *Staphylococcus epidermidis* RP62A | Bacteria | 2 | 2444 | 2,646,310 | TIGR |
| *Streptococcus mutans* UA159 | Bacteria | 1 | 1871 | 2,030,921 | University of Oklahoma |
| *Streptococcus pneumoniae* R6 hex | Bacteria | 1 | 2043 | 2,038,615 | NCBI |
| *Streptococcus pneumoniae* type4 | Bacteria | 1 | 2094 | 2,160,837 | TIGR |
| *Streptococcus pyogenes* M1 GAS | Bacteria | 1 | 1697 | 1,852,441 | University of Oklahoma |
| *Streptococcus pyogenes* MGAS8232 | Bacteria | 1 | 1845 | 1,895,017 | NCBI |
| *Sulfolobus solfataricus* P2 | Archaea | 1 | 2977 | 2,992,245 | NCBI |
| *Sulfolobus tokodaii* | Archaea | 1 | 2826 | 2,694,765 | NCBI |
| *Synechocystis sp.* PCC6803 | Bacteria | 1 | 3169 | 3,573,470 | NCBI |
| *Thermoanaerobacter tengcongensis* MB4T | Bacteria | 1 | 2588 | 2,689,445 | NCBI |
| *Thermoplasma volcanium* | Archaea | 1 | 1499 | 1,584,804 | NCBI |
| *Thermotoga maritima* MSB8 | Bacteria | 1 | 1846 | 1,860,725 | TIGR |
| *Treponema pallidum* Nichols | Bacteria | 1 | 1031 | 1,138,011 | NCBI |
| *Ureaplasma urealyticum* serovar 3 (Uu) | Bacteria | 1 | 611 | 751,719 | NCBI |
| *Vibrio cholerae* N16961 | Bacteria | 2 | 3835 | 4,033,464 | TIGR |
| *Xanthomonas axonopodis citri str.* 306 | Bacteria | 1 | 4312 | 5,175,554 | NCBI |
| *Xanthomonas campestris* ATCC 33913 | Bacteria | 1 | 4181 | 5,076,188 | NCBI |
| *Xylella fastidiosa* | Bacteria | 3 | 2831 | 2,731,750 | NCBI |
| *Yersinia pestis* CO-92 Biovar Orientalis | Bacteria | 4 | 4083 | 4,829,855 | Sanger Center |

From Rosalyn Wilson, Bill Marshall, and James Brown, Bioinformatics, GSK.

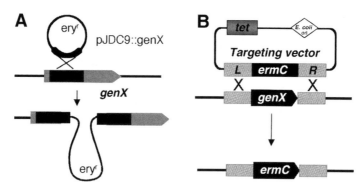

Fig. 1. Gene essentiality testing. (**A**) Plasmid insertion mutagenesis is a rapid method for gene disruption, but results can be misleading due to polarity effects and residual activity from intact gene fragments. (**B**) Allelic replacement mutagenesis results in deletion of the target gene and its replacement with a selective marker. Reproduced with permission from Current Drugs and Payne et al. *(40)*, copyright 2001, Current Drugs.

tial genes. With the availability of genome sequence data, it is now possible to identify every potential gene in the chromosome of a bacterium (*see* Chapter 3). Armed with this information, one can perform a systematic evaluation of the essentiality of each and any predicted gene. Indeed, a consortium of European laboratories is now attempting to evaluate the importance of every gene in the *Bacillus subtilis* genome.

A number of methods are available for knocking out genes in a directed fashion. The simplest form is plasmid insertion mutagenesis *(10)*. In this method, a small defined internal fragment (typically 500 bp) of the target gene is cloned in a nonreplicative plasmid harboring a selective marker that is expressed in the host organism. This plasmid is introduced into the bacterium and grown under selective conditions. Maintenance of the plasmid DNA requires homologous recombination, which results in the insertion of the plasmid sequence into the target gene (**Fig. 1A**). The method is quick and simple and, together with the availability of the sequence of the chromosome, has allowed a rapid genome-wide identification of essential genes in *Streptococcus pneumoniae (11)*. Interpretation of the data however, presents several problems. Because efficient recombination requires hundreds of base pairs of homology, a significant portion of the gene may remain intact after recombination, raising the issue of partial gene activity being expressed from the intact portions of the disrupted open reading frame. Partial activity of an essential gene may be sufficient for cell viability, and thus single crossover insertional inactivations may misrepresent genes that are truly essential. Moreover, insertional mutagenesis requires integration of the entire vector into the chromo-some, often altering the expression of genes downstream of the target (i.e., polarity effects). As a result, a gene may be considered essential when, in fact, the effect derives from a downstream gene vital for cell growth (i.e., false-positive analysis).

An improved knockout approach achieves removal of the gene to ensure its inactivation and replacement with a small resistance determinant designed not to introduce polarity effects *(12,13)*. This allelic replacement method requires a double recombination event using homologous sequences that flank the tar-get gene. The gene is deleted and replaced by a selective marker (**Fig. 1B**). The procedure identifies genes that are likely to be essential, and very effectively defines those genes that are not. The results can be used to focus attention on targets critical for bacterial functions.

Perhaps the best approach to confirm that a gene is essential for cell viabil-ity is to generate a conditional mutant. One method of achieving this is to place the target gene under the control of an inducible promoter. This method pro-vides quantitative information useful for evaluating potential targets for drug discovery because, in contrast to knockout methods that result in the complete loss of activity of the gene product, promoter control systems allow the level of expression to be regulated often in a titratable manner, ranging from no expres-sion (thereby testing the essentiality of the gene) to overexpression (in which the cellular concentration of the target exceeds that normally found in the wild-type strain). In this way, it is possible to determine the level of a particular protein necessary for survival/growth. A number of inducible promoter sys-tems have been developed and used in both nonpathogenic *(14)* and patho-genic *(15,16)* bacteria (**Fig. 2**). Bacterial strains carrying these regulated gene systems can be used to confirm the MOA of an antibacterial agent; if regula-tion of the levels of a particular gene target gives rise to a selective and con-comitant change in bacterial sensitivity to an inhibitor of the target, this provides corroboration of the mode of antibacterial action.

## 2.1.3. Examples of Novel Antibacterial Targets

In the following subsections, we discuss some of the targets progressed via the target-based screening strategy, and focus on the role played by genomics in identifying and progressing these targets. In some of these examples genomics has contributed to the identification of the target, while in others it has provided new insights into an established target area, which has enabled better exploitation of the target.

### 2.1.3.1. Aminoacyl tRNA Synthetases

Aminoacyl tRNA synthetases catalyze the amino acylation of amino acids with their cognate tRNA prior to incorporation of the amino acid into a grow-

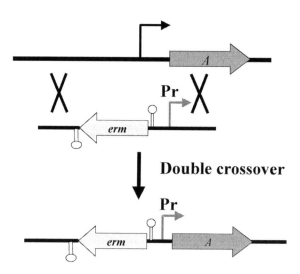

Fig. 2. The use of inducible promoters to evaluate gene essentiality. Gene A is placed under the control of an inducible promoter. In this way, the gene product can be titrated down to a level that cannot support growth, thereby illustrating the essential nature of the gene. Reproduced with permission from Current Drugs and Payne et al. *(40)*, copyright 2001, Current Drugs.

ing peptide chain. Consequently, there are 20 different tRNA synthetases in Gram-negative bacteria (but only 19 in Gram-positive bacteria), and each is essential for the survival of bacterial cells. The potential of these enzymes as antibacterial targets is demonstrated by the antibiotic mupirocin, which is a potent and selective inhibitor of isoleucyl tRNA synthetase (IRS). Prior to 1995 these tRNA synthetases had not been well characterized in different pathogens. Genomics, however, has allowed determination of the sequences of tRNA synthetases from multiple bacterial pathogens, and comparative analysis of these sequences has enabled valid consideration to be given to the spectrum and selectivity of each enzyme. Moreover, this analysis can demonstrate whether a specific enzyme may have use as a broad- or narrow-spectrum target. For example, as there are differences between the methionyl tRNA synthetases (MRS) of Gram-positive and Gram-negative bacteria, this enzyme may be an appropriate target if a selective anti-Gram-positive or anti-Gram-negative agent were required *(17)* (**Fig. 3A**). Conversely, as the glutamyl tRNA synthetases from Gram-positive and Gram-negative bacteria are more similar, this would be a more appropriate target if a broad-spectrum agent were desired *(17)* (**Fig. 3B**).

Structural data for many of these enzymes exist, and tRNA synthetases are highly amenable to HTS. In this example, genomics has expanded the opportu-

**A**

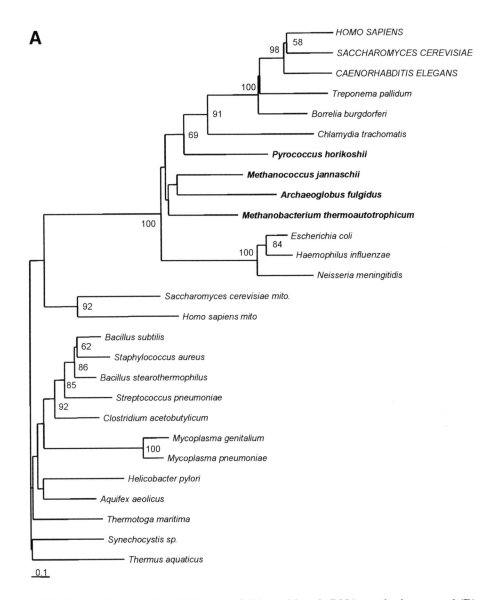

Fig. 3. Neighbor-joining (NJ) trees of **(A)** methionyl-tRNA synthethetase and **(B)** glutaminyl- and glutamyl-tRNA synthetases, labeled as GlnRS and GluRS, respectively. The NJ tree was constructed using the program NEIGHBOR of the PHYLIP 3.57c package (http//evolution.genetics.washington.edu/phylip.html). In order, the numbers at nodes represent the percent occurrence of nodes in 1000 bootstrap resamplings of the NJ tree. Different type faces are used to distinguish species of the Bacteria (italics), Eucarya (upper-case, italics) and Archaea (italics, bold type). Also

*(continued)*

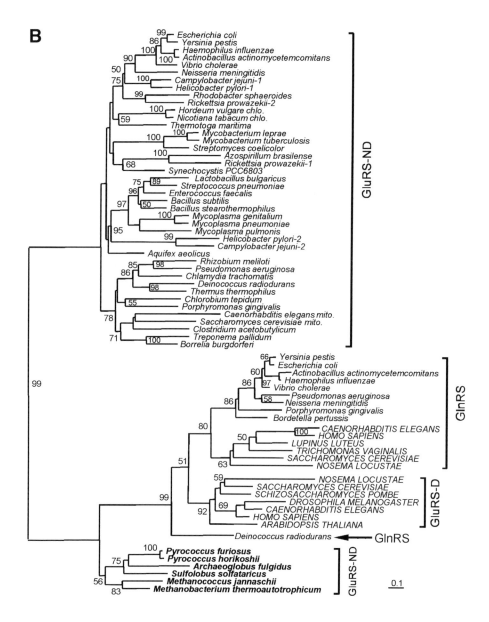

Fig. 3. *(continued)* indicated are nuclear-encoded but mitochondria (mito.) or chloroplast (chlo.) targeted isoforms. The scale bar represents 0.1 expected amino acid replacements per site as estimated by the program PROTDIST using the Dayhoff option *(17)*. Reproduced with permission from Current Drugs and Payne et al. *(17)*, copyright 2000, Current Drugs.

nities demonstrated by mupirocin and IRS to provide 19 novel antibacterial targets all worthy of similar exploitation.

### 2.1.3.2. MURA

MurA, which is involved in an early stage of peptidoglycan synthesis, and fosfomycin, an antibiotic that inhibits this enzyme, have both been known for some 30 yr. However, recent genome analysis has revealed that Gram-positive cocci possess two similar but distinct copies of MurA (MurA1 and MurA2) (**Fig. 4**). Kinetic characterization of both enzymes has failed to demonstrate significant differences between them, and both are inhibited by fosfomycin *(18)*. Single-gene deletions of either MurA1 or MurA2 do not cause a loss in viability, whereas deletion of both genes is lethal. Therefore, it is critical for any potential MurA inhibitor to possess inhibitory activity against both MurA1 and MurA2 to achieve activity against Gram-positive cocci. Clearly, access to complete bacterial genomes enables the identification and characterization of paralogs (*see* Chapter 3) for particular targets that, like MurA, will have a significant impact on inhibitor discovery.

### 2.1.3.3. CHORISMATE BIOSYNTHESIS

Bacteria use the chorismate biosynthetic pathway for the synthesis of aromatic amino acids, *p*-aminobenzoic acid, vitamin K, ubiquinone, and enterochelin (**Fig. 5**). Mammals do not possess this pathway; consequently, it is a highly selective antibacterial target. Moreover, as bacteria cannot obtain the end products from the host, this pathway is essential for bacterial survival. This has been demonstrated in Gram-negative bacteria, where mutations give rise to strains that are auxotrophic in vitro and highly attenuated in vivo.

Prior to the advent of genomics, the vast majority of work on the biosynthetic enzymes involved in this pathway, and antibacterials targeted at them, focused on laboratory strains of nonpathogenic *Escherichia coli.* However, analysis of the genomes of *S. pneumoniae* and *S. aureus* has identified the presence and conservation of the chorismate biosynthetic genes in these Gram-positive pathogens. This has facilitated the characterization and essentiality testing of relevant enzymes. For example, deletion of *aroA,* which encodes 5-enolpyruvylshikimate-3-phosphate (EPSP) synthase, from *S. pneumoniae* results in significant growth attenuation in two different animal models with no recoverable bacteria (Bryant, Ingraham, Chalker, Marra, and Holmes, GlaxoSmithKline, personal communication). Moreover, the *S. pneumoniae* EPSP synthase has been characterized extensively and an HTS format developed. If the data obtained with this enzyme are representative of the other enzymes in this pathway, then chorismate biosynthesis provides a series of broad-spectrum antibacterial targets *(17)*.

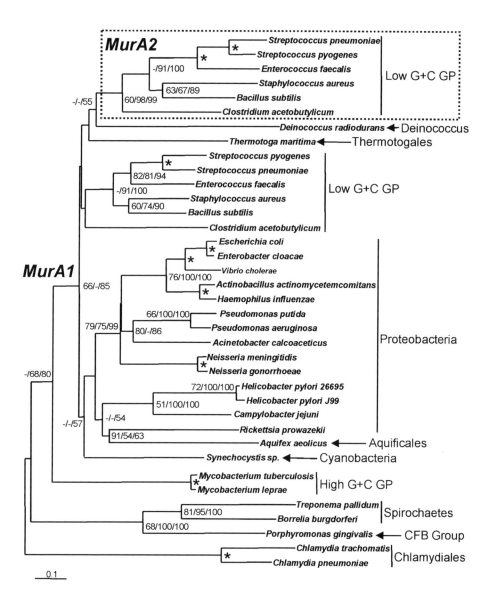

Fig. 4. Phylogenetic tree showing divergence of UDP-*N*-acetylglucosamine enolpyruvyl transferase isoforms, MurA1 and MurA2. (Reproduced from **ref. 18** with permission from American Society for Microbiology.) Tree was constructed using the neighbor-joining (NJ) method as implemented by the program NEIGHBOR of the PHYLIP 3.57c package. The scale bar represents 0.1 expected amino acid replacements per site as estimated by the program PROTDIST using the Dayhoff PAM substitution matrix. Numbers at the branching points represent the percent occurrence in 1000 maximum likelihood (ML) puzzling steps and 1000 random bootstrap replications

## Phosphoenol pyruvate + erythrose 4-phosphate

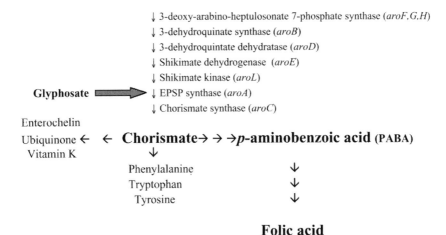

↓ 3-deoxy-arabino-heptulosonate 7-phosphate synthase (*aroF,G,H*)

↓ 3-dehydroquinate synthase (*aroB*)

↓ 3-dehydroquintate dehydratase (*aroD*)

↓ Shikimate dehydrogenase (*aroE*)

↓ Shikimate kinase (*aroL*)

Glyphosate ⟹ ↓ EPSP synthase (*aroA*)

↓ Chorismate synthase (*aroC*)

Enterochelin

Ubiquinone ← ← **Chorismate→ → →*p*-aminobenzoic acid (PABA)**

Vitamin K ↓

Phenylalanine ↓

Tryptophan ↓

Tyrosine ↓

## Folic acid

Fig. 5. Biosynthesis of chorismate and folic acid, *E. coli* gene nomenclature. Reproduced with permission from Current Drugs and Payne et al. *(17)*, copyright 2000, Current Drugs.

### 2.1.3.4. TWO-COMPONENT SIGNAL TRANSDUCTION (TCST) SYSTEMS

TCST systems are typically made up of a histidine kinase and a response regulator (**Fig. 6**). Bacteria depend on these systems to monitor their environment and adapt to their surroundings. A number of TCST systems had been characterized, and some were shown to be involved in bacterial pathogenesis, suggesting their potential role as antibacterial targets. Little was known about the number of different TCST systems that existed in various pathogens or whether the systems were conserved between species. New genome information has revealed entire sets of TCST systems in a wide range of bacterial pathogens, including *Haemophilus influenzae (19), H. pylori (20),* and *S. pneumoniae (21)*. Bioinformatic analysis has allowed detailed comparisons of the homology that exists between the TCST systems within specific pathogens as well as between pathogens. For example, *S. pneumoniae* contains 13 differ-

---

Fig. 4. *(continued)* of maximum parsimony (MP) and NJ methods, respectively. ML and MP analyses were done using the programs PUZZLE v4.0 and PAUP*, respectively. Nodes where values by all three methods were 90% or greater are labeled with an asterisk (*). Values less than 50% are not shown or indicated by a hyphen (-). Major taxonomic groups of bacteria are indicated with abbreviations given for Gram-positive bacteria (GP) and Cytophaga-Flexibacter- Bacteroides (CFB) group.

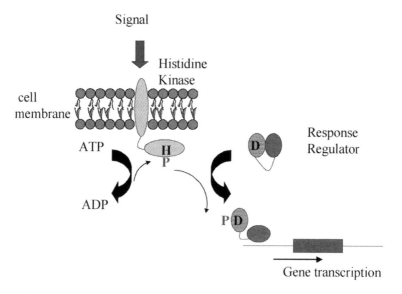

Fig. 6. Scheme to show the mechanism of action of a two-component signal transduction system *(17)*. TCSTS are typically made up of two proteins, a histidine kinase and a response regulator. The histidine kinase generally acts as a sensor to external signals and responds by autophosphorylating on a histidine residue. This phosphate is then transferred to an aspartate residue of the response regulator, which activates the response regulator by causing it to bind to DNA and alter gene transcription. In this way, the bacterium monitors its environment and adapts to its surroundings. Reproduced with permission from Current Drugs and Payne et al. *(17)*, copyright 2000, Current Drugs.

ent TCST systems, several of which are essential for the maintenance of infection; and one (*yycFG*) that is essential for in vitro growth *(21)*. Clearly, a paninhibitor of TCST systems might prove to be a very attractive broad-spectrum antibacterial, as it would target multiple systems whose inhibition would compromise the ability of bacteria to adapt and survive in an infection environment. To date, no such inhibitors have been reported, but the identification of these important regulatory systems now allows the development of physiologically relevant HTS assays. Bacterial genomics has clearly proven instrumental in highlighting inhibition of TCST systems as an extremely compelling and rational antibacterial strategy.

### 2.1.3.5. Cell Division

Bacterial cell division remains one of the least understood aspects of bacterial growth. Septum formation, which is a critical part of this process, involves the dynamic formation and dissolution of a complex protein–protein interaction network. FtsZ plays a central role in this network, by polymerizing into a ring structure at the mid cell, and by recruiting other proteins that play a role in the downstream events in the cell-division cycle. A number of these protein–protein interactions, such as the interaction of FtsA with FtsZ *(22)*, are essential for cell viability and thereby represent potential antibacterial targets. Analysis of the genes encoding FtsA and FtsZ proteins in a variety of bacterial genomes has demonstrated that these proteins are widely distributed and highly conserved. Furthermore, alignment of the sequences of FtsZ from a variety of bacteria demonstrated a conserved 10-amino-acid motif in the C-terminal region of the protein, which was proposed to play an important role in interacting with other cell-division proteins. Detailed mutational analysis of this region of FtsZ in *S. aureus* demonstrated that a single residue (Phe376) is essential for the interaction with FtsA *(23)*. Similar work with *E. coli* FtsZ has confirmed that changing single residues in this region significantly affects binding to FtsA *(24)*. These data suggest that it may be possible to design small molecules that could either enhance or antagonize the interaction of FtsA with FtsZ. Given the importance of the interaction between FtsA and FtsZ, either mechanism of action might result in antibacterial activity *(23)*. An HTS assay for the FtsA/FtsZ interaction has been successfully developed, which we hope will facilitate exploitation of this novel antibacterial strategy *(25)*.

2.1.3.6. ISOPRENOID BIOSYNTHESIS

Isoprenoids, which are ubiquitous in nature, comprise a family of over 23,000 products, each composed of repeating 5-carbon isopentenyl diphosphate (IPP) subunits. Two different pathways for the biosynthesis of IPP have been described in bacteria: the classical "mevalonate" pathway and the more recently identified "nonmevalonate" pathway. Although genetic and biochemical evidence exists for each of the pathways, the entire set of pathway enzymes had not been identified in any one pathogen. Comparative genome analysis has now allowed the identification of the complete set of mevalonate-pathway enzymes in streptococci, staphylococci, and enterococci *(26)*. Access to this genome information has provided the tools needed to perform genetic disruption experiments, with the result that five proteins involved in this pathway (HMG-CoA synthase, HMG-CoA reductase, mevalonate kinase, phosphomevalonate kinase, and mevalonate diphosphate decarboxylase) have been shown to be essential for the in vitro growth of *S. pneumoniae*. The sequences of the genes found in Gram-positive cocci are highly divergent from their mammalian counterparts, and HMG-CoA reductase in *S. aureus* was found to be far less sensitive ($10^{-4}$-fold) than human HMG-CoA reductase to the statin drug fluvastatin. These data raise the possibility of producing agents that selectively inhibit the bacterial HMG-CoA reductase or other bacterial isoprenoid path-

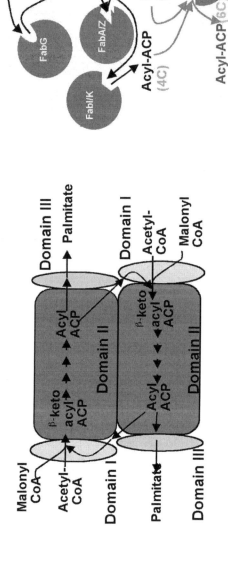

Fig. 7. Comparison of fatty acid biosynthesis in humans and bacteria (reproduced from **ref. 28**). Reprinted with permission from Elsevier Science.

**Table 2**
**Identification of Fab Genes in Key Respiratory and Gram-Positive Pathogens**

| Target | *S. aureus* | *S. pneumoniae* | *E. faecalis* | *H. influenzae* |
|--------|-------------|-----------------|---------------|-----------------|
| FabG | ✓ | ✓ | ✓ | ✓ |
| FabK | X | ✓ | ✓ | X |
| FabI | ✓ | X | ✓ | ✓ |
| FabA | X | X | X | ✓ |
| FabZ | ✓ | ✓ | ✓ | ✓ |
| FabB | X | X | X | ✓ |
| FabF | ✓ | ✓ | ✓ | X |
| FabH | ✓ | ✓ | ✓ | ✓ |
| FabD | ✓ | ✓ | ✓ | ✓ |

✓ = homolog present in genome; X = homolog absent from genome.
Reproduced with permission from **ref. 40**, copyright 2001, Current Drugs.

way enzymes *(27)*. Thus, genomics has helped identify another selective pathway that is a potential target for antibacterial intervention against Gram-positive cocci.

2.1.3.7. FATTY-ACID BIOSYNTHESIS

In mammals, fatty-acid biosynthesis is performed by a single multifunctional polypeptide chain, whereas in bacteria the pathway requires seven discrete enzymes (**Fig. 7**). This key difference implies that bacterial fatty-acid biosynthetic enzymes are also potential selective targets for antibacterial drugs. The majority of the work on this pathway and its component enzymes has focused on laboratory strains of *E. coli*. Early work demonstrated that cerulenin, thiolactomycin, diazaborines, and triclosan all exerted their antibacterial activity via inhibition of one or more of the enzymes in the pathway (reviewed in *28*). Recent analysis of sequenced bacterial genomes has identified the component enzymes in a variety of important clinical pathogens (**Table 2**). The data allow the potential spectrum of each of the component pathway enzymes to be identified. For example, inhibitors of FabK would function as narrow-spectrum antistreptococcal agent, whereas inhibitors of FabH, FabD, and FabG would theoretically show broad-spectrum activity. Genomic analysis has thus enabled the potential of this entire pathway to be exploited as an antibacterial strategy *(28)*.

Enoyl ACP reductase (FabI and FabK) provides an example of the potential of this pathway to provide targets for new antibacterial agents. Furthermore, this example demonstrates the successful progression from target to HTS hit to in vivo proof of concept. When this project was initiated, FabI was thought to be the unique enoyl ACP reductase in all bacteria. However, following the completion of several additional genome sequences, it became apparent that an alternative enoyl ACP reductase (FabK) existed in some pathogens, such as *S. pneumoniae* and *E. faecalis*. Consequently, it was postulated that inhibitors of FabI could be useful as selective agents for *S. aureus,* whereas compounds that inhibited both FabK and FabI could have broad-spectrum activity. Below we present our process and progress toward achieving both of these objectives.

HTS of compounds against the FabI enzyme of *S. aureus* identified a compound designated Compound 1, as a low-potency inhibitor of FabI with no detectable antibacterial activity against *S. aureus* (**Table 3**). Opening the seven-membered ring of Compound 1 and removing the side chain yielded Compound 2 (**Table 3**), which had slightly improved target potency and antibacterial activity. However, Compound 2 was chemically unstable, and further optimization of the right-hand side was performed to yield a chemically stable molecule with the same basic pharmacophore (Compound 3). Compound 3 showed a 32-fold increase in antibacterial activity and further improvements in target potency (**Table 3**). However, the antibacterial spectrum of Compound 3 was still very narrow, being restricted mainly to *S. aureus,* with no inhibitory activity being observed against FabK. Subsequent derivatization of Compound 3 yielded analogs with dramatically enhanced target potency and greater breadth and potency of antibacterial activity. Compound 4 was three orders of magnitude more potent against FabI than the original lead from the HTS screen and possessed exquisite activity against *S. aureus* along with a broader spectrum of antibacterial activity encompassing *H. influenzae, S. pneumoniae,* and *E. faecalis* (**Table 3**). MOA studies confirmed that Compounds 2–4 selectively inhibited acetate incorporation in *S. aureus* and *H. influenzae.* Moreover, all the compounds exhibited raised MIC against *S. aureus* and *H. influenzae* strains overexpressing FabI. These data provided conclusive evidence that the MOA of all these compounds was via inhibition of FabI, and thereby facilitated a robust lead optimization program. Despite Compound 4 inhibiting FabK, as yet no data have been generated to demonstrate that the MOA of Compound 4 against *S. pneumoniae* was via inhibition of FabK. The antistaphylococcal activity of Compound 4 is excellent (MIC$_{90}$ 0.06 µg/mL for 22 multiresistant strains of MRSA). Furthermore, Compound 4 demonstrated excellent in vivo activity in experimental *S. aureus* infections *(29)*.

These data clearly demonstrate the potential of enzymes in the fatty-acid biosynthesis pathway as novel antibacterial targets, and illustrate how a successful cascade from novel genomic target to proof of concept in vivo is achievable.

**Table 3**
**Antibacterial and Enzyme Inhibitory Activity of FabI Inhibitors**

| Compound | IC$_{50}$ (µM) | | | MIC (µg/mL) | | | | |
|---|---|---|---|---|---|---|---|---|
| | S. aureus FabI | H. influenzae FabI | S. pneumoniae FabK | S. aureus (Oxford) | E. faecalis (1) | S. pneumoniae (1629) | H. influenzae (Q1) | M. catarrhalis (1502) |
| 1 | 17.1 | 6.9 | >30 | >64 | >64 | >64 | >64 | 32 |
| 2 | 6.7 | 4.7 | >30 | 16 | >64 | >64 | >64 | 64 |
| 3 | 2.4 | 4.2 | >30 | 0.5 | >64 | >64 | >64 | 4 |
| 4 | 0.047 | 0.13 | 2 | 0.06 | 16 | 8 | 2 | <0.06 |

Reproduced from **ref. 29** with permission from American Society for Microbiology.

## 2.2. Strategy II: Whole-Cell Antibacterial Screening

The explosion in attempts to increase the chemical diversity of antimicrobial agents has meant that the numbers of compounds included in screening collections has increased from tens of thousands to hundreds of thousands. Combinatorial chemistry and new natural product sources and strategies have underpinned this expansion, with new technologies allowing for antibacterial screening in real-time assays. Whole-cell antibacterial screening affords a direct approach to identifying novel antibacterials and is facilitated by advances in techniques for defining the molecular target(s) of novel antibacterial compounds.

Although whole-cell antibacterial screening generally lacks the sensitivity of molecular target screening, it carries the advantage that all potential targets, both known and unknown, are screened simultaneously within a physiological context, and with an in vitro assay that has demonstrated relevance to the clinical endpoint. Additionally, a key advantage of this approach is that the compound already possesses antibacterial activity, which is often exceptionally difficult to engineer into compounds identified from a target-based screen; past experience has shown that leads from target-based screens often possess no significant antibacterial activity. However, a challenge to this approach is that antibacterial activity is frequently encountered as the result of nonspecific effects, often membrane derived. Consequently, assays are required to exclude advancing such compounds and to focus on compounds with truly exploitable, bacterially selective modes of action.

### 2.2.1. MOA Technologies

#### 2.2.1.1. OVEREXPRESSION AND REGULATED GENE EXPRESSION

The relationship between target abundance, especially for single-enzyme targets, and whole-cell sensitivity suggests overexpression constructs could be used as tools to investigate the targets of antibacterial compounds. For example, overexpression of a target in most cases will cause a concomitant increase in MIC. Strategies for obtaining overexpression include both regulated and unregulated plasmid-mediated expression, and genomic essentiality data should allow preparation of overexpression constructs for all potential antibacterial targets. Susceptibility testing of libraries of overexpression constructs together with wild-type strains allows rapid MOA screening for antibacterial compounds against a considerable number of potential targets (Huang and McDevitt, GlaxoSmithKline, personal communication). MOA hypotheses generated by these global analyses can be tested by use of over- or underexpression constructs for specific targets.

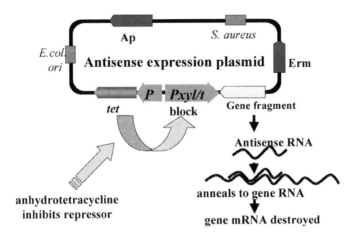

Fig. 8. The use of antisense RNA to evaluate gene essentiality. Gene fragments are cloned, in reverse orientation, downstream of an inducible promoter. Expression of the antisense RNA leads to degradation of the target gene transcript and a consequent reduction in active protein in the cell. This technology has been used to demonstrate the importance of genes important for pathogenesis in animal models of infection. Reproduced with permission from Current Drugs and Payne et al. *(40)*, copyright 2001, Current Drugs.

Inducible promoters can be used to generate libraries of strains expressing antisense RNA to specific gene transcripts. In these experiments, fragments of genes are inserted into a plasmid in the antisense orientation, downstream of a controllable promoter. In the absence of antisense induction, cells are viable, since the normal gene is transcribed and translated. However, in the presence of inducer, antisense RNA is generated, leading to downregulation of expression of the target gene and, if the gene is essential, this will result in decreased growth and/or cell death (**Fig. 8**). Induction of antisense RNA to a gene that encodes an antibacterial target will result in a selective increase in sensitivity to the corresponding antibiotic. Antisense libraries encoding antisense constructs to every gene in a genome can be generated, and this enables rapid global MOA screening of compounds against a considerable number of potential targets. The utility of this procedure for both essentiality testing and MOA screening has been reported *(30–33)*.

2.2.1.2. Expression Analysis

Global transcription profiling using microarrays has been used to identify antibiotic signature expression patterns in *H. influenzae (34)* and *S. aureus*

*(35).* A library of expression profiles has been established for MOA studies in *S. aureus* exposed to ciprofloxacin, triclosan, and mupirocin, with resulting inhibition of DNA replication, fatty-acid biosynthesis, and protein synthesis, respectively *(35).* Cluster analysis of these profiles indicated that the modes of action of the different classes of antibiotics could be distinguished by their unique expression patterns. Therefore, the expression profile of a novel anti-bacterial compound can be compared with the profiles of known antibacterials to indicate its likely mode of action, although the exact molecular target is unlikely to be identified using this technique. A similar approach is possible via the generation of signature patterns using proteomics *(36)* (*see* Chapter 6).

### 2.2.1.3. RESISTANT MUTANT ANALYSIS

Many antibacterials with a specific action have the potential to select target-based resistance, manifested either spontaneously or following mutagenesis *(1).* Hence, investigation of resistant mutants through library generation and complementation has the potential to determine the molecular target of a novel antibacterial. Strides made in molecular cloning technologies, especially in Gram-positive pathogens, such as *S. aureus,* have greatly facilitated this approach.

Once the MOA of an antibacterial compound has been identified, these tools then play the same role as they do in the target-based screening approach, in providing methodologies to confirm the MOA of structural analogs produced during attempts to optimize the properties of the initial lead compound.

### 2.2.1.4. MEMBRANE ASSAYS

A range of membrane assays have been used, including red blood cell lysis and intracellular $K^+$ leakage *(37),* to identify compounds with nonspecific del-eterious effects on membranes as the mechanism of antibacterial activity. The use of dyes sensitive to membrane potential, including flow cytometry applica-tions, also offers approaches to investigating the effect of antibacterials on bac-terial cell membranes *(38;* Gentry, Wilding, and Gwynn, GlaxoSmithKline, personal communication). These assays are critical for early identification of compounds that may kill bacteria via nonspecific, and thus potentially toxic, mechanisms.

### 2.2.1.5. MACROMOLECULAR BIOSYNTHESIS

The effect of antibacterials on the incorporation of radiolabeled precursors into macromolecules has been used in a convenient microtiter format to define whether the primary action of a new antibacterial is on DNA replication, tran-scription, protein synthesis, fatty-acid biosynthesis, or cell wall biosynthesis

*(39)*. Collectively, the gene products of these macromolecular processes represent a significant proportion of the essential genome of pathogens. Identification of one of these processes as the target of a novel antibacterial compound excludes the compound having a nonspecific MOA and enables more focused studies to address the specific targeting.

## 3. Concluding Remarks

The use of genomics has considerably expanded the number of novel and validated antibacterial targets. Soon we will know all the essential genes for key bacterial pathogens, and thus all possible antibacterial targets will be revealed. Furthermore, advances in genetic and genomic technologies means that we now have considerable resources available to identify the mode of action of novel antibacterial compounds identified from screening processes. There is considerable synergy between the whole-cell and target-based screening approaches, as these utilize similar technologies and prosper from the application of genomics (**Fig. 9**). However, a number of challenges still face this therapeutic area, and while genomics has facilitated the identification of new antibacterial strategies, potential targets vary in the ease with which they can be exploited. For example, MurA, FabI, and the aminoacyl tRNA synthetases illustrate the importance of thorough analysis of phylogenetic distribution and conservation as a critical aspect of target selection. In addition, target abundance is an important consideration in target selection, as a highly abundant target may be difficult to inhibit in the physiological environment of the cell. Furthermore, the genes encoding some targets may be prone to high levels of spontaneous mutation, or as yet undiscovered bypass mechanisms may exist, either of which may result in resistance. However, the most substantial challenge is arguably the identification of good quality hits and leads from HTS. It is apparent that bacteria and many of their molecular targets appear to be quite refractory to selective inhibition by small molecules; therefore, it is imperative that screening approaches utilize large collections of novel, chemically diverse, and developable compounds. Natural product strategies such as shuffling the genes of known secondary metabolite pathways to yield new molecules, and application of combinatorial chemistry approaches to design compound libraries especially suited to the parameters common to antibacterial compounds, are both welcome approaches.

New antibacterial agents that inhibit novel targets are required to help overcome the resistance problems that plague present classes of antibiotics. From this discussion, it should be clear that genomics is playing a pivotal role in the discovery of such agents. It is to be hoped that the continued application of this technology will result in the delivery of new antibacterials that are safe, well-tolerated, and efficacious.

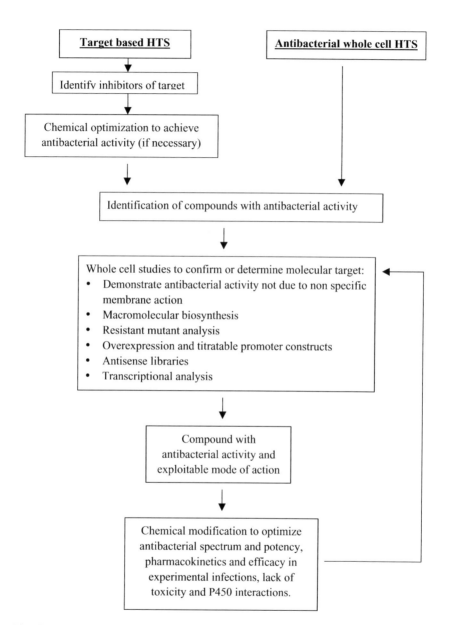

Fig. 9. Interplay of genomics and genomic technologies for enzyme-based and whole cell-based screening strategies.

## References

1. Gale, E. F., Cundliffe, E., Reynolds, P. E., Richmond, M. H., and Waring, M. J. (1981) *The Molecular Basis of Antibiotic Action, 2nd ed.* John Wiley and Sons, London, United Kingdom.
2. Sievert, D. M. (2002) *Staphylococcus aureus* resistant to vancomycin, United States. *Morbidity, Mortality Weekly Reports* **51**, 565–567.
3. Swartz, M. N. (1994) Hospital-acquired infections: diseases with increasingly limited therapies. *Proc. Natl. Acad. Sci. USA* **91**, 2420–2427.
4. Tatusov, R. L., Koonin, E. V., and Lipman, D. J. (1997) A genomic perspective on protein families. *Science* **278**, 631–637.
5. Arigoni, F., Talabot, F., Peitsch, M., Edgerton, M. D., Meldrum, E., Allet, E., et al. (1998) A genome-based approach for the identification of essential bacterial genes. *Nat. Biotechnol.* **16**, 851–856.
6. Venter, J. C., Adams, M. D., Myers, E. W., Li, P. W., Mural, R. J., Sutton, G. G., et al. (2001) The sequence of the human genome. *Science* **291**, 1304–1351.
7. International Human Genome Sequencing Consortium. Initial sequencing and analysis of the human genome. *Nature* **409**, 860–921.
8. Hensel, M., Shea, J. E., Gleeson, C., Jones, M. D., Dalton, E., and Holden, D. W. (1995) Simultaneous identification of bacterial virulence genes by negative selection. *Science* **269**, 400–403.
9. Mei, J. M., Nourbakhsh, F., Ford, C. W., and Holden, D. W. (1997) Identification of *Staphylococcus aureus* virulence genes in a murine model of bacteraemia using signature-tagged mutagenesis. *Mol. Microbiol.* **26**, 399–407.
10. Apfel, C. M., Takacs, B., Fountoulakis, M., Stieger, M., and Keck, W. (1999) Use of genomics to identify bacterial undecaprenyl pyrophosphate synthetase: cloning, expression, and characterization of the essential *uppS* gene. *J. Bacteriol.* **181**, 483–492.
11. Baltz, R. H., Norris, F. H., Matsushima, P., DeHoff, B. S., Rockey, P., Porter, G., et al. (1998) DNA sequence sampling of the *Streptococcus pneumoniae* genome to identify novel targets for antibiotic development. *Microb. Drug Resist.* **4**, 1–9.
12. Hynes, W. L., Hancock, L., and Ferretti, J. J. (1995) Analysis of a second bacteriophage hyaluronidase gene from *Streptococcus pyogenes:* evidence for a third hyaluronidase involved in extracellular enzymatic activity. *Infect. Immun.* **63**, 3015–3020.
13. O'Connell, C., Pattee, P. A., and Foster, T. J. (1993) Sequence and mapping of the *aroA* gene of *Staphylococcus aureus* 8325-4. *J. Gen. Microbiol.* **139**, 1449–1460.
14. Geissendorfer, M. and Hillen, W. (1990) Regulated expression of heterologous genes in *Bacillus subtilis* using the Tn10 encoded tet regulatory elements. *Appl. Microbiol. Biotechnol.* **33**, 657–663.
15. Stieger, M., Wohlgensinger, B., Kamber, M., Rolf, L., and Keck, W. (1999) Integrational plasmids for the tetracycline-regulated expression of genes in *Streptococcus pneumoniae. Gene* **226**, 243–251.

16. Zhang, L., Fan, F., Palmer, L. M., Lonetto, M. A., Petit, C., Voelker, L. L., et al. (2000) Regulated gene expression in *Staphylococcus aureus* for identifying conditional lethal phenotypes and antibiotic mode of action. *Gene* **255**, 297–305.

17. Payne, D. J., Wallis, N. G., Gentry, D. R., and Rosenberg, M. (2000) The impact of genomics on novel antibacterial targets. *Current Opinion in Drug Discovery and Development* **3**, 177–190.

18. Du, W., Brown, J. R., Sylvester, D. R., Huang, J., Chalker, A. F., So, C. Y., et al. (2000) Two active forms of UDP-*N*-acetylglucosamine enolpyruvyl transferase in Gram-positive bacteria. *J. Bact.* **182**, 4146–4152.

19. Fleischmann, R. D., Adams, M. D., White, O., Clayton, R. A., Kirkness, E. F., Kerlavage, A. R., et al. (1995) Whole-genome random sequencing and assembly of *Haemophilus influenzae* Rd. *Science* **269**, 496–512.

20. Tomb, J-F., White, O., Kerlavage, A. R., Clayton, R. A., Sutton, G. G., Fleischmann, R., et al. (1997) The complete genome sequence of the gastric pathogen *Helicobacter pylori*. *Nature* **388**, 539–547.

21. Throup, J. P., Koretke, K. K., Bryant, A. P., Ingraham, K. A., Chalker, A. F., Ge, Y., et al. (2000) Genomic analysis of two-component signal transduction in *Streptococcus pneumoniae*. *Mol. Microbiol.* **35**, 566–576.

22. Addinall, S. G. and Lutkenhaus, J. (1996) FtsA is localized to the septum in an FtsZ-dependent manner. *J. Bact.* **178**, 7167–7172.

23. Yan, K., Pearce, K., and Payne, D. J. (2000) A hydrophobic residue at the extreme C-terminus of FtsZ is critical for the FtsA–FtsZ interaction in *Staphylococcus aureus*. *Biochem. Biophys. Res. Commun.* **270**, 387–392.

24. Ma, X. and Margolin, W. (1999) Genetic and functional analyses of the concerved C-terminal core domain of *Escherichia coli* FtsZ. *J. Bact.* **181**, 7531–7544.

25. Kang Y., Lonsdale, J., Macarron, R., and Payne, D. J. (2002). Biochemical characterization of FtsA–FtsZ binding. *Protein Soc.* **11** (**Suppl. 1**), 16.

26. Wilding, E. I., Brown, J. R., Bryant, A. P., Chalker, A. F., Holmes, D. J., Ingraham, K. A., et al. (2000) Identification, evolution and essentiality of the mevalonate pathway for isopententyl diphosphate biosynthesis in Gram-positive cocci. *J. Bacteriol.* **182**, 4319–4327.

27. Wilding, E. I., Kim, D. Y., Bryant, A. P., Gwynn, M. N., Lunsford, R. D., McDevitt, D., et al. (2000) Essentiality, expression and characterization of the Class II 3-hydroxy-3-methylglutaryl coenzymeA reductase of *Staphylococccus aureus*. *J. Bacteriol.* **182**, 5147–5152.

28. Payne, D. J., Warren, P. V., Holmes, D. J., Ji, Y., and Lonsdale, J. T. (2001) Bacterial fatty acid biosynthesis: a genomics driven target for antibacterial drug discovery. *Drug Discov. Today* **6**, 537–544.

29. Payne, D. J, Miller, W. H., Berry, V., Brosky, J., Burgess, W. J., Chen, E., et al. (2002) Discovery of a novel and potent class of FabI directed antibacterial agents. *Antimicrob. Agents Chemother.* **46**, 3118–3124.

30. Ji, Y., Marra, A., Rosenberg, M., and Woodnutt, G. (1999) Regulated antisense RNA eliminates alpha-toxin virulence in *Staphylococcus aureus* infection. *J. Bacteriol.* **181**, 6585–6590.

31. Ji, Y., Woodnutt, G., Rosenberg, M., and Burnham, M. K. R. (2002) Identification of essential genes in *Staphylococcus aureus* using inducible antisense RNA. Bacterial Pathogenesis, Part C; Identification, regulation and function of virulence factors. *Methods in Enzymology* **358**, 123–128.

32. Ji, Y., Zhang, B., Van Horn, S. F., Warren, P., Woodnutt, G., Burnham, M.K., et al. (2001) Identification of critical staphylococcal genes using conditional phenotypes generated by antisense RNA. *Science* **293**, 2266–2269.

33. Forsyth, R. A., Haselbeck, R. J., Ohlsen, K. L., Yamamoto, R. T., Xu, H., Trawick, J. D., et al. (2002) A genome-wide strategy for the identification of essential genes in *Staphylococcus aureus. Mol. Microbiol.* **43**, 1387–1400.

34. Gmuender, H., Kuratli, K., Di Padova, K., Gray, C. P., Keck, W., and Evers, S. (2001). Gene expression changes triggered by exposure of *Haemophilus influenzae* to novobiocin or ciprofloxacin: combined transcription and translation analysis. *Genome Res.* **11**, 28–42.

35. Chan, P. F., Gagnon, R., Boyle, R., O'Brien, S., Clark, S. M., Javed, R., et al. (2002) Microarray gene expression profiling to study the mode of action of DNA replication class of inhibitors in *Staphylococcus aureus*. Poster A-43, American Society for Microbiology 102nd General Meeting, Salt Lake City, Utah.

36. Evers, S., Di Padova, K., Meyer, M., Langen, H., Fountoulakis, M., Keck, W., et al. (2001) Mechanism-related changes in the gene transcription and protein synthesis patterns of *Haemophilus influenzae* after treatment with transcriptional and translational inhibitors. *Proteomics* **1**, 522–544.

37. Alksne, L. E., Burgio, P., Hu, W., Feld, B., Singh, M. P., Tuckman, M., et al. (2000) Identification and analysis of bacterial protein secretion inhibitors utilizing a SecA–LacZ reporter fusion system. *Antimicrob. Agents Chemother.* **44**, 1418–1427.

38. Novo, D. J., Perlmutter, N. G., Hunt, R. H., and Shapiro, H. M. (2000) Multiparameter flow cytometric analysis of antibiotic effects on membrane potential, membrane permeability, and bacterial counts of *Staphylococcus aureus* and *Micrococcus luteus. Antimicrob. Agents Chemother.* **44**, 827–834.

39. Greenwood, R. C., Nicholas, R. O., Nwanguma, N., VanAller, G. S., Payne, D. J., and Gentry, D. (2001) A microtiter-based macromolecular synthesis assay for assessing the mode of action (MOA) of antibacterial agents against Gram-positive and Gram-negative pathogens. Poster A-24 American Society for Microbiology 101st General Meeting, Orlando, FL.

40. Payne, D., Holmes, D. J., and Rosenberg, M. (2002) Delivering novel targets and antibiotics from genomics. *Curr. Opin. Investig. Drugs* **2**, 1028–1034.

# 12

## Using the Genome to Understand Pathogenicity

### Dawn Field, Jennifer Hughes, and E. Richard Moxon

#### Summary

Genome sequencing, the determination of the complete complement of DNA in an organism, is revolutionizing all aspects of the biological sciences. Genome sequences make available for scientific scrutiny the complete genetic capacity of an organism. With respect to microbes, this means we now have the unprecedented opportunity to investigate the molecular basis of commensal and virulence behavior. We now have genome sequences for a wide range of bacterial pathogens (obligate, facultative, and opportunistic); this has facilitated the discovery of many previously unidentified determinants of pathogenicity and has provided novel insights into what creates a pathogen. In-depth analyses of bacterial genomes are also providing new perspectives on bacterial physiology, molecular adaptation to a preferred niche, and genomic susceptibility to the uptake of foreign DNA, three key factors that can play a significant role in determining whether a species, or a strain, will have pathogenic potential.

**Key Words:** Genome; pathogenicity; comparative genomics; virulence determinants; orphan genes; physiology; ecological niche; adaptation; horizontal gene transfer; bacterial evolution.

## 1. Introduction

Whole-genome sequencing is fueling an information revolution that is changing the face of biology. The ability to determine the complete complement of DNA of a wide range of organisms is allowing us to ask in unprecedented detail, fundamental questions about the molecular basis of life. First, what is the genetic capacity of an organism? Second, what does an organism do with its genetic capacity? Third and very importantly, how susceptible is this genetic capacity to evolutionary change, what types of changes occur, and which are of biological significance?

From: *Methods in Molecular Biology, vol. 266: Genomics, Proteomics, and Clinical Bacteriology: Methods and Reviews*
Edited by: N. Woodford and A. Johnson © Humana Press Inc., Totowa, NJ

With respect to the genomes of bacterial pathogens, we are interested in how answers to these questions about genetic capacity help to explain pathogenicity. In using the genome to understand pathogenicity we are faced with attempting to address the difficult but fundamentally important question of "what creates a pathogen?" *(1–4)*. We cannot hope to understand the intricate details of the molecular basis of the pathogen–host relationship without integrated epidemiological, population-level, phylogenetic, ecological, and, perhaps most importantly, experimental studies (preferably in an animal model *[2]*). These approaches, and the impact that genome sequencing is having on these fields of research, are covered in other chapters. Here we focus on what kinds of questions can be posed and answered using *in silico* studies of whole-genome sequences. To do this, we first give an overview of the current status of pathogen genome-sequencing efforts, and then we present three key examples of how insights derived from the analysis of a large number of bacterial genomes is helping to place the study of pathogens into context. For the remainder of the chapter, we discuss how genomes are shedding light on the numbers and types of virulence determinants in bacteria, and further, on the molecular basis of three aspects of bacterial biology that play an important role in pathogenicity—physiology, adaptation to a preferred niche, and the ability to evolve through the acquisition of foreign DNA.

## 2. The Era of Whole-Genome Sequencing of Pathogens

The first bacterial genome sequence to be completed (in 1995) was that of *Haemophilus influenzae,* a human commensal with the potential to cause a variety of diseases including meningitis *(5)*. The sequencing of this 1.8-million-base-pair genome was a conceptual and technical breakthrough. This milestone was made possible by the shotgun sequencing approach (*see* Chapter 3), which removed the need for painstakingly constructed genetic maps. The shotgun approach relies on sequencing enough clones with unique DNA fragments to increase the probability that all bases in the genome have been sampled past an acceptably high threshold (ideally 100%). The challenge lies in the assembly of hundreds of thousands of overlapping partial DNA fragments of 300–500 nt (followed by the manual closure of the inevitable remaining gaps in the sequence). This advance opened the door for the sequencing of genomes even from species which are poorly characterized, e.g., those of intracellular pathogens that have fastidious growth requirements, like *Treponema pallidum (6)*.

The success of sequencing the *H. influenzae* genome also reflected the application of new computational methods *(5)*, and demonstrated that a bacterial genome could be sequenced in 6 months for less than 50 cents a base *(7)*. In short order, proposals for tens of genomes were funded, and the pace of genome sequencing rapidly increased thereafter (**Fig. 1**). As of this writing, we now

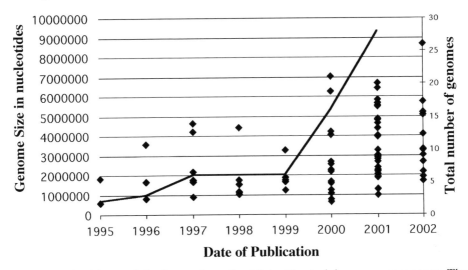

Fig. 1. Rapid growth in the number of published bacterial genome sequences. The first 80 sequenced genomes are plotted according to their year of publication. The primary axis shows genome size in nucleotides (diamonds) and the secondary axis shows the total number of genomes sequenced each year (black line).

have more than 80 complete genome sequences, and the largest sequencing laboratories can finish the raw sequencing of complete bacterial genomes in hours. The Joint Genome Institute (JGI) recently held a "microbial marathon," sequencing 15 bacterial genomes in a single month (http://www.jgi.doe.gov/). In 2001, 28 genomes were published, an average of just over two per month. Not only is the number of completed bacterial genomes increasing, but so too is the size of genome sequences being determined. In 2002, the genome of *Streptomyces coelicolor,* the largest bacterial genome sequenced to date, was published *(8)*. This species has more genes than the eukaryotic yeast *Saccharomyces cerevisiae (9)* and is a key producer of natural antimicrobials *(8)*.

Of the first 80 bacterial genomes, published between 1995 and summer 2002 *(10)*, 16 are for members of the domain Archaea, a vast taxonomic grouping of bacteria with no known pathogenic representatives (**Table 1**). A total of 64 strains of Eubacteria, representing 41 genera, were completed, and 23 of these genera contain species that have the capacity to be human pathogens. Of these, 29 species (37 strains) represent obligate, facultative, and opportunistic pathogens (**Table 2**). These pathogens include the causative agents of some of the most prevalent human diseases (tuberculosis), and agents of historically important pandemics and recurrent epidemics (plague, cholera, typhoid, and typhus). Other species are the etiological agents of nosocomial- and community-acquired infections (*Staphylococcus aureus*) and opportunistic infections

**Table 1**
**The Taxonomic Diversity Found in the First 80 Published Genome Sequences**

|  | Genomes (strains) | Species | Genera |
|---|---|---|---|
| Archaea | 16 | 16 | 11 |
| Eubacteria | 64 | 52 | 41 |
| Eubacterial human pathogens | 37 | 29 | 23 |
| Total | 80 | 68 | 52 |

This collection contains genomes that belong to species of both bacterial domains and represent a total of 68 species from 52 genera. Among Eubacteria, we have genomes from 29 species (37 strains) with the capacity to cause disease in humans (*see* **Table 2**).

(*Pseudomonas aeruginosa*). We are also beginning to sequence pathogens of other species, thereby adding to the value of experimental models of human disease *(11–14)*.

## 2.1. Standard Output of a Bacterial Pathogen Genome-Sequencing Project

The standard workflow of a genome-sequencing project is depicted in **Fig. 2**. The first crucial step is the selection of the target species. Our ability to study pathogenicity is largely dictated by past and future decisions about which organisms to sequence. Earlier sequencing efforts were distinctly biased towards human pathogens *(15)*, but even now the desire to learn more about pathogen biology continues to be the single most significant impetus behind the sequencing of bacterial genomes. Pathogenic species have been chosen for a variety of reasons. For fastidious intracellular obligate pathogens like *T. pallidum* and species of *Chlamydia* in which classical genetics has proved very difficult, genome sequences provide crucial new tools for studying their biology *(6,16)*. The investment in the five chlamydial genomes sequenced thus far shows the perceived importance of this approach. Chlamydial species were also selected because they are deeply separated from other eubacteria *(16)*. *Escherichia coli* K12 was originally sequenced to complement long-term molecular genetic studies of a model laboratory organism *(17)*. It now provides an invaluable basis of comparison for the study of two more recently sequenced pathogenic strains of *E. coli* that are responsible for outbreaks of serious human disease *(18,19)*. The serotype B strain of *Neisseria meningitidis* was sequenced in the hopes that access to a complete genome sequence would accelerate the development of a vaccine *(20–22)*. Although pathogenic, mycoplasmas were primarily selected because they have extremely small genomes *(23)*. As such,

**Table 2**
**Genomes Sequenced From Strains That Have the Capacity to Cause Disease in Humans**

| Pub | Date | Species | Causative agent of |
|-----|------|---------|--------------------|
| 1 | 1995 | *Haemophilus influenzae* | Meningitis, otitis media |
| 2 | 1995 | *Mycoplasma genitalium* | Urethritis |
| 3 | 1996 | *Mycoplasma pneumoniae* | Pneumonia |
| 4 | 1997 | *Helicobacter pylori* | Gastric ulcers, cancer |
| 5 | 1997 | *Borrelia burgdorferi* | Lyme disease |
| 6 | 1998 | *Mycobacterium tuberculosis* | Tuberculosis |
| 7 | 1998 | *Treponema pallidum* | Syphilis |
| 8 | 1999 | *Chlamydia trachomatis* | Genital tract infections, trachoma |
| 9 | 1999 | *Rickettsia prowazekii* | Typhus |
| 10 | 1999 | *Helicobacter pylori* | (As above) |
| 11 | 1999 | *Chlamydia pneumoniae* | Pneumonia and bronchitis |
| 12 | 2000 | *Ureaplasma urealyticum* | Opportunistic infections during pregnancy |
| 13 | 2000 | *Campylobacter jejuni* | Gastroenteritis |
| 14 | 2000 | *Chlamydia pneumoniae* | (As above) |
| 15 | 2000 | *Neisseria meningitidis* | Meningitis and septicemia |
| 16 | 2000 | *Neisseria meningitidis* | (As above) |
| 17 | 2000 | *Chlamydia pneumoniae* | (As above) |
| 18 | 2000 | *Vibrio cholera* | Cholera |
| 19 | 2000 | *Pseudomonas aeruginosa* | Sepsis in immunocompromised hosts |
| 20 | 2001 | *Escherichia coli* | Gastroenteritis and urinary tract infections |
| 21 | 2001 | *Mycobacterium leprae* | Leprosy |
| 22 | 2001 | *Escherichia coli* | (As above) |
| 23 | 2001 | *Pasteurella multocida* | Cellulitis |
| 24 | 2001 | *Streptococcus pyogenes* | TSS, rheumatic fever, scarlet fever, pharyngitis |
| 25 | 2001 | *Staphylococcus aureus* | Boils, abscesses, sepsis |
| 26 | 2001 | *Staphylococcus aureus* | (As above) |
| 27 | 2001 | *Mycobacterium tuberculosis* | (As above) |
| 28 | 2001 | *Streptococcus pneumoniae* | Pneumonia, septicemia, and meningitis |
| 29 | 2001 | *Rickettsia conorii* | Mediterranean spotted fever |
| 30 | 2001 | *Yersinia pestis* | Plague |
| 31 | 2001 | *Salmonella typhi* | Typhoid fever |
| 32 | 2001 | *Salmonella typhimurium* | Gastroenteritis |
| 33 | 2001 | *Listeria monocytogenes* | Listeriosis |
| 34 | 2002 | *Brucella melitensis* | Malta fever |
| 35 | 2002 | *Clostridium perfringens* | Gas gangrene |
| 36 | 2002 | *Fusobacterium nucleatum* | Dental caries |
| 37 | 2002 | *Streptococcus pyogenes* | (As above) |

Many of these species have the capacity to cause a wide range of diseases, and only the most common or most severe human diseases are listed. Entries in gray text represent duplicate strains from a single species. TSS, toxic shock syndrome.

they afford a model with which to study the minimal set of genes required for a free-living, replicating cell (24).

The selection of an index strain is the next critical step. This decision has important implications because there is tremendous genetic and phenotypic diversity within bacterial species (25). Most species that colonize humans and are capable of causing disease also contain strains that are nonpathogenic, or even beneficial (26). Among pathogenic strains in a species there are often a significant variety of distinct serotypes, biovars, or otherwise genetically distinct strains that cause different types of disease, and show different tissue and even host tropisms (27). Finally, the DNA of the genomes of strains within a single species can differ by up to 20% (28), and often it is debatable whether we really understand what a bacterial species is (28,29).

Once the DNA of a genome has been sequenced and assembled, the annotation process begins. The quality of annotation is critical, as it impacts on downstream analyses based on the genome sequence and the quality of our public electronic databases (see Chapter 2). The basic features of this process were established in the groundbreaking analysis of the first complete bacterial genome sequence (5), but sequence annotation remains an active area of research (30,31). The analysis of a final annotated gene list is a first true look at the genetic capacity of an organism, and both the discovery of unexpected genes and the lack of expected genes are valuable outcomes of this process. Also valuable is an appreciation of the number of genes that can be ascribed a putative function vs those whose function remains undefined. The ability to assign functions to a significant number of genes in a genome, especially a large proportion of those highly conserved genes with housekeeping roles, further makes it possible to construct a metabolic and biochemical schematic of a cell. Similar analysis (presence/absence, shared/novel) of the pathways represented in this schematic likewise provides clues about the "lifestyle" of an organism.

Equally important, a schematic of the major landmarks and features of a bacterial genome sequence, which can be constructed from a complete annotation, can have a profound impact on our understanding of biological potential. For example, genes found in operons and clusters often have related functions, while the annotation of the various classes of intergenic regions and repetitive sequences in a genome provides information about how organisms regulate their repertoire of genes (32), how genetic diversity is generated (33), and how long-term evolutionary processes, like the reduction of genome size, proceed (11).

The publication of a genome sequence, no matter how informative, is only a first step towards new and productive lines of investigation. This holds true

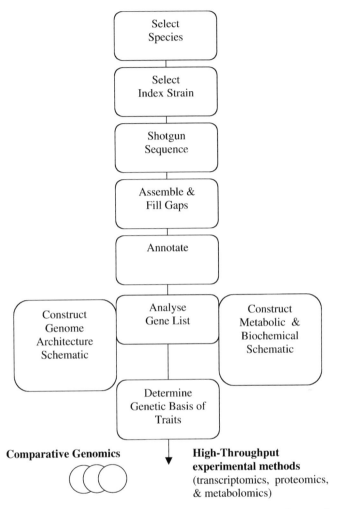

Fig. 2. The standard work flow of a bacterial genome-sequencing project. The first key decisions involve the selection of a species and index strain. All bacterial genome sequences have been completed using the shotgun sequencing method. After assembly and closure of gaps the genome sequence is annotated using computational methods that produce four key results: a list of genes, an overview of genome architecture, an overview of cellular metabolism, and an increased understanding of the molecular basis of key phenotypic traits. Genome sequences are a key step toward comparative genomic, transcriptomic, proteomic, and metabolomic studies.

whether the genome is a first window into the biology of a poorly characterized species *(6)* or is yet another tool, made more powerful when used in parallel with data collected over decades of study of a species like *E. coli (17)*.
When properly analyzed, genome sequences should raise more questions than
they answer. Scrutiny of a genome sequence will elucidate the genetic basis of
particular traits, but the longer-lasting contribution of each new genome is a
list of candidate genes that is made available for the future study of a much
larger variety of phenotypic traits. Powerful comparative genomic (*see* Chapter 4) and experimental postgenomic technologies, direct outgrowths of the
successes of whole-genome sequencing, now provide the promise of allowing
researchers to elucidate the information contained in genome sequences to
determine genotype-phenotype relationships explicitly and explore what
organisms really *do* with their genetic potential *(34,35)*.

## 2.2. General Insights into Bacterial Genomes

The collection of pathogenic and nonpathogenic organisms sequenced over
the past 8 yr represents a wealth of genetic and ecological diversity, and provides excellent starting material for comparative genomic studies. The fact that
we now have genome sequences for a large number of bacteria, half of which
have the capacity to cause disease in humans, has highlighted a core set of
fundamental biological properties of bacterial genomes that place the in-depth
study of bacterial pathogens into proper context.

### 2.2.1. Orphan Genes

Perhaps the most striking general insight to emerge from whole genome
sequencing (of both bacteria and eukaryotes) is the number of genes without
known relatives—also referred to as "orphans" *(36)* or "ORFans" *(37)*. Whole-
genome sequences are of particular value because they contribute a copy of all
the genes found in a genome, and not just the most intensively studied genes.
We already have around 225,000 ORFs from bacterial genome sequences
alone, and yet, despite this wealth of data, there continue to be large numbers
of genes that are known only from a single species (**Fig. 3**). Some authors have
suggested that the compilation of this long list of unshared, orphan genes is the
biggest surprise to come from genome sequencing thus far *(38)*. Orphans are
not a phenomenon restricted to species for which only distantly related genome
sequences are available, but may also be found in strains within a species. For
example, when two pathogenic strains of *E. coli* were compared with strain
K12, a total of 1632 ORFs were found, 1387 of them strain-specific *(18,19)*.
With respect to pathogens, the detection of orphans and taxonomically
restricted genes is of special interest because some may be responsible for their
virulence behavior *(39,40)*.

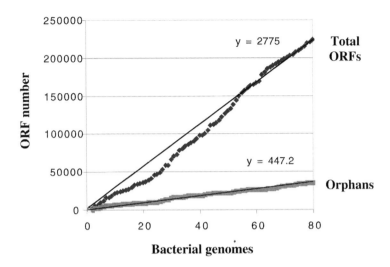

Fig. 3. The total number of ORFs and orphans in bacterial genomes. For each of the first 80 published genomes (shown in order of publication) the cumulative totals of both ORFs and orphans are plotted. The slope is shown for each linear trendline, showing that, on average, each new genome sequence has contributed around 2700 new genes, of which 447 are orphans. Data taken from individual genome reports. Data on orphans (as compared with all known genes) are available for 64 genomes as taken from primary genome report publications.

## 2.2.2. Gene Number Is a Function of Genome Size

Another striking observation that has emerged from the sequencing of tens of genomes is the relationship between genome size and coding capacity (**Fig. 4**) *(38)*. We have genome sequences that represent a significant portion of the total range of genome sizes observed in bacteria *(41)*. From the trendline in **Fig. 4** it can be seen that on average there are 1000 genes per million base pairs of a bacterial genome sequence. This relationship holds true for the smallest genomes of pathogenic mycoplasmas and the symbiont *Buchnera aphidacola*, and also for the largest sequenced genomes of generalist soil bacteria like *P. aeruginosa*. It is due to the fact that overall gene length (approx 1000 bp) and the proportion of noncoding DNA (85–94% of the total genome is usually coding DNA) is generally conserved across bacterial taxa.

The intracellular pathogen *Mycobacterium leprae,* though, is a striking exception *(11,38)*. This species deviates significantly because it is in the process of shedding large numbers of genes *(11)*. Caught in the process of genetic decay, *M. leprae* represents a rare glimpse into the biological phenomenon of "reductive genome evolution"—a key discovery to come from the genome

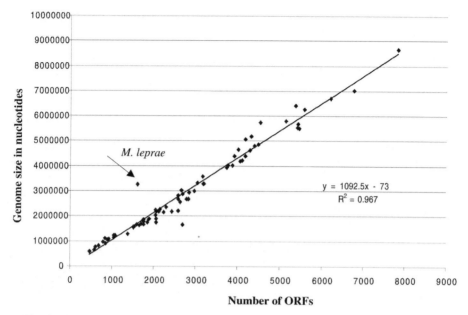

Fig. 4. The strict relationship between ORFs and genome size. The number of ORFs in each of the first 80 published bacterial genomes is plotted against genome size. The trendline fits a straight line through this plot, showing the very strong correlation between ORF number and genome size across a wider range of taxonomically diverse lineages. *M. leprae* is an exception owing to the large number of pseudogenes in this genome.

sequencing of highly host-adapted intracellular pathogens *(11,42–44)*. Only 49.5% of the *M. leprae* genome encodes for functional proteins, while 27% contains recognizable genes that have decayed into nonfunctional pseudogenes through the accumulation of mutations. The remaining 23.5% is purely noncoding *(11)*. When the 1,116 pseudogenes found in the genome are included, *M. leprae* nonetheless fits the expected ratio of ORFs to genome size *(38)*. The larger genome of the closely related *Mycobacterium tuberculosis* contains functional copies of many of these pseudogenes and has a standard ratio of ORFs to genome size *(11)*.

## 2.2.3. The Evolutionary Potential of Horizontal Gene Transfer Is Revealed in Detail

Genome sequencing has further underscored the high frequency and bio-logical importance of the common ability of bacteria to acquire foreign DNA *(45)*. Horizontally acquired sequences can be detected in a single genome in association with mobile elements or through significant departures from genome-wide average values for features such as G+C content, dinucleotide frequencies, and codon composition *(45,46)*. Comparative genomic analysis has further provided ample evidence of significant genetic transfer between strains within a species *(18,19)* to the level of taxa that belong to different bacterial domains *(47)*.

Transfer events can involve the acquisition of large numbers of small sequences over time (e.g., mediated by phage *[18,19,48]*) or the acquisition of large sequences that, when they carry a set of genes involved in virulence, are termed *pathogenicity islands* (PAIs). This feature of bacterial genomes was originally described in *E. coli* in the latter half of the 1980s *(49)*. Subsequently, in part owing to the analysis of whole genomes, pathogenicity islands have been discovered to be one of the major "themes" of bacterial pathogenesis, due both to their prevalence in bacterial populations and the importance of the traits they confer on their host genomes *(50)*. The confirmation of high rates of horizontal transfer between bacterial genomes, by contrast, is also encouraging a growing appreciation of the concept of the *genetic backbone,* comprising a core set of indigenous genes that remain stable over time *(28)*. Genetic backbones are now actively being defined by homology comparisons for various evolutionary groups including serogroups *(48)*, species *(18,19)*, and even bacterial domains *(51)*.

## 3. Exploring and Understanding the Genetic Capacity Required to Become a Pathogen

The challenge in analyzing the genome and genetic capacity of any bacterial pathogen lies in distinguishing the genes responsible for disease from all the other genes in its genome *(52)*. Access to the wealth of data provided by complete genome sequences underscores two important facts—first, the inherent difficulty of defining what a virulence determinant is, and second, that pathogenicity is often due to a complex combination of genes working together *(1–3)*. Accordingly, it has been suggested that there are three classes of proteins responsible for virulence *(1)*. These are true virulence factors, which are found only in pathogens *(50)*, genes that regulate "true virulence genes," and "life-style" genes, such as those that are required to colonize the host *(1)*.

Questions like "How many genes are required for virulence?" can be answered only by experimentation *(2)*, but the analysis of a genome can play an important role by providing a list of candidate virulence determinants. Given a draft list of candidate virulence determinants we can ask further questions, like "What proportion of the total coding capacity of the genome do these genes represent?" and "What classes of genes are involved?" Answers to these questions differ among pathogens with different lifestyles. Experimental studies have demonstrated that in enteric bacteria like *E. coli,* becoming a pathogen can be as simple as acquiring a single region of DNA (by horizontal transfer), which is integrated into a "permissible" genetic background full of genes that allow (benign) colonization of the human host *(3)*. The intracellular enteric pathogen *Salmonella enterica* requires at least 60 genes to be a pathogen, and many of these genes are shared with nonpathogenic strains or have house-keeping roles *(3)*.

To put the types and frequencies of virulence determinants in a genome into proper context, we need to understand an organism's physiology, ecological niche, and susceptibility to the acquisition of foreign DNA. Physiology and preferred niche are intimately linked—biochemical capacities evolve in response to the environment (in the case of pathogens, the host), and physiology determines where and how an organism can live. The probability of acquiring foreign DNA and the niche are also intimately related. The ecological niche of a bacterium should determine the frequency at which a pathogen has a chance to acquire foreign DNA and the probability that this DNA will be from another species, genus, or even a more distantly related taxon. Pathogens that live inside host cells are expected to have limited opportunities to acquire foreign DNA, while enteric and mucosal pathogens, which live in complex microbial communities, should experience frequent opportunities for genetic exchange.

### 3.1. Using Genome Sequences to Discover New Determinants of Virulence

Identifying the determinants of virulence not only provides essential clues to the biology and evolution of an organism, it also paves the way for the development of novel vaccines *(20)* and diagnostics *(53)*. The complex progression of steps in the behavior of a pathogen can be described in five steps—adherence, invasion, multiplication within the host, interference with host defenses, and damage to the host *(2,54)*. The corresponding categories of genes share many "themes," or mechanisms of action *(50)*. These are *adhesins,* like pili/fimbrae; *invasins,* which allow bacterial cells to move to their preferred location within the host; *pabulins,* or genes that help the bacterium scavenge energy and essential nutrients like iron; *evasins,* or factors like capsule and surface proteins that provide mechanisms of surviving host defenses; and *toxins,* or factors that damage the host and the secretion systems that deliver these toxins into host tissues. Genome sequencing not only provides an overview of all the virulence factors in an organism; it also gives insights into the relative genetic investment in each class of factor. This in turn provides ecological insights into the host–pathogen relationship and the relative adaptive strategies of ecologically different pathogens.

The genome of every pathogen sequenced thus far has provided a significant list of newly discovered candidate genes ready for further empirical characterization. The contribution of genomics to our overall understanding of pathogenicity and virulence is especially large when the target organism is a fastidious grower or an organism for which genetic tools have not been developed. For example, prior to the sequencing of the *T. pallidum* genome, no virulence factors were known in this species. In the annotation of this sequence, a total of 70 candidate virulence factors were identified *(6)*. One interesting exception that highlights how little we know about certain molecular mechanisms of virulence came from the genome analysis of *Borrelia burgdorferi,* the spirochete agent of Lyme disease, which goes through a tick–rodent–human life cycle *(55)*. In this case, the surprise was how few virulence factors could be found. Instead of a wide range of recognizable genes involved in virulence or host–parasite interactions, the genome revealed a tremendous repertoire of lipoproteins unique to *Borrelia* species. Even more interestingly, 63 of the total 105 lipoproteins are carried on the 17 plasmids contained within the genome. The loci encoding these largely unknown lipoproteins make up 8% of the genome, compared with only 1.3% in *Helicobacter pylori,* attesting to their biological importance *(55)*.

There are three methods of detecting novel virulence factors within genome sequences. One can search for genes homologous to those encoding known virulence factors in other species, detect virulence factors *de novo* based on their unusual DNA characteristics, or identify genes found only in pathogens. The majority of virulence factors are found using homology-based methods. While this approach is extremely effective for well-studied taxonomic groups containing well-studied pathogens, it is obviously less useful for undersampled taxa, and can shed no light on virulence genes for which there are no known copies in public databases. Similarity analysis can also be used to characterize sequences (paralogs) within the same genome, including those that are not found in public databases. For example, completion of the *M. genitalium* genome allowed the identification of all copies of the *MgPa* operon, a key virulence determinant of this species *(23)*. Analysis of the *M. tuberculosis* genome led to the discovery of the large PE and PPE genes families, which comprise 10% of the total coding capacity of this species and are suspected to play a role in virulence and antigenic variation *(56)*.

Methods that are not reliant on the detection of homology include the detection of horizontally transferred DNA sequences *(45)*, the identification of genes encoding surface-exposed peptides by sequence motifs related to their secretion or surface binding *(21,57)*, and the detection of sequences potentially involved in generating antigenic variation by their repetitive nature or their

membership of large gene families *(56)*. One of the most striking examples of
the power of detection methods that are not homology based involves searches
for simple sequence repeats (SSRs). Also known as microsatellites in eukaryotes, these hypervariable, short tandem repeats are prone to expansion and con-
traction through the gain or loss of repeat units via replication slippage *(58)*.
When found in promoters and coding regions, mutations in these repeats can
cause "phase variation" of the genes with which they are associated. The ability to vary phase, or undergo rapid switching between distinct phenotypic
states, is a trait shared by a wide range of pathogenic bacteria, and such loci
have been termed "contingency loci," as they generate random but useful
genetic variation that plays an important role in adaptation *(59)*. Phase variation through SSRs often produces antigenic variation, but it can also facilitate
adaptation to different microniches within the host *(33)*.

A wide variety of molecular mechanisms drive the generation of mutations
at contingency loci *(59)*, but SSRs lend themselves extremely well to discovery by
*in silico* analyses *(60)*. Prior to the publication of the *H. influenzae* genome in
1995, only a few such SSR-containing contingency loci were known. However, a computational survey of the fully sequenced genome revealed nine
tetranucleotide repeats positioned in coding regions such that they had the
potential to be translational molecular switches *(60)*. A knockout mutant lacking one of these genes, *lgtC,* had reduced virulence *(60)*. All of these repeats,
plus others not found in the index genome (Rd), have now been empirically
demonstrated to cause phase variation of genes responsible for lipopolysaccharide biosynthesis, adhesion, and iron scavenging *(61)*, and the power of this
*in silico* method has been applied to other genomes. Most notably, we now
appreciate that the genomes of *H. pylori (62)*, *Campylobacter jejuni (63)*, *N.
meningitidis (22,64)*, and *Streptococcus pneumoniae (65)* are each characterized by sets of between 10 and 35 contingency loci containing SSRs.

The third method of identifying genes involved in virulence is perhaps the
least specific, but offers a chance to find potential virulence determinants for
which we lack any *a priori* knowledge upon which we can build detection
methods. This method requires two or more genome sequences and relies on
using subtractive methods to detect unshared genes or noncoding regions
between genomes *(40)*. This approach can now be applied to very closely
related genomes, since there are two or more genomes available for a total of
eight pathogenic species (**Table 3**). *E. coli* is one species for which we have the
genomes of both pathogenic and nonpathogenic strains, and comparisons have
revealed that as much as 25% of the genome of pathogenic strains is not shared
by strain K12. It is interesting to note that these regions contain more genes for
degradation than for biosynthetic capacity *(17–19)*. The concept of these "electronic filtering experiments" can now be extended to the comparison of

**Table 3**
**Sequencing of Multiple Strains from a Single Species**

| Species | Genomes | Nonpathogenic | Pathogenic |
|---|---|---|---|
| *Chlamydia trachomatis* | 2 | 0 | 2 |
| *Chlamydia pneumoniae* | 3 | 0 | 3 |
| *Escherichia coli* | 3 | 1 | 2 |
| *Helicobacter pylori* | 2 | 0 | 2 |
| *Mycobacterium tuberculosis* | 3 | 0 | 3 |
| *Neisseria meningitidis* | 2 | 0 | 2 |
| *Staphylococcus aureus* | 2 | 0 | 2 |
| *Streptococcus pyogenes* | 2 | 0 | 2 |
| *Streptococcus pneumoniae* | 2 | 1 | 1 |

There are already nine species of bacteria capable of causing disease in humans for which we have the genome sequence of more than one strain. We have both pathogenic and nonpathogenic strains from both *E. coli* and *S. pneumoniae*.

unsequenced genomes. Microarrays (*see* Chapter 10) containing large numbers of bacterial genes or complete genomes are now being developed for *genomotyping,* or the profiling the genetic similarities and dissimilarities between collections of strains *(39,48,65–68)*. Such technologies promise to provide unprecedented levels of information about the nature and significance of the extensive population-level genomic diversity that is known to characterize bacterial species.

## 3.2. Genome-Based Insights into the Physiology of Pathogens

Complete genome sequences provide a catalog not only of genes, but also of pathways that, when combined into functional categories, explain the metabolism and biosynthetic capacities of an organism. We now know that the tight relationship between genome size and total numbers of genes (**Fig. 4**) also translates into relative functional complexity *(32)*. Since the publication of the first bacterial genome sequence, annotated genes have been assigned to functional categories, an approach originally pioneered by Riley in a seminal study of *E. coli (69)*. Adherence to a standard classification schema allows a familiar schema to be produced for each genome (**Fig. 5**). The types and relative numbers of gene categories in a genome can be read to explain various biological attributes of a cell. Such analysis can be used both to confirm the molecular basis of observed traits as well as to discover new ones. Even more importantly, this cellular blueprint can be compared across genomes to gain insight into how different bacteria are built to occupy their specific ecological niches.

Knowledge of the unique metabolic and genetic features of pathogenic strains and species may make a large impact on human health by facilitating the development of cheaper, faster diagnostic methods (*see* Chapters 7 and 8). Analysis of pathways and metabolic capacity might also reveal clues to the culturing of fastidious pathogens. Finally, an understanding of the genes required for essential metabolic tasks that are not shared with eukaryotes (especially humans) is the key to developing potential targets for novel antimicrobials (*see* Chapter 11).

The stylized blueprint of the physiology and metabolism of a generalized cell is shown in **Fig. 5.** This schema is vastly simplified compared to those reported in genome analysis papers, but is a composite of the essential and optional features of a bacterial cell. Core features shared with even the smallest of genomes (*M. genitalium*) include DNA replication, transcription and translation, some form of energy production or energy scavenging, and the ability to obtain key metabolites either through biosynthesis or by scavenging through the use of transporters with various affinities. Cells with larger genomes and more genes have increased functional diversity, both in terms of the overall number of unique pathways and with respect to genetic redundancy *(32,70)*. Optional features include such things as genes for motility and chemotaxis, secretion systems, and drug efflux systems.

The cellular blueprint of a bacterial cell reflects its lifestyle and, for commensal and virulent bacteria, the nature of the relationship with their host. Highly specialized, obligate intracellular pathogens have small genome sizes, and therefore low numbers of genes, and lack some or most biosynthetic pathways, having evolved over time in association with their hosts to scavenge rather than produce key metabolites. Fascinatingly, it is now known that there are many evolutionary routes to becoming an intracellular pathogen, and different pathways can be lost—in varying orders—during evolution of a reduced genome *(38)*. The number of pathways (e.g., the ability to synthesize amino acids, purines, or pyrimidines) lost reflects the degree of dependence a bacterium has on its host. Genome analysis shows in intricate detail the genetic trade-offs associated with adaptation to a given ecological niche—if a cell does not have a biosynthetic pathway for an essential molecule, it must have at least one transporter/scavenging mechanism in its place. Moreover, the mechanism a bacterium evolves for such scavenging can determine the type and extent of damage done to host cells. In some cases, robbing from the host environment is not a passive but an aggressive activity that involves special enzymes (toxins) that dissolve host tissues to release component parts, as in the case of the flesh-eating bacterium *Clostridium perfringens (71)*. In combination with appropriate transporters, such systems can remove the need for entire biosynthetic pathways, which are then lost over evolutionary time, gene by gene.

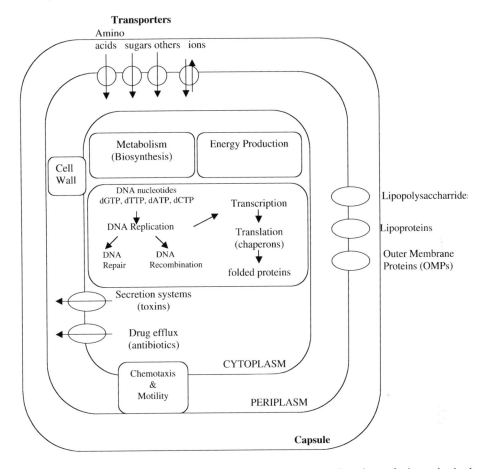

Fig. 5. A stylized schema of a bacterial cell. This composite view of a hypothetical cell shows all the major features of bacterial cells, all of which can be reconstructed from the annotation of a complete genome sequence. Essential features include DNA replication, transcription, and translation, and energy production. Other features are optional. Secretion systems are found only in Gram-negative bacteria. Capsule is a virulence determinant only in certain species, and many of these species, like *H. influenzae,* have both capsulated and noncapsulated strains. Possession of different combinations and numbers of these components reflects different physiologies and, for pathogens, different relationships with the host.

Larger genomes encode a more extensive genetic capacity that includes both larger numbers of regulatory sequences and more extensive and often redundant pathways for synthesizing and acquiring nutrients and other molecules. The cosmopolitan, generalist, soil bacterium *P. aeruginosa* has the genetic capacity to become an opportunistic pathogen because the functional diversity and complex regulatory mechanisms in this large genome provide this species with resistance to disinfectants and antimicrobials *(32)*.

The analysis of the relative proportion of genes in each functional category is of special interest, especially when the lack of key metabolic capacities explains particular dependencies of pathogens on their host *(42)*. Also informative are the abundances and types of surface molecules that mediate interactions with the host. For example, the genome of *T. pallidum* is marked by a near absence of genes that encode outer-membrane proteins *(6)* while the *M. tuberculosis* genome is unique for the unusually large portion of the coding capacity dedicated to the production of enzymes involved in lipogenesis and lipolysis *(56)*.

### 3.3. Genome-Based Insights into the Preferred Niche of Bacteria Capable of Causing Disease

Genome sequences are shedding new light on how bacteria survive within the host. These insights include the identification of genes, pathways, and even global features of genome sequences that help to explain the ability of bacteria to invade and colonize specific locations in the host. For example, the ability of *H. pylori* to colonize the human stomach is due to the enzyme urease, but completion of the genome sequence also revealed that there are twice as many basic amino acids (lysine and arginine) in the proteins of *H. pylori* than in other bacteria *(62)*. This unusual amino acid composition provides an additional adaptation to life in an acidic environment. It is also possible to discover how species adapt to microniches within humans, and why certain species fail to colonize certain niches. Analysis of the genome of the bacterium *Fusobacterium nucleatum,* which colonizes the oral cavity, revealed that the restricted ecological niche of this species is due to the lack of genes that encode choline-binding proteins *(72)*. This is in contrast to other common commensals like *H. influenzae* or *S. pneumoniae,* which have these genes and can therefore colonize the nasopharynx.

Genomes also provide insights into how pathogens modify their environments—a process that can contribute significantly to host-cell damage and death. The genome sequence of the flesh-eating bacterium *C. perfringens* provides an extreme example of the ability of bacteria to change and utilize the resources in their chosen environments *(71)*. This anaerobe uses its genetic capacity when colonizing wounds to create a preferred anaerobic environment by making carbon dioxide, the gas that leads to gangrene, through anaerobic fermentation *(71)*. The genome sequence revealed the limited biosynthetic capacities of *C. perfringens,* and its repertoire of more than 200 transporters. These features place the toxins and enzymes it uses to obtain nutrients by degrading host tissues into proper context *(71)*.

## 3.4. How Genetic Capacity Is "Engineered" to Evolve

One of the most fascinating results to emerge from genome sequencing is an improved understanding of bacterial genome architecture, or the macro- and microlevel organization of genes, elements that fill noncoding regions, repetitive sequences, and compositional patterns within a genome (**Fig. 6**). A direct outcome of this knowledge is our growing appreciation of the fluidity of genome sequences (*see* Chapter 5). Genome analysis is revealing both the nature and extent of molecular mechanisms that both promote mutation (vertical inheritance) and gene transfer (horizontal inheritance). For example, the presence of SSRs near virulence determinant genes in mucosal pathogens suggests the potential for clones to generate quickly genetic variation of biological significance through mutation *(33,58,60)*. More striking, perhaps, is the extent to which certain genomes are "engineered" to rearrange the order and composition of their genes and noncoding regions by the presence of plasmids, bacteriophage, transposons, and other classes of repetitive DNAs, like uptake sequences (discovered as a result of the sequencing of the *H. influenzae* genome *[73]*). In contrast, the genomes of other organisms, particularly those of intracellular pathogens and symbionts with very small genomes, are not *(45,74)*. Through genome sequencing, we now know that the small genomes of intracellular pathogens have lost most if not all of the mobile and self-propagating mediators of genome fluidity found in some other bacteria. In contrast, these elements can represent a significant proportion of the larger genome sequences of free-living, cosmopolitan bacteria, and represent a tremendous capacity for evolutionary change *(45,74)*.

The ability to acquire foreign DNA can impact on both the short- and long-term phenotypic evolution of pathogens. While the true frequency of horizontal transfer and the percentage of horizontally transferred sequences is debated *(46)*, it is clear that genetic transfer can lead to biological innovation *(45)*. Traits such as antibiotic resistance, which protect bacteria against noxious chemicals they may encounter in certain environments, can circulate in bacterial populations and jump across evolutionarily distinct lineages to form new combinations of genes. In pathogenic strains, these new combinations can give rise to clones with changed pathogenic potential; and in nonpathogenic strains, the repeated uptake of foreign DNA can even give rise to new species of pathogens.

Species of bacteria capable of causing disease in humans have repeatedly evolved in the tree of life. Many species contain both pathogenic and nonpathogenic strains. Both these facts demonstrate the ease with which bacteria can apparently evolve strategies that result in virulence and damage to the host. In many cases, this ease can now be shown, through the comparison of appro-

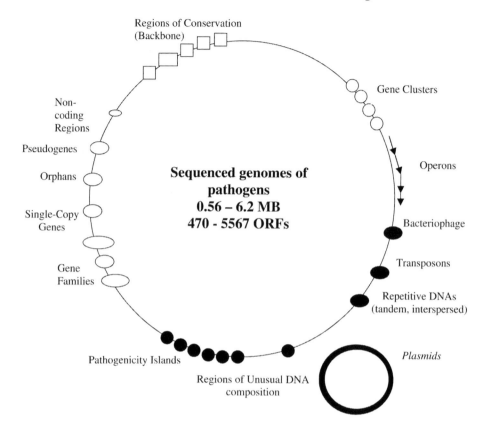

Fig. 6. A stylized schema of a bacterial genome. This composite view of a hypothetical genome shows the core features of bacterial genomes. The abundances of these
features vary among genomes, and some features are absent from particular genomes.
A large portion of each genome (80% or more) will belong to the genetic backbone of
a species. Features in black are those genetic elements and sequences most susceptible
to genetic transfer between genomes.

priate sets of genome sequences, to be the direct result of the ability to acquire
DNA horizontally. The primary source of genetic diversity between the
sequenced genomes of the pathogenic strains of *Streptococcus pyogenes (48)*,
and between genomes of virulent and avirulent *E. coli (17–19)*, are phage and
phage-like sequences that have inserted themselves and additional short segments of foreign DNA into the genetic backbone of these species. In contrast,
the *Vibrio cholerae* genome provides an unprecedented opportunity to study
how a free-living aquatic organism emerged to become a pathogen of humans
through the lateral acquisition of a collection of virulence factors, such as the
filamentous CTX bacteriophage that encodes the cholera toxin, and the TCP

pathogenicity island *(27)*. Analysis of this genome supports the view that horizontal transfer has not only been a means for strains of this species to expand and reorganize micro-sections of their genetic capacity, but has also provided a mechanism for this species to acquire the genetic capacity contained on a second chromosome. The composition of genes on the second, smaller chromosome of *Vibrio cholerae* suggests that it is derived from a former captured megaplasmid *(27)*.

With complete genome sequences we can not only ask what percentage of a genome sequence may have been acquired by lateral gene transfer, but we can also investigate the types of genes that are involved *(45)*. With respect to pathogens, genome analysis is revealing that many of the genes acquired by lateral transfer are found in only a subset of strains in a population, and that this set of genes contains a disproportionate number of orphans and genes with unknown function. However, of those genes that are characterized, many encode traditional virulence factors *(39,48,65–68)*.

## 4. The Future

The past 8 yr have seen the field of bacterial genomics rapidly progress toward becoming a mature field of study. This success has revolutionized the study of pathogens and pathogenesis. As of this writing, over 80 complete bacterial genomes have been published, half of which belong to strains that cause disease. More than a hundred further genomes are in draft form, or are in the process of being sequenced *(10)*, and many more are held privately, for example by Integrated Genomics (http://www.integratedgenomics.com/). We have seen the publication of the genome sequence of more than one pathogenic strain in a single report *(75)* and the simultaneous publication of the genome sequence of a pathogen and a closely related nonpathogen *(76)*. Increasingly, authors of genome reports are choosing to present the combined results of a genome-sequencing project and high-throughput experimental analyses (e.g., microarray studies) *(48,65)*. These facts attest to both the raised estimation of the value of genome sequencing and the reduced costs and improved experimental and bioinformatic technologies now available to the research community. They also suggest a tremendous potential for advancement in future years.

The ability to integrate information derived from genomes with other sources of information will be central to our future ability to exploit bacterial genomes. The sequencing of pathogen genomes has proved beyond doubt the value of using genomes to discover new determinants of virulence. It has also proved the power of using genomes to elucidate the molecular basis of physiology, adaptation to niche, and the mechanisms by which genomes change over both short and long evolutionary time spans. In order to build on this knowledge, we must (1) find new ways to characterize the vast number of genes without known

**Table 4**
**Internet Resources for Examining Genomic Resources Available for Pathogens**

Electronic resources

**Get a full list of complete and in progress genomes at GOLD:**
http://www.integratedgenomics.com/gold/

**Access all complete bacterial genome sequences:**
http://www.ncbi.nlm.nih.gov/PMGifs/Genomes/eub_g.html

**Access the taxonomy of all sequenced pathogens:**
http://www.ncbi.nlm.nih.gov/PMGifs/Genomes/new_micr.html
(also *see* http://www.cme.msu.edu/bergeys/)

**Find individual genes among all known public DNA or protein sequences:**
http://www.ncbi.nlm.nih.gov/Entrez http://www.ebi.ac.uk/srs

**Access curated lists of individual proteins in Swissprot:**
http://www.ebi.ac.uk/swissprot/

**Conduct similarity searches against finished and unfinished microbial genomes:**
http://www.ncbi.nlm.nih.gov/cgi-bin/Entrez/genom_table.cgi

***See* features shared by bacterial pathogen proteomes:**
http://www.ebi.ac.uk/proteome/

***See* lists of known virulence factors in the Prints Database:**
http://www.jener.ac.uk/BacBix3/Pprints.htm

Internet resources have grown quickly in number and complexity because the completion of the first bacterial genome sequence. For specific information on particular pathogens it is best to find a database dedicated to that pathogen. International genome sequencing centers like TIGR (http://www.tigr.org) and the Sanger Centre (http://www.sanger.ac.uk) have developed individual databases for each sequenced genome.

function, especially orphans; (2) continue to look for global patterns that help define unusual features of specific sequences, strains, and species; (3) more effectively translate the results of genome sequencing into hypotheses that can be tested by experiment; (4) explore the extent and significance of population-level genetic diversity *(26)*; and (5) place each genome into its proper "organismal" context by understanding its specific ecology and evolution *(77)*. Perhaps most importantly, this heterogeneous and complex information must be captured in next-generation databases that will improve the scope, quality, and interconnectivity of our current invaluable electronic resources (**Table 4**) *(78)*. With this integrative approach, based on a sound knowledge of genome sequences, we will be able to explore in more detail than ever before what makes a pathogen. This will occur through an increased understanding of the genetic capacity of a wider range of pathogens, a greater appreciation of what pathogens do with their genetic capacity, and a deeper knowledge of exactly how this capacity is engineered to change over time and how much of this change is biologically significant.

## References

1. Wassenaar, T. M. and Gaastra, W. (2001) Bacterial virulence: can we draw the line? *FEMS Microbiol. Lett.* **201,** 1–7.
2. Moxon, R. and Tang, C. (2000) Challenge of investigating biologically relevant functions of virulence factors in bacterial pathogens. *Philos. Trans. R. Soc. Lond. B. Biol. Sci.* **355,** 643–656.
3. Groisman, E. A. and Ochman, H. (1997) How *Salmonella* became a pathogen. *Trends Microbiol.* **5,** 343–349.
4. Groisman, E. A. and Ochman, H. (1994) How to become a pathogen. *Trends Microbiol.* **2,** 289–294.
5. Fleischmann, R. D., Adams, M. D., White, O., Clayton, R. A., Kirkness, E. F., Kerlavage, A. R., et al. (1995) Whole-genome random sequencing and assembly of *Haemophilus influenzae* Rd. *Science* **269,** 496–512.
6. Fraser, C. M., Norris, S. J., Weinstock, G. M., White, O., Sutton, G. G., Dodson, R., et al. (1998) Complete genome sequence of *Treponema pallidum,* the syphilis spirochete. *Science* **281,** 375–388.
7. Nowak, R. (1995) Bacterial genome sequence bagged. *Science* **269,** 468–470.
8. Bentley, S. D., Chater, K. F., Cerdeno-Tarraga, A. M., Challis, G. L., Thomson, N. R., James, K. D., et al. (2002) Complete genome sequence of the model actinomycete *Streptomyces coelicolor* A3(2). *Nature* **417,** 141–147.
9. Goffeau, A., Barrell, B. G., Bussey, H., Davis, R. W., Dujon, B., Feldmann, H., et al. (1996) Life with 6000 genes. *Science* **546,** 63–67.
10. Bernal, A., Ear, U., and Kyrpides, N. (2001) Genomes OnLine Database (GOLD): a monitor of genome projects world-wide. *Nucleic Acids Res.* **29,** 126–127.
11. Cole, S. T., Eiglmeier, K., Parkhill, J., James, K. D., Thomson, N. R., Wheeler, P. R., et al. (2001) Massive gene decay in the leprosy bacillus. *Nature* **409,** 1007–1011.
12. Read, T. D., Brunham, R. C., Shen, C., Gill, S. R., Heidelberg, J. F., White, O., et al. (2000) Genome sequences of *Chlamydia trachomatis* MoPn and *Chlamydia pneumoniae* AR39. *Nucleic Acids Res.* **28,** 1397–1406.
13. Chambaud, I., Heilig, R., Ferris, S., Barbe, V., Samson, D., Galisson, F., et al. (2001) The complete genome sequence of the murine respiratory pathogen *Mycoplasma pulmonis. Nucleic Acids Res.* **29,** 2145–2153.
14. McClelland, M., Sanderson, K. E., Spieth, J., Clifton, S. W., Latreille, P., Courtney, L., et al. (2001) Complete genome sequence of *Salmonella enterica* serovar Typhimurium LT2. *Nature* **413,** 852–856.
15. Field, D., Hood, D., and Moxon, R. (1999) Contribution of genomics to bacterial pathogenesis. *Curr. Opin. Genet. Dev.* **9,** 700–703.

16. Kalman, S., Mitchell, W., Marathe, R., Lammel, C., Fan, J., Hyman, R. W., et al. (1999) Comparative genomes of *Chlamydia pneumoniae* and *C. trachomatis. Nat. Genet.* **21,** 385–389.

17. Blattner, F. R., Plunkett, G., 3rd, Bloch, C. A., Perna, N. T., Burland, V., Riley, M., et al. (1997) The complete genome sequence of *Escherichia coli* K-12. *Science* **277,** 1453–1474.

18. Perna, N. T., Plunkett, G., 3rd, Burland, V., Mau, B., Glasner, J. D., Rose, D. J., et al. (2001) Genome sequence of enterohaemorrhagic *Escherichia coli* O157:H7. *Nature* **409,** 529–533.

19. Hayashi, T., Makino, K., Ohnishi, M., Kurokawa, K., Ishii, K., Yokoyama, K., et al. (2001) Complete genome sequence of enterohemorrhagic *Escherichia coli* O157:H7 and genomic comparison with a laboratory strain K-12. *DNA Res.* **8,** 11–22.

20. Moxon, R. and Rappuoli, R. (2002) Bacterial pathogen genomics and vaccines. *Br. Med. Bull.* **62,** 45–58.

21. Pizza, M., Scarlato, V., Masignani, V., Giuliani, M. M., Arico, B., Comanducci, M., et al. (2000) Identification of vaccine candidates against serogroup B meningococcus by whole-genome sequencing. *Science* **287,** 1816–1820.

22. Tettelin, H., Saunders, N. J., Heidelberg, J., Jeffries, A. C., Nelson, K. E., Eisen, J. A., et al. (2000) Complete genome sequence of *Neisseria meningitidis* serogroup B strain MC58. *Science* **287,** 1809–1815.

23. Fraser, C. M., Gocayne, J. D., White, O., Adams, M. D., Clayton, R. A., Fleischmann, R. D., et al. (1995) The minimal gene complement of *Mycoplasma genitalium. Science* **270,** 397–403.

24. Hutchison, C. A., Peterson, S. N., Gill, S. R., Cline, R. T., White, O., Fraser, C. M., et al. (1999) Global transposon mutagenesis and a minimal Mycoplasma genome. *Science* **286,** 2165–2169.

25. Fleischmann, R. D., Alland, D., Eisen, J. A., Carpenter, L., White, O., Peterson, J., et al. (2002) Whole-genome comparison of *Mycobacterium tuberculosis* clinical and laboratory strains. *J. Bact.* **184,** 5479–5490.

26. Joyce, E. A., Chan, K., Salama, N. R., and Falkow, S. (2002) Redefining bacterial populations: a post-genomic reformation. *Nat. Rev. Genet.* **3,** 462–473.

27. Heidelberg, J. F., Eisen, J. A., Nelson, W. C., Clayton, R. A., Gwinn, M. L., Dodson, R. J., et al. (2000) DNA sequence of both chromosomes of the cholera pathogen *Vibrio cholerae. Nature* **406,** 477–483.

28. Lan, R. and Reeves, P. R. (2000) Intraspecies variation in bacterial genomes: the need for a species genome concept. *Trends Microbiol.* **8,** 396–401.

29. Lan, R. and Reeves, P. R. (2001) When does a clone deserve a name? A perspective on bacterial species based on population genetics. *Trends Microbiol.* **9,** 419–424.

30. Miller, W. (2001) Comparison of genomic DNA sequences: solved and unsolved problems. *Bioinformatics* **17,** 391–397.

31. Overbeek, R., Fonstein, M., D'Souza, M., Pusch, G. D., and Maltsev, N. (1999) The use of gene clusters to infer functional coupling. *Proc. Natl. Acad. Sci. USA* **96,** 2896–2901.

32. Stover, C. K., Pham, X. Q., Erwin, A. L., Mizoguchi, S. D., Warrener, P., Hickey, M. J., et al. (2000) Complete genome sequence of *Pseudomonas aeruginosa* PA01, an opportunistic pathogen. *Nature* **406,** 959–964.
33. Bayliss, C. D., Field, D., and Moxon, E. R. (2001) The simple sequence contingency loci of *Haemophilus influenzae* and *Neisseria meningitidis. J. Clin. Invest.* **107,** 657–662.
34. Schilling, C. H., Covert, M. W., Famili, I., Church, G. M., Edwards, J. S., and Palsson, B. O. (2002) Genome-scale metabolic model of *Helicobacter pylori* 26695. *J. Bacteriol.* **184,** 4582–4593.
35. Edwards, J. S. and Palsson, B. O. (2000) The *Escherichia coli* MG1655 *in silico* metabolic genotype: its definition, characteristics, and capabilities. *Proc. Natl. Acad. Sci. USA* **97,** 5528–5533.
36. Oliver, S. G. (1996) From DNA sequence to biological function. *Nature* **379,** 597–600.
37. Fischer, D. and Eisenberg, D. (1999) Finding families for genomic ORFans. *Bioinformatics* **15,** 759–762.
38. Doolittle, R. F. (2002) Biodiversity: microbial genomes multiply. *Nature* **416,** 697–700.
39. Salama, N., Guillemin, K., McDaniel, T. K., Sherlock, G., Tompkins, L., and Falkow, S. (2000) A whole-genome microarray reveals genetic diversity among *Helicobacter pylori* strains. *Proc. Natl. Acad. Sci. USA* **97,** 14,668–14,673.
40. Huynen, M. A., Diaz-Lazcoz, Y., and Bork, P. (1997) Differential genome display. *Trends Genet.* **13,** 389–390.
41. Casjens, S. (1998) The diverse and dynamic structure of bacterial genomes. *Annu. Rev. Genet.* **32,** 339–377.
42. Tamas, I., Klasson, L. M., Sandstrom, J. P., and Andersson, S. G. (2001) Mutualists and parasites: how to paint yourself into a (metabolic) corner. *FEBS Lett.* **498,** 135–139.
43. Andersson, J. O. and Andersson, S. G. (1999) Insights into the evolutionary process of genome degradation. *Curr. Opin. Genet. Dev.* **9,** 664–671.
44. Andersson, S. G., Zomorodipour, A., Andersson, J. O., Sicheritz-Ponten, T., Alsmark, U. C., Podowski, R. M., et al. (1998) The genome sequence of *Rickettsia prowazekii* and the origin of mitochondria. *Nature* **396,** 133–140.
45. Ochman, H., Lawrence, J. G., and Groisman, E. A. (2000) Lateral gene transfer and the nature of bacterial innovation. *Nature* **405,** 299–304.
46. Eisen, J. A. (2000) Horizontal gene transfer among microbial genomes: new insights from complete genome analysis. *Curr. Opin. Genet. Dev.* **10,** 606–611.
47. Nelson, K. E., Clayton, R. A., Gill, S. R., Gwinn, M. L., Dodson, R. J., Haft, D. H., et al. (1999) Evidence for lateral gene transfer between Archaea and bacteria from genome sequence of *Thermotoga maritime. Nature* **399,** 323–329.
48. Smoot, J. C., Barbian, K. D., Van Gompel, J. J., Smoot, L. M., Chaussee, M. S., Sylva, G. L., et al. (2002) Genome sequence and comparative microarray analysis of serotype M18 group A Streptococcus strains associated with acute rheumatic fever outbreaks. *Proc. Natl. Acad. Sci. USA* **99,** 4668–4673.

49. Knapp, S., Hacker, J., Jarchau, T., and Goebel, W. (1986) Large, unstable inserts in the chromosome affect virulence properties of uropathogenic *Escherichia coli* O6 strain 536. *J. Bacteriol.* **168**, 22–30.
50. Finlay, B. B. and Falkow, S. (1997) Common themes in microbial pathogenicity revisited. *Microbiol. Mol. Biol. Rev.* **61**, 136–169.
51. Graham, D. E., Overbeek, R., Olsen, G. J., and Woese, C. R. (2000) An archaeal genomic signature. *Proc. Natl. Acad. Sci. USA* **97**, 3304–3308.
52. Weinstock, G. M. (2000) Genomics and bacterial pathogenesis. *Emerg. Infect. Dis.* **6**, 496–504.
53. Chizhikov, V., Rasooly, A., Chumakov, K., and Levy, D. D. (2001) Microarray analysis of microbial virulence factors. *Appl. Environ. Microbiol.* **67**, 3258–3263.
54. Mitchell, T. J. (1998) Molecular basis of virulence. *Arch. Dis. Child.* **78**, 197–199; discussion 99–200.
55. Fraser, C. M., Casjens, S., Huang, W. M., Sutton, G. G., Clayton, R., Lathigra, R., et al. (1997) Genomic sequence of a Lyme disease spirochaete, *Borrelia burgdorferi. Nature* **390**, 580–586.
56. Cole, S. T., Brosch, R., Parkhill, J., Garnier, T., Churcher, C., Harris, D., et al. (1998) Deciphering the biology of *Mycobacterium tuberculosis* from the complete genome sequence. *Nature* **393**, 537–544.
57. Wizemann, T. M., Heinrichs, J. H., Adamou, J. E., Erwin, A. L., Kunsch, C., Choi, G. H., et al. (2001) Use of a whole genome approach to identify vaccine molecules affording protection against *Streptococcus pneumoniae* infection. *Infect. Immun.* **69**, 1593–1598.
58. De Bolle, X., Bayliss, C. D., Field, D., van de Ven, T., Saunders, N. J., Hood, D. W., et al. (2000) The length of a tetranucleotide repeat tract in *Haemophilus influenzae* determines the phase variation rate of a gene with homology to type III DNA methyltransferases. *Mol. Microbiol.* **35**, 211–222.
59. Moxon, E. R., Rainey, P. B., Nowak, M. A., and Lenski, R. E. (1994) Adaptive evolution of highly mutable loci in pathogenic bacteria. *Curr. Biol.* **4**, 24–33.
60. Hood, D. W., Deadman, M. E., Jennings, M. P., Bisercic, M., Fleischmann, R. D., Venter, J. C., et al. (1996) DNA repeats identify novel virulence genes in *Haemophilus influenzae. Proc. Natl. Acad. Sci. USA* **93**, 11,121–11,125.
61. Weiser, J. N. (2000) The generation of diversity by *Haemophilus influenzae. Trends Microbiol.* **8**, 433–435.
62. Tomb, J. F., White, O., Kerlavage, A. R., Clayton, R. A., Sutton, G. G., Fleischmann, R. D., et al. (1997) The complete genome sequence of the gastric pathogen *Helicobacter pylori. Nature* **388**, 539–547.
63. Parkhill, J., Wren, B. W., Mungall, K., Ketley, J. M., Churcher, C., Basham, D., et al. (2000) The genome sequence of the food-borne pathogen *Campylobacter jejuni* reveals hypervariable sequences. *Nature* **403**, 665–668.
64. Parkhill, J., Achtman, M., James, K. D., Bentley, S. D., Churcher, C., Klee, S. R., et al. (2000) Complete DNA sequence of a serogroup A strain of *Neisseria meningitidis* Z2491. *Nature* **404**, 502–506.

65. Tettelin, H., Nelson, K. E., Paulsen, I. T., Eisen, J. A., Read, T. D., Peterson, S., et al. (2001) Complete genome sequence of a virulent isolate of *Streptococcus pneumoniae*. *Science* **293**, 498–506.
66. Dziejman, M., Balon, E., Boyd, D., Fraser, C. M., Heidelberg, J. F., and Mekalanos, J. J. (2002) Comparative genomic analysis of *Vibrio cholerae:* genes that correlate with cholera endemic and pandemic disease. *Proc. Natl. Acad. Sci. USA* **99**, 1556–1561.
67. Dorrell, N., Mangan, J. A., Laing, K. G., Hinds, J., Linton, D., Al-Ghusein, H., et al. (2001) Whole genome comparison of *Campylobacter jejuni* human isolates using a low-cost microarray reveals extensive genetic diversity. *Genome Res.* **11**, 1706–1715.
68. Behr, M. A., Wilson, M. A., Gill, W. P., Salamon, H., Schoolnik, G. K., Rane, S., et al. (1999) Comparative genomics of BCG vaccines by whole-genome DNA microarray. *Science* **284**, 1520–1523.
69. Riley, M. (1993) Functions of the gene products of *Escherichia coli. Microbiol. Rev.* **57**, 862–952.
70. Fraser, C. M., Eisen, J., Fleischmann, R. D., Ketchum, K. A., and Peterson, S. (2000) Comparative genomics and understanding of microbial biology. *Emerg. Infect. Dis.* **6**, 505–512.
71. Shimizu, T., Ohtani, K., Hirakawa, H., Ohshima, K., Yamashita, A., Shiba, T., et al. (2002) Complete genome sequence of *Clostridium perfringens,* an anaerobic flesh-eater. *Proc. Natl. Acad. Sci. USA* **99**, 996–1001.
72. Kapatral, V., Anderson, I., Ivanova, N., Reznik, G., Los, T., Lykidis, A., et al. (2002) Genome sequence and analysis of the oral bacterium *Fusobacterium nucleatum* strain ATCC 25586. *J. Bact.* **184**, 2005–2018.
73. Smith, H. O., Tomb, J. F., Dougherty, B. A., Fleischmann, R. D., and Venter, J. C. (1995) Frequency and distribution of DNA uptake signal sequences in the *Haemophilus influenzae* Rd genome. *Science* **269**, 538–540.
74. Frank, A. C., Amiri, H., and Andersson, S. G. (2002) Genome deterioration: loss of repeated sequences and accumulation of junk DNA. *Genetica* **115**, 1–12.
75. Kuroda, M., Ohta, T., Uchiyama, I., Baba, T., Yuzawa, H., Kobayashi, I., et al. (2001) Whole genome sequencing of meticillin-resistant *Staphylococcus aureus. Lancet* **357**, 1225–1240.
76. Glaser, P., Frangeul, L., Buchrieser, C., Rusniok, C., Amend, A., Baquero, F., et al. (2001) Comparative genomics of *Listeria* species. *Science* **294**, 849–852.
77. Thompson, J. N. (1999) The evolution of species interactions. *Science* **284**, 2116–2118.
78. Stein, L. (2002) Creating a bioinformatics nation. *Nature* **417**, 119–120.

# 13

## Identification of Novel Pathogenicity Genes by PCR Signature-Tagged Mutagenesis and Related Technologies

**Dario E. Lehoux, François Sanschagrin, Irena Kukavica-Ibrulj, Eric Potvin, and Roger C. Levesque**

### Summary

Microbial pathogens possess a repertoire of virulence determinants that make unique contributions to bacterial fitness during infection. In this chapter, we focus on the recent progress and adaptations of signature-tagged mutagenesis (STM) by PCR instead of hybridization. This is a PCR-based STM mutation-based screening method using a population of bacterial mutants for the simultaneous identification of multiple virulence genes in microbial pathogens by negative selection. Modifications of STM developed in our laboratory have been applied to *Pseudomonas aeruginosa* PAO1. Screening of a collection of 6912 STM mutants in the rat chronic lung model of infection identified 214 *P. aeruginosa* STM mutants defective in virulence. For further studies, and to illustrate better the strategies that need to be utilized, we present detailed analysis of nine selected STM mutants. The data obtained indicate that in vivo, defects in virulence give a wide variety of phenotypes: defects in known virulence factors have been found, thereby validating the method; defects have also been found in orthologs with predicted functions, and in some genes whose functions cannot be predicted from databases. A general strategy and a simple scenario is discussed using the nine STM mutants selected for further characterization. PCR-based STM represent a genomics-based method for in vivo high-throughput screening of new virulence factors.

**Key Words:** Microbial genomics; infection; virulence; PCR-based in vivo profiling; essential genes.

## 1. Introduction

Microbial pathogens possess a repertoire of virulence determinants, each of which makes a unique contribution to bacterial fitness during an infection. Understanding expression of the genes involved in infection remains the Holy

From: *Methods in Molecular Biology, vol. 266: Genomics, Proteomics, and Clinical Bacteriology: Methods and Reviews*
Edited by: N. Woodford and A. Johnson © Humana Press Inc., Totowa, NJ

Grail of research in bacterial virulence. The development of molecular biological technologies, including recent progress in microbial genomics and bioinformatics, coupled with molecular genetic approaches, has facilitated the dissection of a plethora of bacterial regulatory systems implicated in virulence *(1)*. However, the complexities of bacterial gene expression in vivo during infection cannot be addressed solely by in vitro experiments and must be coupled to in vivo analysis. The infected host represents a complex and dynamic environment, which may vary during the infection process. One may assume then, that bacterial pathogens modulate gene expression in response to the conditions encountered in vivo, via bacterial sensors responding to a variety of host factors and stimuli. In addition, discrete cellular steps presumably permit the controlled expression of genes essential for initiation and maintenance of infection. To gain overall understanding of an infection, we must determine the proportion of the bacterial genome expressed in vivo and the subset of genes essential for infection. This is a formidable task requiring several complementary techniques, such as in vivo expression technology (IVET), signature-tagged mutagenesis (STM), elegant bioinformatics analysis, functional genomic analysis via proteomics using two-dimensional gel electrophoresis (2DE) differential display (*see* Chapter 6), DNA and protein arrays, and chip technology (*see* Chapter 10) *(1)*. There is also an urgent need for new technologies for analysis of genes expressed in vivo, as the current tools have significant limitations.

In this chapter, we focus on the recent progress and adaptations of STM, a technology that has become accessible to an increasing number of laboratories, especially as screening can now be done rapidly by automated high-throughput robotics using PCR instead of hybridization. The design of tags has been simplified, and several different miniTn5s, each with a unique phenotypic selection, can now be utilized. These modifications of STM, developed in our laboratory, have been applied to the study of *Pseudomonas aeruginosa* PAO1, the 6.3-Mb genome of which has been completely sequenced *(2)*. The strategies of PCR-based STM are described using the *P. aeruginosa* STM project ongoing in our laboratory as an example. We also discuss the significance of in vivo screening.

STM is a mutation-based screening method that uses a population of bacterial mutants for the identification of multiple virulence genes of microbial pathogens by negative selection *(3)*. The technique depends upon in vivo selection of virulent organisms, while those mutants whose virulence genes are inactivated will not persist at infection foci in models of infection. This elegant technique, developed by Holden and colleagues *(3)*, allows a relatively rapid, unbiased search for virulence genes, using animal hosts to select against strains carrying mutations in genes affecting virulence, among a mixed population of mutants. To avoid the typical labor-intensive screening of individual mutants in the first steps, each mutant is tagged with a different and unique DNA signature; the power of STM is that it allows large numbers of different mutants to be screened simultaneously, in the same animal host.

In the classical STM protocol *(3)*, a comparative hybridization technique is used, and employs as mutagens a collection of transposons, each modified by the incorporation of a different DNA tag sequence. The concept is that when the tagged mutagens integrate into the bacterial chromosome, each individual mutant can be distinguished from every other, on the basis of their unique molecular tags. The first tag collections were designed as short DNA segments containing 40-bp variable regions flanked by invariant arms that facilitated the coamplification and radioactive labeling of the central portions by PCR, and which were subsequently used as hybridization probes *(3)*.

Since STM is an en masse screening technique, mutants are usually kept in arrays, using microtiter plates. Colony or DNA dot blots are prepared from these mutants, and compared by hybridization to DNA prepared when the same pool of strains has been passed through the animal host. Technically, PCR is used to prepare labeled probes representing the tags present in the inoculum (input) and recovered from the host (output). Hybridization of the tags from the input and output pools to the colony or dot blot permits the identification of mutants that failed to grow in vivo, because these tags will not be present in the output pools. These mutants can then be identified and recovered from the original arrays, and the nucleotide sequence of DNA flanking the mutation site can be determined. Hence, STM has and will benefit immensely from the sequence data of all bacterial genomics projects.

In the original method, tags were incorporated into a mini-Tn5 transposon, and their suitability was checked prior to use, by hybridization of amplified labeled tags to DNA colony blots of mutants. Mutants from the in vitro pool whose tags failed to yield clear signals on autoradiograms were discarded, while those that gave good signals were assembled into new pools for screening in animals *(3)*. This careful analysis was done prior to STM, so as to diminish the inherent problems of hybridization caused by problematic tags. The STM method was subsequently modified to avoid the prescreening process, where a series of Tn917 transposons were selected prior to mutagenesis, on the basis of efficient tag amplification and labeling, and lack of cross-hybridization to other tags *(4)*. Since in this case the identity of each tag is known, each modified transposon can then be used to generate an infinite number of mutants. It can also be screened using plasmid or tag DNA rather than colony blots, which increases the sensitivity and specificity of the assay. An additional modi-

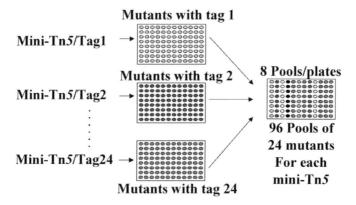

Fig. 1. General strategy for the construction of arrayed libraries using 24 signature tags and three mini-Tn5 derivatives. In each of the 72 defined libraries, the 96 selected mutants have the same tag, but it is inserted at different locations in the bacterial chromosome. After pooling, the end result is one library containing 96 pools of 72 mutants (3 × 24 mutants with unique tag for each mini-Tn5).

fication of the technique introduced 96 tags at a disrupted *URA3* locus of *Candida glabrata (5)*. Different mutants were created by insertional mutagenesis using a plasmid that simultaneously complemented the *URA3* mutation, and the DNA flanking the mutations was cloned by plasmid rescue in *Escherichia coli*. Variations of the original STM method using various transposons have subsequently been applied to many different bacterial pathogens *(6–8)*. Certain transposons are inappropriate tools for genetic analysis of some species of bacteria. For example, Tn5 and its derivatives do not transpose in Gram-positive bacteria, or very poorly in certain Gram-negative bacteria, such as the Pasteurellaceae. However, the recent development of new transposons for genetic analysis, and development of an in vitro transposition assay, should permit the application of STM to a large variety of organisms *(7)*. Suitable properties for transposons to be used in STM include ease of manipulation in vitro, high frequency of transposition, random insertion in the genome of the host, and controlled frequency of insertions as with mini-Tn5s. Obviously, STM cannot provide information about genes that are essential for survival per se, but are inactivated by insertional mutagenesis. Conversely, an insertion within a non-essential gene that is crucial for virulence would not be expected to affect the growth rate of cells in vitro *(9)*.

We present here a general approach to PCR-based STM that can be adapted to any organism of interest.

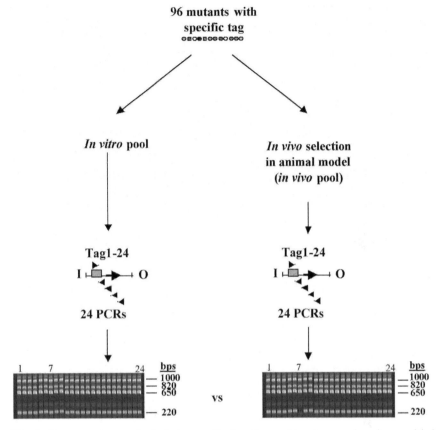

Fig. 2. Comparative analysis between the in vitro and in vivo pools using multiplex PCR. An aliquot is kept as the in vitro pool, and a second aliquot from the same preparation is used for passage into an animal model for in vivo negative selection. After determined time points of infection, bacteria are recovered from animal organs and constitute the in vivo pool. The in vitro and in vivo pools of bacteria are used to prepare DNA in 24 PCR reactions using the 21-mers 1 to 24 in **Table 1** and the Km, Tc, and gfp primers.

## 2. General Strategy for PCR-Based STM

As in the original STM method (3), PCR-based STM is divided into two major steps. The first in vitro step involves the construction of specific and defined DNA tags, which are used for the preparation of a library of tagged mutants (**Fig. 1**). This first phase comprises: the design of oligonucleotides for use as DNA tags; the cloning of these tags into a transposable element that will produce insertional mutations in the organism to be studied; introduction of the

**Table 1**
**Nucleotide Sequences of Signature Tags Used in PCR-Based STM**

| Tag number | Nucleotide sequence (5'–3') |
| --- | --- |
| 1 | GTACCGCGCTTAAACGTTCAG |
| 2 | GTACCGCGCTTAAATAGCCTG |
| 3 | GTACCGCGCTTAAAAGTCTCG |
| 4 | GTACCGCGCTTAATAACGTGG |
| 5 | GTACCGCGCTTAAACTGGTAG |
| 6 | GTACCGCGCTTAAGCATGTTG |
| 7 | GTACCGCGCTTAATGTAACCG |
| 8 | GTACCGCGCTTAAAATCTCGG |
| 9 | GTACCGCGCTTAATAGGCAAG |
| 10 | GTACCGCGCTTAACAATCGTG |
| 11 | GTACCGCGCTTAATCAAGACG |
| 12 | GTACCGCGCTTAACTAGTAGG |
| 13 | CTTGCGGCGTATTACGTTCAG |
| 14 | CTTGCGGCGTATTATAGCCTG |
| 15 | CTTGCGGCGTATTAAGTCTCG |
| 16 | CTTGCGGCGTATTTAACGTGG |
| 17 | CTTGCGGCGTATTACTGGTAG |
| 18 | CTTGCGGCGTATTGCATGTTG |
| 19 | CTTGCGGCGTATTTGTAACCG |
| 20 | CTTGCGGCGTATTAATCTCGG |
| 21 | CTTGCGGCGTATTTAGGCAAG |
| 22 | CTTGCGGCGTATTCAATCGTG |
| 23 | CTTGCGGCGTATTTCAAGACG |
| 24 | CTTGCGGCGTATTCTAGTAGG |

Each 21-mer has a $T_m$ of 64°C and permits PCR amplification in one step when the three primer combinations are used for multiplex screening. Two sets of consensus 5'-ends comprising the first 13 nucleotides have higher ΔGs for optimizing PCR. Twelve variable 3'-ends define tag specificity and allow amplification of specific DNA fragments. The set of 24 21-mers representing the complementary DNA strand in each tag are not represented and can be deduced from the sequences present.

transposon containing specific tags into the organism; a reliable method to select the recipients; and the assembly of an array of tagged mutants. The second step requires an animal or cell model for in vivo screening of the library (**Fig. 2**). In this second phase, experimental design needs to take account of the power and limitations of the models used, and the number of rounds of STM screening and the number of animals or cell cultures to be utilized. STM also involves a systematic characterization of mutants selected in vivo. Crucial data

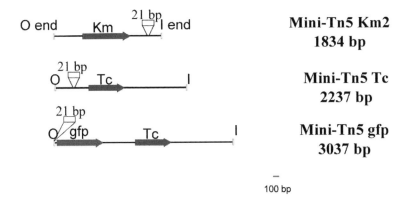

Fig. 3. Maps of the mini-Tn5Km2, mini-Tn5Tet, and mini-Tn5Tetgfp transposons used. The elements are represented by thick black lines; inverted repeats are indicated as open vertical boxes; and genes are indicated by arrows. Abbreviations: I and O, inverted repeat ends; Km, kanamycin; Tc, tetracycline; gfp, green fluorescent protein.

include the DNA sequence around the site of the insertion mutation (which may be obtained by cloning the transposon marker or by RT-PCR) and confirmation that the tagged mutation is the cause of virulence attenuation. The final phase, and the ultimate goal of STM, is to characterize genes shown to play a significant role in virulence, and to define their function and role in pathogenicity. Here, we describe the rationale for PCR-based STM and demonstrate typical results using *P. aeruginosa* as a model.

### 2.1. Cloning of Signature Tags into Transposable Elements

The PCR-based STM method that we have developed uses 24 pairs of complementary 21-bp oligomers (**Table 1**), the first twelve pairs corresponding to ones described by Lehoux et al. *(10)*. These pairs are annealed to generate 24 double-stranded DNA tags, which are then cloned into each of the three plasmid-located mini-Tn5 derivatives: mini-Tn5Km2, mini-Tn5Tc, and mini-Tn5TcGFP (**Fig. 3**) *(11–15)*. These Tn5-derived mini-transposons contain genes encoding kanamycin resistance (Km), tetracycline resistance (Tc), and green fluorescent protein (GFP), respectively, have unique cloning sites for tag insertion, and are flanked by 19-bp terminal repeats (the I and the O ends) *(16,17)*. The transposons are located on the pUT plasmid, a R6K-based suicide delivery plasmid *(15)*. This procedure yields 72 (24 tags × 3 mini-Tn5 vectors) tagged transposons.

There is a precedent for limiting the number of tags used. When *Salmonella* Typhimurium was inoculated into the peritoneal cavities of mice, pools of 96 different mutants gave reproducible hybridization signals after 3 d of infec-

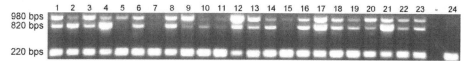

Fig. 4. Agarose gel electrophoresis separation of multiplex PCR amplified products obtained from the in vivo pool of 72 tagged-transposon mutants. Lane 1 to 24, 10 μL of a multiplex PCR reaction using pUTKana2, tetR1, and pUTgfpR2 primers with Tag 1 to 24 respectively. The lane indicated as – is the negative control. The amplified PCR products from tagged mini-Tn5 Km2, *gfp*, and Tc are 220, 820, and 980 bp, respectively. Note the absence of amplification in lanes 5, 7, 15, and 24, where one or two DNA bands are not visible. Rigorous analysis by PCR-based STM implies that the weak signals in lanes 2, 9, 10, 11, and 15 would be considered as positive amplifications and would not be tested further. The clones in lanes 5, 7, 15, and 24 would be further tested by individual PCR and an additional round of in vivo STM screening. In this specific example, no STM mini-Tn5Km mutants were obtained.

tion, whereas pools comprising 192 mutants did not *(3)*. Problems with the complexity of the pool tags have also been described for *Vibrio cholerae,* where it was necessary to reduce the complexity of an orally inoculated pool to 48 mutants, in order to obtain reproducible results *(8)*.

## 2.2. Creating Libraries of Tagged Mutants

Each of the 72 signature-tagged transposons is replicated in *E. coli,* and is then inserted separately into the chromosome of the host bacterium that is to be investigated, to generate insertional mutants. This is usually achieved by conjugation at high frequency, with transconjugants (i.e., mutants) selected using the appropriate mini-Tn5 markers, Km, Tc, or Tc combined with GFP. Ninety-six transconjugants from each mating are arrayed as libraries in 96-well plates. There are, therefore, 72 libraries. Each mutant in a defined library has the same tag, but with the likelihood that it is inserted at a different location in the chromosome from that of other mutants in the same pool.

For the final STM working scheme, one mutant from each of the 72 libraries is then picked to form 96 pools of 72 unique tagged mutants (**Fig. 1**). The 24 signature tags can be used for specific DNA amplifications in multiplex PCR reactions, and tagged products from arrayed bacterial clone pools can be compared as DNA products of a specific length, when separated by agarose gel electrophoresis (**Fig. 4**).

## 2.3. Screening of STM Mutants in Animal Models

STM usually involves the use of an animal model or cell line to mimic aspects of a particular infection, although these models cannot necessarily reproduce faithfully all the conditions associated with the natural disease. Several parameters need to be considered, including the number of different mutant strains to be used in a pool, the route of administration, the size of the inoculum, and the incubation period. An additional consideration is the use of different animal hosts for screening the same pool of STM mutants.

The inoculum size required to initiate infection will determine the complexity of the mutant pool. On the one hand, the pool must contain sufficient cell numbers to initiate infection reliably. On the other, the inoculum size must not be too high, or the host's immune system may be overwhelmed, resulting in the growth of mutants which would not otherwise be detected, and the possibility that animals might die of shock *(6)*. Other important parameters in STM include the route of inoculation and the time-course of a particular infection. Also, certain gene products that are important, directly or indirectly, for initiation or maintenance of infection, may be niche-dependent or expressed specifically in only certain host tissues. If the duration of the infection in STM in vivo selection is short, genes important for establishment of the infection will be found, while if the duration is long, genes important for maintenance of infection may be identified *(6,18)*.

### 2.4. Characterization of Mutant Strains

When a potential STM mutant has been obtained, it is essential to confirm rapidly that the tagged mutant is the cause of attenuation, even after two rounds of in vivo screening. Our recent approach is to do this first, even before the actual DNA sequence flanking the insertion point is known. The degree of virulence attenuation is a prerequisite, and we use the more sensitive and increasingly popular competitive index (CI) and growth index (GI) tests *(18,21,22)*. The analysis is done with an STM mutant strain combined as single or mixed infections with the wild-type parent strain to describe the kinetics of growth, and in certain cases to identify the time and body site where the virulence defect is apparent. In a mixed infection, the ability of the strains to initiate infection or colonize the host provides a measure of their relative virulence. The CI and GI can then be defined as the output ratio of mutant to wild-type bacteria divided by the input ratio of mutant to wild-type bacteria. The CI is thus a quantitative value for the degree of attenuation of a mutant strain, with the CI of a wild-type strain versus a fully virulent derivative being approx 1.0 *(18)*.

### 3. The Use of STM for Studying the Pathogenicity of *Pseudomonas aeruginosa*

As STM is increasingly used and parameters are better defined in different models of infection, several routes of inoculation and different animal models may be used for studying the same organism. We present here the protocol that we have used with *P. aeruginosa.*

### 3.1. A Rat Model of Chronic Lung Infection

For studying *P. aeruginosa,* we have used a rat model of chronic lung infection, which was adapted for this work *(19).* Female Sprague-Dawley rats (140–160 g) are inoculated into the left lobe of the lungs with a suspension of agar beads containing $10^5$ bacterial cells (the in vitro pool); after 7 d, rats are sacrificed, and homogenized lung samples are plated on medium supplemented with chloramphenicol (5 µg/mL). To identify mutants not recovered after the in vivo passage, screening is done by multiplex PCR using bacterial colonies recovered from the lung homogenates. The 21-mers numbered 1 to 24 in **Table 1** were used as a first primer in combination with a primer selected in the Km, Tc, or GFP gene from each transposon. Colony PCR amplification products obtained from the in vivo pool are compared with the in vitro pool, and mutants that give positive results from the in vitro pool, but are absent from the in vivo pool, are kept for further analysis, involving identification and a second round of in vivo screening. Technically, this is confirmed as absence of a PCR product (**Fig. 3**, lanes 5 and 7, and so on).

In our studies, 96 pools of 72 mutants, forming a collection of 6912 mutants, were maintained in the rat, and caused chronic lung infection. From these 6912 mutants, we identified 214 attenuated mutants whose tag did not give a PCR amplification product from the in vivo pool. Mutants attenuated in vivo were further tested by assessing their growth on minimal media (M9 plates). From the collection of 214 mutants, 5 auxotrophic mutants were identified and not studied further. We arrayed 14 new groups with the 214 mutants from the first screening, and 29 *P. aeruginosa* STM mutants previously screened once to confirm the STM attenuated phenotype *(20–22).* When necessary, we completed a group with the wild-type strain PAO1 in order to maintain 72 clones in each group. The in vivo selection and detection of mutants were done as described above. From the 214 mutants initially identified, we retained 42 highly attenuated mutants, whose tags did not give any PCR amplification product from the in vivo pool. These results showed that to identify and obtain the most significant and highly attenuated mutants, a second round of in vivo screening is a prerequisite.

The genomic DNA from these *P. aeruginosa* STM mutants was isolated, digested with *Pst*I, and cloned into pTZ18R (Amersham Pharmacia Biotech) (**Fig. 5**). The recombinant plasmids were introduced into *E. coli* DH5α by electroporation. Selection of recombinant plasmids was based on the cloning

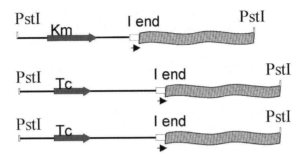

Fig. 5. General scheme for cloning genomic DNA encoding the inactivated gene via selection of the antibiotic resistance genes inserted in STM mutants. The genomic DNA of *P. aeruginosa* is digested completely with *Pst*I and cloned DNA fragments are selected on the basis of encoding either Tc or Km resistance. The cloned genomic DNA is sequenced using primers annealing in the I end of each transposon DNA fragment (I-end sequencing primer: 5'–AGA TCT GAT CAA GAG ACA G–3').

of the Km, Tc, or Tc-Gfp marker; plasmid DNA was prepared and the sequences flanking the transposon markers were analyzed to determine the insertion endpoints.

To determine the GIs of the *P. aeruginosa* STM mutants, the in vitro growth rates of the parent PAO1 strain and mutants derived from it were measured in the rat model of lung infection. Bacteria were recovered 1- and 7-d postinfection from two animals, by direct plating of lung tissue homogenates on medium supplemented with chloramphenicol. Animals used as controls were injected with strains PAO1 and PAO909 (a purine auxotroph of PAO1). For example, nine mutants found to be attenuated are listed in **Table 2**. Among the nine in vivo attenuated mutants, growth rates varied from low to high as reflected by the index values (**Table 2**); the least attenuated mutant had an index of 0.294. Mutants that were considered to be clearly attenuated had a growth rate index of 0.1 or less.

An additional confirmation that we use in combination with the GI to show that the tagged mutation is the cause of attenuation of virulence, is complementation with a functional allele. In this case, the wild-type gene of the STM mutant identified is cloned as a PCR product and introduced in the STM mutant. In a first step, we use RT-PCR to demonstrate in vitro expression and proceed with a GI analysis of the wild-type, the STM mutant, and the mutant strain expressing the complementing allele. The obvious result expected will show that the mutation and the virulence phenotype are linked. Gene analysis can indicate whether the inactivated gene is part of an operon, and that STM

**Table 2**
**In Vitro and In Vivo Growth Index of _P. aeruginosa_ STM Mutants**

| Strains/(genes inactivated) | In vitro GI[a] | Relative GI[b] |
|---|---|---|
| PAO1 | 1.0 | 1.0 |
| PAO909 | 1.1 | 0.0036 |
| G20T2 | 1.5 | 0.0006 |
| (Clp protease, ClpB, Y. enterocolitica) | | |
| G13T12 | 1.2 | 0.025 |
| (Molybdate, binding precursor, ModA, P. aeruginosa) | | |
| G19T12 | 2.0 | 0.01 |
| (Unknown) | | |
| G10T7 | 1.7 | 0.176 |
| (Unknown) | | |
| G30T12 | 1.3 | 0.154 |
| (Hypothetical 64 kDa protein, YfaA, E. coli) | | |
| G56T2 | 2.4 | 0.25 |
| (Hypothetical 51 kDa protein, YgdH, E. coli) | | |
| G18T2 | 1.7 | 0.294 |
| (Transporter protein, AcrB, D, F, Vc1663, V. cholerae) | | |
| G94T2 | 1.5 | 0.0267 |
| (ABC transporter ATP,YbjZ, E. coli) | | |
| G38T4 | 5.8 | 0.034 |
| (Heat shock protein, HtpG, P. aeruginosa) | | |

The disrupted open-reading frames (ORF) in each STM mutant were: PAO090 (G20T2); PA1863 (G13T12); Pa5441 (G19T12); PAO082 (G10T7); PA4491 (G30T12); PA4115 (G56T2); PAO0158(G18T2); PA0073 (G94T2); PA1596(G38T4). Identity values were obtained by comparing the amino acid sequences deduced from the complete ORF.

[a]The in vitro growth index (GI) is the ratio of the growth rate of the mutant compared with that of the wild-type strain PAO1 in rich BHI broth.

[b]The relative GI is the in vivo growth index divided by the in vitro growth index. PAO909 is a purine auxotrophic strain used as a negative control.

attenuation of virulence may be due to a polar effect. In this case, inactivation of the downstream gene(s) is essential to confirm their implication in the decrease of virulence by further testing with the CI tests.

To demonstrate allelic complementation of STM mutants, we will use as an example the G18T2 mutant (PAO0158) described in **Table 2** _(21)_. Mutant

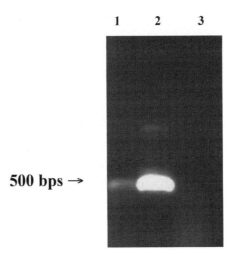

Fig. 6. Agarose gel electrophoresis of RT-PCR products confirming expression of *modA* from the complementing allele. Lanes: 1, wild-type; 2, *modA* STM mutant complemented; 3, STM mutant. The intense 500-bp band is owing to high-level expression from the *lacZ* constitutive promoter in pTZ18R.

G18T2 corresponds to an insertion in an ORF (PA0158) encoding for a protein of 1,016 residues and having 32% identity with a putative ORF VC1663 of *V. cholerae*, 31% identity with AcrD of *Rickettsia prowazekii* and CzrA of *P. aeruginosa*, and 45% identity with a membrane protein which is a putative cation efflux protein (HP0969) of *Helicobacter pylori*, designated Hef. The protein PA0158 has been classified as a probable multidrug efflux system component encoding for a probable RND efflux transporter. In *H. pylori*, *hefF* encodes an RND pump protein homolog. The role and function of RND multiple-drug efflux systems in intrinsic antibiotic resistance has been demonstrated in several human pathogens, including *E. coli*, *Haemophilus influenzae*, *Neisseria gonorrhoeae*, and *P. aeruginosa*. The corresponding mutant has been shown to be mildly attenuated, with a GI value of 0.294 in **Table 2**. Several pseudomonads possess an intrinsic resistance to many front-line antibiotics, due mainly to their low outer membrane permeability and active efflux of antibiotics. We have found by STM that this efflux system may be implicated in virulence in vivo; this system could presumably be active against the host defense system by pumping out toxic molecules (data not shown).

As proof of concept, we have cloned and expressed the *modA* gene **(Table 2)** as determined by RT-PCR **(Fig. 6)**, and shown complementation of ModA in vitro and in vivo using the mutant G13T12 (PA1863). The complemented strain G13T12/pMON-PA1863 was used and compared with the G13T12 strain. We observed that the complemented strain was able to grow 3.4 times faster in vivo than the strain containing the *modA* insertional mutation (GI of 0.24 compared with 0.07, respectively) (data not shown). These results showed that the complementation with *modA* gave a significant recovery to the wild-type phenotype. This indicated that the disruption of *modA* in the mutant G13T12 was responsible for the attenuated phenotype obtained by STM, and confirmed by the in vivo GI experiments. Indeed, insertional inactivation of *modA* gave a significant attenuation of virulence in vivo, but attempts to restore the mutant G13T12 to the wild-type phenotype were only partly successful (GI of 0.6 instead of 1.0).

## 4. Conclusion

The methods described in this chapter have been the first important step in using STM for the elucidation of which genes are essential in *P. aeruginosa* for the maintenance of an infection in the rat lung model. At this stage of the project, we have shown that PCR-based STM can identify typical virulence factors. Indeed, we have isolated *P. aeruginosa* STM mutants with defects in alginate biosynthesis, pilin formation, biofilm production, elastase and pro-tease production, and twitching motility; these have all been shown to have some role in maintenance of infection in vivo *(22)*. In the last few years, hybridization-based STM has been widely used, and has given remarkable insights in the virulence of bacterial pathogens, such as *Salmonella, Vibrio, Streptococcus,* and *Staphylococcus.* The PCR-based STM approach that we present here should be a useful addition to the repertoire of technologies for in vivo functional genomics and the identification of function of new virulence factors.

## Acknowledgments

We express our gratitude to R. W. Hancock, UBC, Vancouver, BC, Canada, for his constant encouragement, support, and suggestions. Work on STM in R.C.L.'s laboratory is funded by the Canadian Institute for Health Research as part of the CIHR Genomics Program, and by the Canadian Bacterial Diseases Network via the Canadian Centers of Excellence. R.C.L. is a Scholar of Exceptional Merit from Le Fonds de la Recherche en Santé du Québec.

## References

1. Handfield, M. and Levesque, R. C. (1999) Strategies for isolation of in vivo expressed genes from bacteria. *FEMS Microbiol. Rev.* **23,** 69–91.
2. Stover, C. K., Pham, X. Q., Erwin, A. L., Mizoguchi, S. D., Warrener, P., Hickey, M. J., et al. (2000) Complete genome sequence of *Pseudomonas aeruginosa* PA01, an opportunistic pathogen. *Nature* **406,** 959–964.

3. Hensel, M., Shea, J. E., Gleeson, C., Jones, M. D., Dalton, E., and Holden, D. W. (1995) Simultaneous identification of bacterial virulence genes by negative selection. *Science* **269,** 400–403.

4. Mei, J.-M., Nourbakhsh, F., Ford, C. W., and Holden, D. W. (1997) Identification of *Staphylococcus aureus* virulence genes in a murine model of bacteraemia using signature-tagged mutagenesis. *Mol. Microbiol.* **26,** 399–407.

5. Cormack, B. P., Ghori, N., and Falkow, S. (1999) An adhesin of the yeast pathogen *Candida glabrata* mediating adherence to human epithelial cells. *Science* **285,** 578–582.

6. Lehoux, D. E. and Levesque, R. C. (2000) Detection of genes essential in specific niches by signature tagged mutagenesis. *Curr. Opin. Biotechnol.* **11,** 434–439.

7. Craig, N. L., Craigie, R., Gellert, M., and Lambowitz, A. M. (2002) *Mobile DNA II,* ASM Press, Washington, DC.

8. Chiang, S. and Mekalanos, J. (1998) Use of signature-tagged transposon mutagenesis to identify *Vibrio cholerae* genes critical for colonization. *Mol. Microbiol.* **27,** 797–805.

9. Lehoux, D. E. and Levesque, R. C. (2001) Discovering essential and infection-related genes. *Curr. Opin. Microbiol.* **4,** 515–519.

10. Lehoux, D. E., Sanschagrin, F., and Levesque, R. C. (1999) Defined oligonucleotide tag pools and PCR screening in signature-tagged mutagenesis of essential genes from bacteria. *Biotechniques* **26,** 473–478.

11. De Lorenzo, V., Herrero, M., Jakubzik, U., and Timmis, K. E. (1990) Mini-Tn5 transposons derivatives for insertion mutagenesis, promoter probing, and chromosomal insertion of cloned DNA in Gram-negative eubacteria. *J. Bacteriol.* **172,** 6568–6572.

12. Herrero, M., de Lorenzo, V., and Timmis, K. N. (1990) Transposon vectors containing non-antibiotic resistance selection markers for cloning and stable chromosomal insertion of foreign genes in Gram-negative bacteria. *J. Bacteriol.* **172,** 6557–6567.

13. Matthysse, A. G., Stretton, S., Dandie, C., McClure, N. C., and Goodman, A. E. (1996) Construction of GFP vectors for use in Gram-negative bacteria other than *Escherichia coli. FEMS Microbiol. Lett.* **145,** 87–94.

14. Fellay, R., Frey, J., and Krish, H. (1987) Interposon mutagenesis of soil and water bacteria: a family of DNA fragments designed for in vitro insertional mutagenesis of Gram-negative bacteria. *Gene* **52,** 147–154.

15. Simon, R., Priefer, U., and Pühler, A. (1983) A broad-host range mobilization system for in vivo genetic engineering transposon mutagenesis in Gram-negative bacteria. *Bio/Technology* **1,** 784–791.

16. Sambrook, J. and Russel, D. (2001) *Molecular Cloning: A Laboratory Manual.* Cold Spring Harbor Laboratory, Cold Spring Harbor, NY.

17. De Bruijn, F. J. and Rossbach, S. (1994) Transposon mutagenesis, in *Methods for General and Molecular Bacteriology,* Gerhardt, P., Murray, R. G. E., and Krieg, N. (eds), ASM Press, Washington, DC, pp. 387–405.

18. Unsworth, K. E. and Holden, D. (2000) Identification and analysis of bacterial virulence genes in vivo. *Phil. Trans. R. Soc. Lond. B* **355,** 613–622.
19. Cash, H. A., Woods, D. E., McCullough, B., Johanson, W. G., Jr., and Bass, J. A. (1979) A rat model of chronic respiratory infection with *Pseudomonas aeruginosa. Am. Rev. Respir. Dis.* **119,** 453–459.
20. Darwin, A. J. and Miller, V. J. (1999) Identification of *Yersinia enterocolitica* genes affecting survival in an animal host using signature-tagged mutagenesis. *Mol. Microbiol.* **32,** 51–62.
21. Lehoux, D. E., Sanschagrin, F., and Levesque, R. C. (2002) Identification of in vivo essential genes from *Pseudomonas aeruginosa* by PCR-based signature-tagged mutagenesis. *FEMS Microbiol. Lett.* **210,** 73–80.
22. Potvin, E., Lehoux, D. E., Kukavica-Ibrulj, I., Richard, K. L., Sanschagrin, F., Lau, G. W., and Levesque, R. C. (2003) In vivo functional genomics of Pseudomonas aeruginosa for high-throughput screening of new virulence factors and antibacterial targets. *Environ. Microbiol.* **5,** 1294–1308.

# 14

## Discovering New Pathogens

*Culture-Resistant Bacteria*

**Andrew J. Lawson**

### Summary

Recent advances in gene-amplification technology and molecular phylogenetics have provided the means of detecting and classifying bacteria directly from their natural habitats without the need for culture. These techniques have revolutionized environmental microbiology, and it is now apparent that the global diversity of microorganisms is much greater than previously thought. In the context of clinical microbiology, this molecular-based approach has facilitated the characterization of culture-resistant bacteria associated with human disease. Examples include *Helicobacter heilmannii,* a cause of gastritis, *Tropheryma whippeli* (the agent of Whipple's disease), and the agents of human ehrlichiosis and bacillary angiomatosis. Molecular-based techniques also provide a means of investigating complex bacterial flora within the human ecosystem, such as feces and dental plaque, without the bias of culture-based isolation. This has given a new perspective to the study of polymicrobial infections such as gingivitis, and offers the potential for the detection and identification of novel bacterial pathogens from among complex and numerous endogenous microbial flora.

**Key Words:** Culture-resistant bacteria; 16S rRNA; 16S rDNA; polymerase chain reaction; phylogenetic analysis.

## 1. Introduction

In the past 15 yr, our knowledge of the extent of microbial diversity has been revolutionized thanks to two key discoveries: the development of the polymerase chain reaction (PCR) for rapid and specific amplification of DNA sequences (*1*), and the establishment of a phylogenetic classification system

From: *Methods in Molecular Biology, vol. 266: Genomics, Proteomics, and Clinical Bacteriology:
Methods and Reviews*
Edited by: N. Woodford and A. Johnson © Humana Press Inc., Totowa, NJ

based on the sequence of genes encoding the RNA component of the small-subunit (SSU) ribosome *(2)*. These breakthroughs have provided the means of identifying and classifying microorganisms in a variety of ecological niches, without the need for cultivation. The impact of these developments has been most apparent among environmental microbiologists, who had long been aware of what was termed "the great plate-count anomaly"; that is, the difference between numbers of bacteria estimated by plate counts of cultivable bacteria, and numbers estimated by direct microscopic counts *(3)*. All in all, it has been estimated that between 0.4 and 5% of all extant bacterial species have been cultured and characterized *(3–5)*. The proportion of a bacterial community that cannot be cultured depends on the ecosystem studied, and may vary from >99% for saltwater to 85% in activated sludge *(4)*. Of all microbial ecosystems, the human body is undoubtedly the most extensively studied, yet our current knowledge of the presence and diversity of microorganisms associated with health and disease is almost entirely derived from cultivation-based methodologies. Given the importance that is placed on microbial identification once humans develop signs of infection, it seems unlikely that the proportion of uncultivated bacteria present in the human microflora is as high as that for environmental niches. Nevertheless, there are examples of pathological conditions associated with bacteria that have resisted isolation by existing culture methods, but which have been successfully characterized by molecular means. The aim of this review is to describe the methodologies employed in the nonculture analysis and characterization of bacterial pathogens.

## 2. What Are Culture-Resistant Bacteria?

The term *culture-resistant* was coined by Wilson *(6)* to describe bacteria that could not be cultured by current methods and which were generally referred to as *unculturable* or *uncultured*. Neither of the latter terms is satisfactory, as some bacteria originally so described have since been cultivated. *Culture-resistant* therefore seems to be a more accurate term.

All bacteria are intrinsically capable of growing and multiplying. When we are unable to induce them to do so in the laboratory, it reflects our inability to fulfill the physiological needs of the microorganism in vitro. In clinical bacteriology, it has long been recognized that many bacterial pathogens are fastidious in their requirements, but may be cultured once the appropriate conditions are provided. In diagnostic clinical bacteriology, the pressures of time, workload, and the necessity of identifying known microbial pathogens result in a somewhat narrow "window of cultivation." Clinical samples with complex microbial floras (such as feces) are often cultured on selective media designed for the isolation of specific pathogens. Moreover, most samples are cultured aerobically at 37°C for 18 to 24 h, with only a subset of specimens being incubated for longer, or at different temperatures, or in different atmospheres. While it is understandable that the function of a clinical bacteriology laboratory is to isolate known bacterial pathogens, this approach can lead to the misconception that the microorganisms isolated by routine laboratory culture methods represent the numerically dominant or significant species in that sample.

An example of this narrow window of cultivation was provided by the discovery of *Helicobacter pylori.* For many years, spiral or curved Gram-negative bacteria had been observed in gastric biopsies of patients suffering from gastritis and gastric ulcers. Efforts to culture these bacteria were unsuccessful or grew only contaminating bacteria. In the early 1980s, Marshall and Warren were attempting to isolate these spiral Gram-negative bacteria using plates incubated for 48 h in a micro-aerobic atmosphere. Results were unsuccessful until a set of plates left by mistake for six days over the Easter holiday yielded growth of novel campylobacter-like bacteria *(7)*. Subsequently, the bacteria, classified as *H. pylori,* became recognized as a cause of gastritis and gastric ulceration, and are associated with the development of gastric cancer. It has been estimated that *H. pylori* infects 50% of the human population worldwide, ranking it as one of the most prevalent bacterial pathogens *(8)*.

## 3. Molecular Detection and Characterization of Culture-Resistant Bacteria

### 3.1. Phylogenetic Analysis of Bacteria

The seminal work of Woese and colleagues *(2,9)* on the comparative analysis of SSU ribosomal RNA (rRNA; 16S in prokaryotes and 18S in eukaryotes) provided, for the first time, an objective framework for determining the evolutionary relationships between organisms, and gave a quantitative measure of diversity, expressed as sequence divergence on a phylogenetic tree (*see* Chapter 16). Woese found that cellular organisms were not simply divided into eukaryote and prokaryote cell types, as had previously been supposed, but rather into three primary domains; the prokaryotic, termed *Bacteria* and *Archaea*; and a single eukaryotic domain, *Eucarya (2)*. Woese also defined 11 major lineages, termed *phyla* or *divisions,* within the bacterial domain, on the basis of 16S rRNA sequences obtained from cultivated microorganisms. This 16S rRNA-based classification of bacteria has largely replaced existing schemes based on traditional phenotypic characterization.

### 3.2. 16S rRNA

Genes encoding rRNA are ideal targets for phylogenetic analysis, as they are present in all organisms and, due to the functional constraints of the ribosome, contain highly conserved regions (being almost invariable for all bacteria). Some variation is possible in other regions of the genes, and mutations accumulate at a slow but constant rate, allowing differences to evolve between the sequences found in different bacterial divisions, genera, and species. Of the genes in the ribosomal operon, the gene encoding 16S rRNA (sometimes termed 16S rDNA) has become the most widely used in the study of bacterial phylogeny because it contains more information than the 5S rRNA, and is more easily sequenced and less prone to contain insertion sequences than 23S rRNA. Eight variable regions have been identified in 16S rRNA, which are termed V1 to V3 and V5 to V9 (region V4 occurs only in eukaryotic 18S rRNA) (**Fig. 1**). Any microorganism can be characterized, provided that the 16S rDNA sequence can be determined. Ribosomal RNA sequence information can be retrieved from both cultured and culture-resistant bacteria present in mixed bacterial communities or infected tissue.

### 3.3. Broad-Range PCR

In the early years of rRNA-based analysis, rRNA sequences were obtained from RNase T1 oligonucleotide catalogs of bulk rRNA, and later from reverse transcriptase sequencing of this RNA *(10)*. Subsequently, PCR has been used to amplify rRNA genes for sequencing. This has been achieved by using a broad-range PCR approach, using primers designed to target regions of the 16S rRNA gene that are universal to members of the bacterial domain, but distinct from eukaryotic 18S rRNA. The design of the PCR assay requires only some knowledge of the nucleotide sequences that flank the target DNA *(1)*. In broad-range PCR, the primers may be designed to target regions of the 16S rDNA that are universal (i.e., present in both Bacteria and Archaea) or specific for the bacterial domain or a specific division or genera within that domain. The resulting amplified DNA contains internal regions of variable sequence that form the basis for specific phylogenetic analysis.

To characterize a culture-resistant, unpurified bacterial pathogen from infected human tissue by PCR, total genomic DNA must first be extracted using a suitable protocol. The nature of the extraction protocol used will depend on the specimen type; for example, feces and blood are known to contain substances that are inhibitory to PCR *(11,12)*. The DNA sequences of the amplified products can be determined either directly or by first cloning the products in a recombinant plasmid vector. Direct sequencing generates a consensus sequence more readily and is less labor intensive, but should be used only with

---

Fig. 1. (*opposite page*) Secondary structure of the small subunit (16S) rRNA of bacteria. Adapted from Neefs et al. *(71)*. The 5' terminus is marked by a filled circle and the 3' terminus by an arrowhead. Helices bearing a single number are common to both Bacteria and Eucarya. Numbers preceded by P are Bacterial-specific helices. Con-

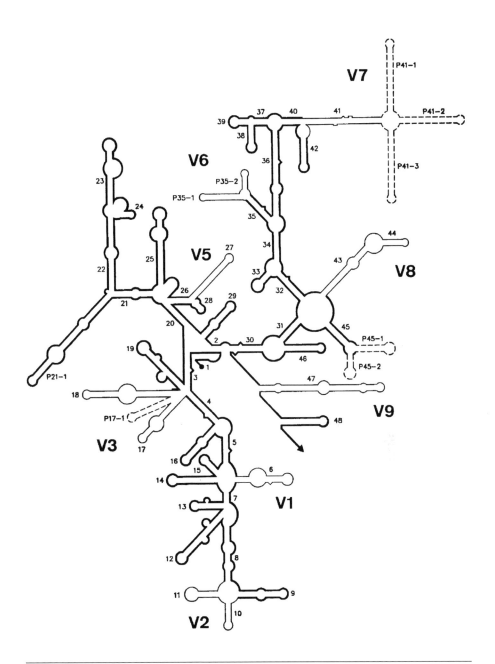

Fig. 1. *(continued)* served areas are drawn in bold lines, areas of variable sequence or length in thin lines. Variable regions numbered V1 to V9 are distinguished (region V4 is absent in prokaryotes). Helices drawn in broken lines are present only in a small number of known structures. Archaeal sequences form similar structures to those of Bacteria except for helix 35, which is unbranched as in the Eucarya.

samples from normally "sterile" sites (i.e., sites with little or no commensal flora). Sequencing cloned products is more laborious, but is essential in examining material that is heavily contaminated with commensal bacteria, where it will reveal sequence heterogeneity and give some measure of the diversity of bacteria present in the sample. This technique can be adapted to polymicrobial environments; in these cases, the PCR products are cloned to create a recombinant DNA library prior to sequencing of individual 16S rRNA genes. Alternatively, PCR primers can be designed to be specific for a single division, genus, or, in some cases, species of bacteria (**Fig. 2**).

### *3.4. Phylogenetic Analysis*

New consensus sequences are aligned with known bacterial 16S rRNA sequences in the current genetic databases *(13–15)* and are analyzed for relatedness using the BLAST utility (*see* Chapter 2) *(16)*. Further PCR primers designed to target specifically the variable regions of 16S rRNA genes from suspected culture-resistant bacteria may be used to investigate further samples for the presence of the same or related bacteria. In cases of novel 16S rDNA sequences that show little relatedness to existing sequences in the available databases, phylogenetic trees can be constructed to allow the evolutionary relationships of the pathogen to be deduced.

### *3.5. Microbial Diversity*

At the time of writing (March 2003), the list of approved bacterial names contains 21 divisions comprising 1146 genera and 6093 species *(17)*. This is a listing of *fully characterized* bacteria that have been cultured successfully. A survey of bacterial diversity conducted in 1998 from the approx 8000 bacterial 16S rRNA gene sequences then available inferred the presence of 36 bacterial divisions *(18)*, with possible evidence for as many as 40 divisions, based on lineages indicated by single environmental isolates. A more recent study of bacterial diversity at the divisional level used only near-full length 16S rRNA sequences (over 1300 nucleotides), and provided a more conservative estimate of 35 division-level lineages, of which 22 have one or more cultivated representatives, and 13 are known only from culture-resistant environmental samples *(19)*.

### *3.6. Candidatus Status*

The vast majority of culture-resistant bacteria are known from little more than their 16S rRNA sequences deposited in genetic databases such as GenBank and the European Molecular Biology Laboratory (EMBL). In many cases these 16S rRNA sequences are incomplete, containing just a few hundred

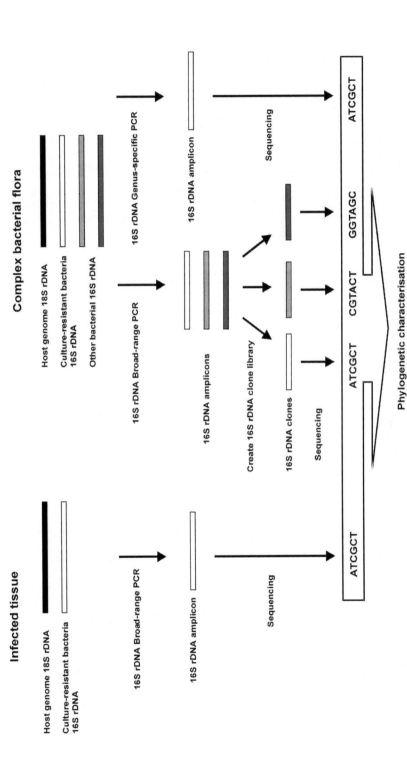

Fig. 2. Summary of methods for PCR-based retrieval of bacterial 16S rDNA sequences from infected tissue and complex bacterial floras. In infected tissue, 16S rDNA-specific broad-range PCR is used to amplify sequences from culture-resistant bacteria from a background of host DNA. In samples from sites with complex bacterial flora, 16S rDNA-specific broad-range PCR is used to generate heterogeneous 16S rDNA amplicons, which are used to create a 16S rDNA clone library from which individual 16S rDNA clones can be sequenced. Alternatively, the 16S rDNA PCR can be designed to amplify only a subset of 16S rDNA from a mixed bacterial population. In this case the PCR assay has been designed to amplify 16S rDNA from a specific bacterial genus.

nucleotides, and have little accompanying information, save details of the source of isolation and phylogenetic association.

For culture-resistant bacteria that have been more closely studied, and for which more information is available, the International Committee on Systematic Bacteriology has created a *Candidatus* category to be used to record formally the properties of putative taxa of prokaryotes that are otherwise insufficiently characterized for formal description *(20,21)*. There are currently 44 officially recognized taxa with *Candidatus* status *(17)*. These mostly include endosymbiotic organisms found in association with nematodes, insects, and plants, and some obligately intracellular parasites of plants and animals. As yet there is only one *Candidatus* associated with human disease, "*Candidatus* Helicobacter suis" *(22)* (*see* **Subheading 5.1.**).

## 4. Disadvantages of Culture-Independent Studies

Culture-independent studies, based on obtaining sequences of 16S rRNA derived from a sample by PCR and cloning, have the potential to improve greatly our understanding of the microbial world. Nevertheless, such molecular methodologies are prone to certain shortcomings and disadvantages.

### *4.1. PCR Amplification*

The PCR reaction itself may be a source of bias and inconsistency when investigating microbial diversity and culture-resistant bacteria, as differential amplification of rDNA from mixed populations may occur in certain cases. For example, it has been shown that bacterial DNA with a high G+C content amplifies relatively poorly *(12,23)*. The number of amplification cycles used can also affect the range of amplified 16S rDNA detected in microbial diversity studies. The greater the number of amplification cycles used, the less diversity apparent in cloned rDNA libraries, as more numerous and more readily amplified rDNAs are copied preferentially at each cycle. Typically, phylogenetic diversity studies employ 9 or 10 cycles of PCR *(24)*. The appearance of PCR artifacts is also a potential risk in the analysis of complex microbial populations, as they suggest the presence of bacteria that do not exist. Deletion mutations may arise due to the formation of stable secondary structures in the template, and point mutations may arise due to misincorporation of nucleotides by DNA polymerases *(12)*. The most insidious PCR artifacts involve the formation of recombinant or chimeric sequences during PCR amplification. Chimera formation is thought to occur when a prematurely-terminated amplicon re-anneals to a foreign DNA strand and is copied to completion in the following PCR cycles. This results in a sequence composed of two or more parent sequences that, when analyzed comparatively with other 16S rDNA sequences, suggest the presence of a nonexistent organism *(25)*. A recent survey of public databases detected a significant number of chimeric 16S rDNA sequences from environmental samples *(25)*. Chimeric 16S rDNA sequences can be readily detected by analysis programs such as CHIMERA_CHECK *(26)* and BELLEROPHON *(25)*.

## 4.2. Establishing Causation

The detection of a DNA sequence in samples from an infected tissue or body site does not prove that the bacterium whose presence is inferred from the sequence is the actual cause of that infection. Traditionally, disease causality has been established by the use of Koch's postulates, which state that (1) the microorganism must be isolated from all cases of disease, (2) the microorganism must produce the characteristic disease pattern when inoculated into a susceptible animal species, and (3) the microorganism must be re-isolated from the experimentally induced disease *(27,28)*. Obviously without a purified or cultivated microorganism, these criteria are difficult to fulfill. The only alternative is to build a compelling body of circumstantial evidence to establish a causal relationship between the culture-resistant bacterium inferred from amplified 16S rDNA, and the disease under investigation. For example, DNA sequences belonging to the putative pathogen should be present in most cases of the infection and absent from hosts or tissues without disease. The significance of an amplified 16S rDNA sequence can be increased by specific hybridization to visible microorganisms in infected tissue samples, or by *in situ* PCR.

## 5. Examples of Culture-Resistant Bacteria

## 5.1. "Helicobacter heilmannii"

Following the successful isolation and characterization of *H. pylori* from human gastric biopsies in 1984 *(7)*, workers reported the microscopic observation of a large, Gram-negative, tightly-coiled spiral bacterium in the gastric mucosa of a small percentage of gastritis cases, that remained resistant to all attempts at culture *(29–32)*. This new bacterium was named "*Gastrospirillum hominis*" *(31)*. Infected patients typically complained of upper-abdominal discomfort, and a diagnosis of chronic active gastritis was confirmed following endoscopic investigation. Treatment with bismuth salts resulted in eradication of the organisms, which coincided with resolution of the symptoms of chronic active gastritis *(32)*, thus implicating the bacterium as the etiological agent of the infection. Subsequently, the 16S rDNA sequence of "*G. hominis*" was detected in specific pathogen-free mice fed homogenized gastric biopsy samples, and the presence of "*G. hominis*" in mouse stomach was confirmed by electron microscopy *(33)*. Comparison with other 16S rDNA sequences showed that "*G. hominis*" was a member of the genus *Helicobacter*, showing closest phylogenetic relatedness to *H. felis, H. bizzozeronii,* and *H. salomonis,* these last being readily cultivable species isolated from the gastric mucosa of cats and dogs *(33)*. In light of this new phylogenetic information, the new name *"Helicobacter heilmannii"* was proposed (reports of similar organisms in animals seemed to suggest that the species designation *hominis* was inappropriate) *(34)*. Although the isolation of *"H. heilmannii"* from human gastric mucosa using conventional culture techniques was reported in 1996 *(35)*, it was subsequently shown that the isolate was actually *H. bizzozeronii (36)*.

Other culture-resistant *Helicobacter*-like organisms from the gastrointestinal tract of animals have been described and accorded *Candidatus* status, namely *"Candidatus* Helicobacter suis" (formerly *"Gastrospirillum suis"* from pigs) *(22)*, and *"Candidatus* Helicobacter bovis" from the abomasum of cattle *(37)*. However, the 16S rRNA gene sequence of *"Candidatus* Helicobacter suis" shares over 99% homology with the corresponding sequence determined for *"H. heilmannii."* It has since been proposed that *"Candidatus* Helicobacter suis" and *"H. heilmannii"* should be referred to as *"Candidatus* Helicobacter heilmannii," to reflect the more frequent nomenclature usage and the nonspecific host range of this organism *(38)*.

### 5.2. Whipple's Disease and Tropheryma whippeli

Whipple's disease, or intestinal lipodystrophy, first described by the American pathologist George Hoyt Whipple in 1907, is an uncommon systemic infectious disease primarily affecting middle-aged Caucasian males *(39)*. Patients typically present with weight loss, arthralgia, diarrhea, and abdominal pain and, less often, central nervous system or cardiac manifestations that may prove fatal if untreated *(39,40)*. The disease is commonly diagnosed by small bowel biopsy, where the appearance of the sample is characterized by inclusions in the lamina propria staining with periodic-acid-Schiff, which represent the causative bacteria *(40)*. In his initial description of the disease, Whipple reported the presence of rod-shaped microorganisms in the vacuoles of macrophages, but did not link them to the causation of disease *(40)*. Thus, the bacterial etiology of this disease was not suspected until 1952, when a case was successfully treated with chloramphenicol. It was not until 1961 that microscopic examination confirmed the presence of Gram-positive bacilli associated with areas of affected small-intestine mucosa *(39)*; however, the organism resisted all attempts at conventional culture. In 1991, Wilson et al. *(41)* used broad-range primers to amplify bacterial 16S rDNA directly from the intestinal tissue of an infected individual, and suggested that the bacillus was most closely related to the actinomycetes. This work was subsequently confirmed with tissues from several other patients, and the provisional name *"Tropheryma whippeli"* was proposed *(42)*. The rDNA sequence of this organism forms the basis of a specific diagnostic PCR test, and further studies have demonstrated the presence of "*T. whippeli*" DNA in 25/38 sewage samples, suggesting that the organism, or a closely related species, was either a widely distributed commensal or an environmental microorganism *(43)*.

In 1997, the Whipple's disease bacterium was cultured from a cardiac valve, using human macrophages inactivated with interleukin-4, but the isolate could not be subcultured *(44)*. Subsequently, it was propagated on a human fibroblast cell line, employing a prolonged incubation period, with a reported generation time of 18 d *(45)*, and the new genus, *Tropheryma*, and species, *T. whipplei,* were characterized and formally recognized *(46)*. The genome of *T. whipplei* has recently been sequenced *(47)* and lacks genes for key biosynthetic pathways and energy-generating metabolism, which may account for its host-adapted lifestyle. However, the genome does encode a large family of surface proteins, which suggests that immune evasion and host interactions play an important part in its life cycle. Available evidence suggests that *T. whipplei* is a widely distributed human pathogen, and it may be that Whipple's disease is just one manifestation of its ability to infect humans.

### 5.3. Human Ehrlichiosis

Ehrlichiosis is an arthropod-borne veterinary disease caused by several bacterial species of the genus *Ehrlichia*, which primarily affects dogs, cattle, sheep, goats, and horses. In 1987 an apparent case of human ehrlichiosis was reported, involving a middle-aged man who developed a febrile illness after being bitten by a tick during a trip to Arkansas *(48)*. Initially it was assumed that he was suffering from Rocky Mountain spotted fever, caused by the tick-borne bacterium *Rickettsia rickettsii*, which is known to infect humans. Serological tests for Rocky Mountain spotted fever were negative, but did provide evidence of infection with *Ehrlichia canis,* which causes ehrlichiosis in dogs *(48)*. Retrospective serological analyses demonstrated further unrecognized cases of human ehrlichiosis, and analysis of blood smears by electron microscopy showed the culture-resistant bacteria forming intracytoplasmic inclusions within leucocytes *(27)*. In 1991, broad-range PCR with bacterium-specific primers was used to amplify 16S rDNA from the blood of a patient from Fort Chaffee, Arkansas *(49)*. Phylogenetic analysis of the retrieved 16S rRNA gene sequence showed that the organism from which the DNA was sequenced was closely related to, but not identical with, *E. canis*. This novel organism was provisionally named "*Ehrlichia chaffeensis*," and has since been detected in other clinical samples *(6,50)* and cultivated in a canine macrophage cell line *(51)*. In the case of "*E. chaffeensis,*" the nature of the clinical illness, the presence of leukocytic inclusions, and serological cross-reactivity with *E. canis* are all suggestive of a causal role for "*E. chaffeensis*" in disease.

## 5.4. Bacillary Angiomatosis and Bartonella henselae

Bacillary angiomatosis is an angioproliferative, AIDS-related disease involving the skin and numerous internal organs *(52)*. The presence of clumps of pleomorphic bacilli in infected tissues, and a positive response to antibiotic therapy, strongly suggested a bacterial etiology for this condition, but the bacterium remained resistant to culture. Broad-range PCR for 16S rRNA was used to amplify bacterial DNA from frozen tissue samples from an immunocompromised patient with disseminated visceral bacillary angiomatosis. The resulting phylogenetic analysis identified the pleomorphic bacilli as a proteobacterium *(53)*. Its closest relative was *"Rochalimaea quintana,"* which causes trench fever and is also associated with bacillary angiomatosis *(54)*. This newly described bacterium was subsequently isolated and named *"R. henselae" (50,55,56)*, but was thereafter reclassified (along with *"R. quintana"*), as a member of the genus *Bartonella*, and renamed *Bartonella henselae (57)*. All *Bartonella* species will grow on blood agar, but are culture-resistant in that they typically take 12–14 d to produce visible colonies on primary isolation. On subculture, colonies usually appear after 3–5 d *(54)*.

## 6. Investigating Complex Bacterial Flora

The endogenous microbial flora of the human body is extensive and varied, reflecting the wide variety of sites available for colonization, with distinct bacterial populations inhabiting the skin, the genitourinary tract, the oral cavity, and various niches within the gastrointestinal tract. It has been estimated that the human body is colonized with 10 viable bacteria for every cell in the human body: $10^{14}$ bacterial cells compared with $10^{13}$ human cells *(58)*. The endogenous bacterial flora is thought to play a beneficial role in nutrition and the physiological development of the host, in addition to stimulating the immune system and providing resistance to colonization by pathogens. Nevertheless, the endogenous flora may also contain organisms capable of acting as opportunistic pathogens.

## 6.1. Dental Caries and Periodontal Disease

In excess of 300 bacterial species have been identified by culture from dental plaque, a polymicrobial matrix that is responsible for two of the world's most prevalent infectious diseases, dental caries and periodontal disease *(59)*. Certain species, such as *Porphyromonas gingivalis, Bacteroides forsythus,* and *Treponema denticola* have been identified as putative periodontal pathogens because they are detected in higher levels from diseased rather than healthy periodontal and gingival samples. However, only 75% of the bacteria visible by microscopy in subgingival plaque can be cultured on nonselective media, and it is possible that the uncultured portion of the microflora may contain pathogens *(28)*. Several studies have used broad-range PCR to amplify bacterial 16S rDNA directly from subgingival plaque in order to create clone libraries from the resulting heterogeneous population of 16S rRNA amplicons *(60–62)*. Analysis of the sequences of these clones has demonstrated that a significant proportion of the resident human bacterial flora remains poorly characterized, with between 40% and 50% of the directly amplified 16S rDNA sequences showing no close homology with 16S rDNA sequences in public databases *(60–62)*. Comparison of the bacterial flora of healthy subjects and that of subjects with periodontal disease showed that some 16S rDNA sequences from culture-resistant bacteria were associated with disease *(61,62)*. Subsequently, Sakamoto and colleagues designed PCR assays specific for five distinct 16S rDNA sequences (called *phylotypes*) retrieved from cases of periodontal disease *(63)*. In a study of 45 patients with periodontitis and 18 healthy subjects, one particular phylotype, designated AP24, was significantly associated with plaque samples from subjects with periodontitis, suggesting that this phylotype represents the 16S rDNA sequence of a culture-resistant bacterium involved in periodontal disease *(63)*. Interestingly, phylotype AP24 showed a close phylogenetic relationship to the clone DA014 reported by Paster et al. *(62)* in an earlier study of subgingival plaque.

### 6.2. The Lower Gastrointestinal Tract

The majority of the estimated $10^{14}$ bacterial cells that comprise the endogenous microbial flora of humans are found in the lower gastrointestinal tract *(58)*. Extensive efforts have been made to cultivate the bacteria found in human feces, making the microflora of the lower gastrointestinal tract one of the most widely studied bacterial ecosystems. Despite a large degree of individual variation, the microflora of the lower gastrointestinal tract consists primarily of strictly anaerobic species coexisting with smaller populations of facultative organisms *(23)*. Nevertheless, the recent application of 16S rDNA PCR-based methodologies has demonstrated that between 50% and 90% of the lower gastrointestinal tract microflora is as yet uncultivated *(23,58,64)*. Hence, the human lower gastrointestinal tract contains many undiscovered bacterial lineages. Molecular studies of the bacterial diversity of the gastrointestinal tract have focused on the microflora of healthy individuals. A recent extensive study of infectious intestinal disease in England determined that microbial etiologies could be established in only approx 45% of cases of apparent gastrointestinal infection *(65)*. While it is likely that some of the unattributed cases of gastroenteritis represent viral infection, it may well be that some are caused by culture-resistant bacteria. It is noteworthy that *Campylobacter jejuni*, now widely acknowledged as the single most important bacterial cause of human gastroenteritis worldwide *(66)*, was itself regarded as culture-resistant prior to the

1970s. It was not until the development of effective isolation techniques, such as filtration *(67)* and antibiotic-containing selective media *(68)*, that the true significance of this bacterium became apparent. More recently, the application of a 16S rDNA PCR assay specific for genus *Campylobacter* detected sequences from a novel *Campylobacter* species in the feces of both healthy humans and those with diarrhea. This novel and initially culture-resistant organism was subsequently isolated and named *C. hominis (69,70)*.

## 7. Conclusions

The combination of PCR and molecular phylogeny in the form of broad-range 16S rDNA-based analyses has helped to elucidate the nature of several previously uncharacterized culture-resistant bacterial pathogens. Using these methodologies, clinical bacteriologists now have the means to explore complex bacterial floras, and potentially to detect as yet undescribed pathogens. The potential usefulness of this and related molecular phylogenetic approaches is considerable, as they offer a new perspective on the diversity of microbial life in human health and disease.

## References

1. Mullis, K. B. and Faloona, F. A. (1987) Specific synthesis of DNA in vitro via a polymerase-catalyzed chain reaction. *Methods Enzymol.* **155,** 335–350.
2. Woese, C. R. (1987) Bacterial evolution. *Microbiol. Rev.* **51,** 221–271.
3. Staley, J. T. and Konopka, A. (1985) Measurement of in situ activities of nonphotosynthetic microorganisms in aquatic and terrestrial habitats. *Annu. Rev. Microbiol.* **39,** 321–346.
4. Amann, R. I., Ludwig, W., and Schleifer, K. H. (1995) Phylogenetic identification and in situ detection of individual microbial cells without cultivation. *Microbiol. Rev.* **59,** 143–169.
5. Relman, D. A. (1998) Detection and identification of previously unrecognized microbial pathogens. *Emerg. Infect. Dis.* **4,** 382–389.
6. Wilson, K. H. (1994) Detection of culture-resistant bacterial pathogens by amplification and sequencing of ribosomal DNA. *Clin. Infect. Dis.* **18,** 958–962.
7. Marshall, B. J. and Warren, J. R. (1984) Unidentified curved bacilli in the stomach of patients with gastritis and peptic ulceration. *Lancet* **1,** 1311–1315.
8. Owen, R. J. (1998) Helicobacter—species classification and identification. *Br. Med. Bull.* **54,** 17–30.
9. Woese, C. R. and Fox, G. E. (1977) Phylogenetic structure of the prokaryotic domain: the primary kingdoms. *Proc. Natl. Acad. Sci. USA* **74,** 5088–5090.
10. Lane, D. J., Pace, B., Olsen, G. J., Stahl, D. A., Sogin, M. L., and Pace, N. R. (1985) Rapid determination of 16S ribosomal RNA sequences for phylogenetic analyses. *Proc. Natl. Acad. Sci. USA* **82,** 6955–6959.

11. Widjojoatmodjo, M. N., Fluit, A. C., Torensma, R., Verdonk, G. P., and Verhoef, J. (1992) The magnetic immuno polymerase chain reaction assay for direct detection of salmonellae in fecal samples. *J. Clin. Microbiol.* **30**, 3195–3199.

12. von Wintzingerode, F., Gobel, U. B., and Stackebrandt, E. (1997) Determination of microbial diversity in environmental samples: pitfalls of PCR-based rRNA analysis. *FEMS Microbiol. Rev.* **21**, 213–229.

13. Gutell, R. R., Weiser, B., Woese, C. R., and Noller, H. F. (1985) Comparative anatomy of 16-S-like ribosomal RNA. *Prog. Nucleic Acid Res. Mol. Biol.* **32**, 155–216.

14. Olsen, G. J. (1987) Earliest phylogenetic branchings: comparing rRNA-based evolutionary trees inferred with various techniques. *Cold Spring Harb. Symp. Quant. Biol.* **52**, 825–837.

15. Olsen, G. J., Larsen, N., and Woese, C. R. (1991) The ribosomal RNA database project. *Nucleic Acids Res.* **19**, 2017–2021.

16. Altschul, S. F., Gish, W., Miller, W., Myers, E. W., and Lipman, D. J. (1990) Basic local alignment search tool. *J. Mol. Biol.* **215**, 403–410.

17. List of bacterial names with standing in nomenclature. http://www.bacterio.cict.fr.

18. Hugenholtz, P., Goebel, B. M., and Pace, N. R. (1998) Impact of culture-independent studies on the emerging phylogenetic view of bacterial diversity. *J. Bacteriol.* **180**, 4765–4774.

19. Hugenholtz, P. (2002) Exploring prokaryotic diversity in the genomic era. *Genome Biol.* **3**, 0003.1–0003.8.

20. Murray, R. G. and Schleifer, K. H. (1994) Taxonomic notes: a proposal for recording the properties of putative taxa of procaryotes. *Int. J. of Syst. Bacteriol.* **44**, 174–176.

21. Murray, R. G. and Stackebrandt, E. (1995) Taxonomic note: implementation of the provisional status *Candidatus* for incompletely described procaryotes. *Int. J. Syst. Bacteriol.* **45**, 186–187.

22. De Groote, D., van Doorn, L. J., Ducatelle, R., Verschuuren, A., Haesebrouck, F., Quint, W. G., et al. (1999) "*Candidatus* Helicobacter suis," a gastric helicobacter from pigs, and its phylogenetic relatedness to other gastrospirilla. *Int. J. Syst. Bacteriol.* **49**, 1769–1777.

23. Wilson, K. H. and Blitchington, R. B. (1996) Human colonic biota studied by ribosomal DNA sequence analysis. *Appl. Environ. Microbiol.* **62**, 2273–2278.

24. Bonnet, R., Suau, A., Dore, J., Gibson, G. R., and Collins, M. D. (2002) Differences in rDNA libraries of faecal bacteria derived from 10- and 25-cycle PCRs. *Int. J. Syst. Evol. Microbiol.* **52**, 757–763.

25. Hugenholtzt, P. and Huber, T. (2003) Chimeric 16S rDNA sequences of diverse origin are accumulating in the public databases. *Int. J. Syst. Evol. Microbiol.* **53**, 289–293.

26. Maidak, B. L., Cole, J. R., Lilburn, T. G., Parker, C. T., Jr., Saxman, P. R., Farris, R. J., et al. (2001) The RDP-II (Ribosomal Database Project). *Nucleic Acids Res.* **29**, 173–174.

27. Fredricks, D. N. and Relman, D. A. (1996) Sequence-based identification of microbial pathogens: a reconsideration of Koch's postulates. *Clin. Microbiol. Rev.* **9,** 18–33.
28. Wilson, M. J., Weightman, A. J., and Wade, W. G. (1997) Applications of molecular ecology in the characterization of uncultured microorganisms associated with human disease. *Rev. Med. Microbiol.* **8,** 91–101.
29. Dent, J. C., McNulty, C. A., Uff, J. C., Wilkinson, S. P., and Gear, M. W. (1987) Spiral organisms in the gastric antrum. *Lancet* **2,** 96.
30. Dye, K. R., Marshall, B. J., Frierson, H. F., Jr., Guerrant, R. L., and McCallum, R. W. (1989) Ultrastructure of another spiral organism associated with human gastritis. *Dig. Dis. Sci.* **34,** 1787–1791.
31. McNulty, C. A., Dent, J. C., Curry, A., Uff, J. S., Ford, G. A., Gear, M. W. et al. (1989) New spiral bacterium in gastric mucosa. *J. Clin. Pathol.* **42,** 585–591.
32. Heilmann, K. L. and Borchard, F. (1991) Gastritis due to spiral shaped bacteria other than *Helicobacter pylori:* clinical, histological, and ultrastructural findings. *Gut* **32,** 137–140.
33. Solnick, J. V., O'Rourke, J., Lee, A., Paster, B. J., Dewhirst, F. E., and Tompkins, L. S. (1993) An uncultured gastric spiral organism is a newly identified Helicobacter in humans. *J. Infect. Dis.* **168,** 379–385.
34. Stolte, M., Wellens, E., Bethke, B., Ritter, M., and Eidt, H. (1994) *Helicobacter heilmannii* (formerly *Gastrospirillum hominis*) gastritis: an infection transmitted by animals? *Scand. J. Gastroenterol.* **29,** 1061–1064.
35. Andersen, L. P., Norgaard, A., Holck, S., Blom, J., and Elsborg, L. (1996) Isolation of a "Helicobacter heilmanii"–like organism from the human stomach. *Eur. J. Clin. Microbiol. Infect. Dis.* **15,** 95–96.
36. Jalava, K., On, S. L., Harrington, C. S., Andersen, L. P., Hanninen, M. L., and Vandamme, P. (2001) A cultured strain of "Helicobacter heilmannii," a human gastric pathogen, identified as *H. bizzozeronii:* evidence for zoonotic potential of Helicobacter. *Emerg. Infect. Dis.* **7,** 1036–1038.
37. De Groote, D., van Doorn, L. J., Ducatelle, R., Verschuuren, A., Tilmant, K., Quint, W. G., et al. (1999) Phylogenetic characterization of "*Candidatus* Helicobacter bovis," a new gastric helicobacter in cattle. *Int. J. Syst. Bacteriol.* **49,** 1707–1715.
38. On, S. L. W. (2002) International Committee on Systematics of Prokaryotes: Subcommittee on the taxonomy of Campylobacter and related bacteria. Minutes of the meetings, 2 and 4 September 2001, Freiburg, Germany. *Int. J. Syst. Evol. Microbiol.* **52,** 2339–2341.
39. Dutly, F. and Altwegg, M. (2001) Whipple's disease and "Tropheryma whippelii." *Clin. Microbiol. Rev.* **14,** 561–583.
40. Marth, T. and Raoult, D. (2003) Whipple's disease. *Lancet* **361,** 239–246.
41. Wilson, K. H., Blitchington, R., Frothingham, R., and Wilson, J. A. (1991) Phylogeny of the Whipple's-disease-associated bacterium. *Lancet* **338,** 474–475.
42. Relman, D. A., Schmidt, T. M., MacDermott, R. P., and Falkow, S. (1992) Identification of the uncultured bacillus of Whipple's disease. *N. Engl. J. Med.* **327,** 293–301.

43. Maiwald, M., Schuhmacher, F., Ditton, H. J., and von Herbay, A. (1998) Environmental occurrence of the Whipple's disease bacterium (*Tropheryma whippelii*). *Appl. Environ. Microbiol.* **64,** 760–762.

44. Schoedon, G., Goldenberger, D., Forrer, R., Gunz, A., Dutly, F., Hochli, M., et al. (1997) Deactivation of macrophages with interleukin-4 is the key to the isolation of *Tropheryma whippelii*. *J. Infect. Dis.* **176,** 672–677.

45. Raoult, D., La Scola, B., Lecocq, P., Lepidi, H., and Fournier, P. E. (2001) Culture and immunological detection of *Tropheryma whippelii* from the duodenum of a patient with Whipple disease. *JAMA* **285,** 1039–1043.

46. La Scola, B., Fenollar, F., Fournier, P. E., Altwegg, M., Mallet, M. N., and Raoult, D. (2001) Description of *Tropheryma whipplei* gen. nov., sp. nov., the Whipple's disease bacillus. *Int. J. Syst. Bacteriol.* **51,** 1471–1479.

47. Bentley, S. D., Maiwald, M., Murphy, L. D., Pallen, M. J., Yeats, C. A., Dover, L. G., et al. (2003) Sequencing and analysis of the genome of the Whipple's disease bacterium *Tropheryma whipplei. Lancet* **361,** 637–644.

48. Maeda, K., Markowitz, N., Hawley, R. C., Ristic, M., Cox, D., and McDade, J. E. (1987) Human infection with *Ehrlichia canis,* a leukocytic rickettsia. *N. Engl. J. Med.* **316,** 853–856.

49. Anderson, B. E., Dawson, J. E., Jones, D. C., and Wilson, K. H. (1991) *Ehrlichia chaffeensis,* a new species associated with human ehrlichiosis. *J. Clin. Microbiol.* **29,** 2838–2842.

50. Relman, D. A. (1993) The identification of uncultured microbial pathogens. *J. Infect. Dis.* **168,** 1–8.

51. Dawson, J. E., Anderson, B. E., Fishbein, D. B., Sanchez, J. L., Goldsmith, C. S., Wilson, K. H., et al. (1991) Isolation and characterization of an *Ehrlichia* sp. from a patient diagnosed with human ehrlichiosis. *J. Clin. Microbiol.* **29,** 2741–2745.

52. Stoler, M. H., Bonfiglio, T. A., Steigbigel, R. T., and Pereira, M. (1983) An atypical subcutaneous infection associated with acquired immune deficiency syndrome. *Am. J. Clin. Pathol.* **80,** 714–718.

53. Relman, D. A., Loutit, J. S., Schmidt, T. M., Falkow, S., and Tompkins, L. S. (1990) The agent of bacillary angiomatosis. An approach to the identification of uncultured pathogens. *N. Engl. J. Med.* **323,** 1573–1580.

54. Jacomo, V., Kelly, P. J., and Raoult, D. (2002) Natural history of Bartonella infections (an exception to Koch's postulate). *Clin. Diagn. Lab. Immunol.* **9,** 8–18.

55. Slater, L. N., Welch, D. F., Hensel, D., and Coody, D. W. (1990) A newly recognized fastidious Gram-negative pathogen as a cause of fever and bacteremia. *N. Engl. J. Med.* **323,** 1587–1593.

56. Regnery, R. L., Anderson, B. E., Clarridge, J. E., 3rd, Rodriguez-Barradas, M. C., Jones, D. C., and Carr, J. H. (1992) Characterization of a novel *Rochalimaea* species, *R. henselae* sp. nov., isolated from blood of a febrile, human immunodeficiency virus-positive patient. *J. Clin. Microbiol.* **30,** 265–274.

57. Brenner, D. J., O'Connor, S. P., Winkler, H. H., and Steigerwalt, A. G. (1993) Proposals to unify the genera Bartonella and Rochalimaea, with descriptions of *Bartonella quintana* comb. nov., *Bartonella vinsonii* comb. nov., *Bartonella*

*henselae* comb. nov., and *Bartonella elizabethae* comb. nov., and to remove the family Bartonellaceae from the order Rickettsiales. *Int. J. Syst. Bacteriol.* **43,** 777–786.

58. Berg, R. D. (1996) The indigenous gastrointestinal microflora. *Trends Microbiol.* **4,** 430–435.

59. Moore, W. E. (1987) Microbiology of periodontal disease. *J. Periodontal Res.* **22,** 335–341.

60. Kroes, I., Lepp, P. W., and Relman, D. A. (1999) Bacterial diversity within the human subgingival crevice. *Proc. Natl. Acad. Sci. USA* **96,** 14,547–14,552.

61. Sakamoto, M., Umeda, M., Ishikawa, I., and Benno, Y. (2000) Comparison of the oral bacterial flora in saliva from a healthy subject and two periodontitis patients by sequence analysis of 16S rDNA libraries. *Microbiol. Immunol.* **44,** 643–652.

62. Paster, B. J., Boches, S. K., Galvin, J. L., Ericson, R. E., Lau, C. N., Levanos, V. A., et al. (2001) Bacterial diversity in human subgingival plaque. *J. Bacteriol.* **183,** 3770–3783.

63. Sakamoto, M., Huang, Y., Umeda, M., Ishikawa, I., and Benno, Y. (2002) Detection of novel oral phylotypes associated with periodontitis. *FEMS Microbiol. Lett.* **217,** 65–69.

64. Suau, A., Bonnet, R., Sutren, M., Godon, J. J., Gibson, G. R., Collins, M. D., et al. (1999) Direct analysis of genes encoding 16S rRNA from complex communities reveals many novel molecular species within the human gut. *Appl. Environ. Microbiol.* **65,** 4799–4807.

65. Wheeler, J. G., Sethi, D., Cowden, J. M., Wall, P. G., Rodrigues, L. C., Tompkins, D. S., et al. (1999) Study of infectious intestinal disease in England: rates in the community, presenting to general practice, and reported to national surveillance. The Infectious Intestinal Disease Study Executive. *BMJ* **318,** 1046–1050.

66. Skirrow, M. B. (1994) Diseases due to Campylobacter, Helicobacter and related bacteria. *J. Comp. Pathol.* **111,** 113–149.

67. Butzler, J. P., Dekeyser, P., Detrain, M., and Dehaen, F. (1973) Related vibrio in stools. *J. Pediatr.* **82,** 493–495.

68. Skirrow, M. B. (1977) Campylobacter enteritis: a "new" disease. *Br. Med. J.* **2,** 9–11.

69. Lawson, A. J., Linton, D., and Stanley, J. (1998) 16S rRNA gene sequences of "*Candidatus* Campylobacter hominis," a novel uncultivated species, are found in the gastrointestinal tract of healthy humans. *Microbiology* **144,** 2063–2071.

70. Lawson, A. J., On, S. L., Logan, J. M., and Stanley, J. (2001) *Campylobacter hominis* sp. nov., from the human gastrointestinal tract. *Int. J. Syst. Evol. Microbiol.* **51,** 651–660.

71. Neefs, J. M., Van de Peer, Y., Hendriks, L., and De Wachter, R. (1990) Compilation of small ribosomal subunit RNA sequences. *Nucleic Acids Res.* **18,** 2237–2317.

# 15

## Exploring the Concept of Clonality in Bacteria

### Brian G. Spratt

### Summary

Isolates of bacterial species that are indistinguishable in genotype are assigned as a clone, with the implication that they are descended from the same recent ancestor. Clones are difficult to define with precision since bacteria are not truly asexual, and recombinational replacements result in diversification of the ancestral genotype of a clone, to produce a cluster of increasingly diverse genotypes (a clonal complex). The rate at which clonal diversification occurs depends on the extent of recombination, which varies among bacteria, so that some species have rather stable clones (e.g., *Salmonella enterica*), whereas in other species (e.g., *Helicobacter pylori*) clones may be so transient that they cannot readily be discerned. Clones and clonal complexes need to be assigned by indexing genetic variation that is selectively neutral, and currently this is best achieved using multilocus sequence typing. Some species of bacterial pathogens are very diverse, whereas others are genetically uniform, and some are, in essence, a single clone of a mother species that has been raised to species status due to the distinctiveness of the disease it causes (e.g., *Yersinia pestis, Salmonella typhi,* or *Burkholderia mallei*). The population structures of bacteria depend on the rate of recombination, and comparative measures of the extent of recombination during clonal diversification can be obtained from multilocus sequence typing data, as can measures of the longer-term impact of recombination. These studies show a wide range of recombination rates among bacterial species, and indicate that recombination in many bacteria has been sufficiently extensive that a reliable evolutionary history of the species cannot be inferred.

**Key Words:** Population structure; recombination; multilocus sequence typing; clonal complex.

## 1. Introduction

The realization over 30 yr ago that particular serovars of *Escherichia coli* were strongly associated with disease led to the bacterial clone concept. In

From: *Methods in Molecular Biology, vol. 266: Genomics, Proteomics, and Clinical Bacteriology: Methods and Reviews*
Edited by: N. Woodford and A. Johnson © Humana Press Inc., Totowa, NJ

subsequent decades it has become well established that all isolates of pathogenic species are not equal, and that in a typical pathogenic species there are a small number of genotypes (clones) that are greatly over-represented among those recovered from disease or from a particular type of disease.

The recognition that particular clones are much more strongly associated with disease than others was first established using serological methods, but most of our current knowledge of the genetic structure of bacterial populations, and of the association of particular genotypes with disease, is due to the introduction of improved methods for distinguishing the diversity of genotypes that are present within most bacterial species. In particular, it was the introduction of multilocus enzyme electrophoresis (MLEE) into microbiology by Milkman *(1)* and Selander and Levin *(2)* that provided the key experimental tool that led to a greatly increased understanding of bacterial populations. MLEE indirectly assesses genetic variation at multiple housekeeping loci by examining the electrophoretic mobility of their gene products on starch gels *(3)*. By numbering the different mobility variants at each locus, and by equating these with different alleles at the underlying genetic locus, a strain can be characterized by a string of integers that correspond to the alleles at each of the multiple housekeeping loci. MLEE was introduced into microbiology by eukaryotic population geneticists who had an evolutionary perspective that had been lacking from earlier studies of bacterial populations. They stressed the limitations of serological methods, which take advantage of extensive variation in bacterial components that interact with the host immune system, and the importance of using neutral variation for strain characterization and for understanding the evolutionary relationships among isolates of bacterial species.

MLEE is highly discriminatory, and the identification in the early studies of isolates in *E. coli* populations that were indistinguishable in genotype, although recovered several decades apart, was a surprise, and was taken to imply that the alleles at different genetic loci in the population are not being shuffled by recombination *(2)*. Thus, the sexual processes of conjugation and transduction that had been studied so extensively in the laboratory were considered to be relatively ineffective at promoting genetic exchange and genetic diversification among *E. coli* strains in nature. From these pioneering studies of *E. coli* emerged the view that bacterial pathogens have clonal population structures, which arise as a consequence of low rates of genetic exchange in nature *(4–8)*. This view has changed significantly over the last decade, and there is now a general recognition that the extent of genetic recombination among bacterial species varies, such that there is a spectrum of population structures, from highly clonal through to essentially nonclonal *(9)*.

Recently, MLEE has increasingly been replaced by multilocus sequence typing (MLST *[10]*), and this chapter focuses on the use of the latter technique for precise strain characterization, and on some of the insights that both MLEE and MLST can provide about the nature and evolution of bacterial populations, and the concept of clonality. For more detailed accounts of some of the topics covered in this chapter there are a number of recent review articles *(11–16)*.

## 2. Recombination and Bacterial Population Structures

The sequences of the genomes of bacterial pathogens have provided strong evidence for a major impact of recombination over evolutionary timescales *(17)*. Many genes that are common to all isolates of a species appear to have been acquired from distantly related species, and there is evidence for more recent gene acquisition (and gene loss), because different isolates of the same species differ substantially in gene content *(17–19)*.

Acquisition of novel genes may be of huge significance in the evolution of pathogens, since gene acquisition may, occasionally, produce a strain within a commensal species that has an improved mode of transmission to new hosts— e.g., production of a cough, vomiting, or diarrhea—which we observe as an ability to cause disease *(18,19)*. However, the type of recombination that occurs at a frequency that can impact on population structure, and the rate of diversification of clones, which is the focus of this chapter, involves extensive sequence similarity (and thus occurs typically between strains of the same or very closely related species) and is mediated by homologous recombination.

Recombination in bacteria is, of course, very different from that in typical eukaryotes, where reassortment of the alleles from each parent occurs in every generation. In bacteria, recombination is not linked to reproduction and involves the replacement of a small region of the genome (typically, from a few hundred to a few thousand base pairs) of a recipient bacterium by the corresponding region from another isolate of the same or a closely related species *(20)*. To produce genetic change there must be both a sexual mechanism and opportunities for donor and recipient bacteria to meet. Transformation (in species that are naturally transformable) and phage-mediated transduction are probably the main mechanisms in nature.

The rate of recombination is a key parameter in bacterial population genetics. In the total absence of recombination, all genetic variation within housekeeping genes arises by mutational events, and genetic novelties (e.g., adaptive mutations) occurring in one strain are not able to spread horizontally into other strains. The absence of recombination has another important consequence; fitter variants that arise at intervals will increase in frequency in the population and, in the absence of recombination, may rise to high frequency in the population, leading to a loss of other genotypes (periodic selection; *[21]*). The continual purging of genotypes by repeated rounds of periodic selection leads to a population where there are fewer genotypes, and clones are much more apparent, than in populations where recombination is frequent.

If rates of recombination are very high, those fitter variants that arise will increase in frequency in the population but, while doing so, will diversify rapidly as a result of the accumulation of frequent recombinational replacements. The rapid diversification of any emerging fitter genotypes can result in an absence of observable clones within the population, and a nonclonal population, such as that of *Helicobacter pylori (22)*, in which all isolates from epidemiologically unlinked sources may be different in genotype, as assessed using MLEE or MLST.

Populations in which recombination is either absent or extremely rare, or very frequent, probably represent the minority of situations, and most bacterial species are likely to have intermediate levels of recombination, and population structures that are between these highly clonal and nonclonal extremes.

### 3. Characterizing Bacterial Populations Using Neutral Variation

As emphasized by Selander and colleagues, population structures and the relatedness between isolates of a bacterial species should be assessed by examining genes for which the genetic variation within the population is likely to be selectively neutral, or nearly so *(3,8)*. Genes that encode enzymes involved in essential metabolic processes (housekeeping genes) are therefore used in both MLEE and MLST, rather than, for example, genes that encode cell-surface components where strong selection for variation may be imposed by the host immune response. Thus, while genes involved in pathogenicity and virulence are of crucial importance, they cannot be used for defining population structures. Rather, the evolutionary behavior of these genes has to be mapped onto the population structure and relatedness between isolates, as inferred from the analysis of genetic variation at neutral loci.

MLEE has a number of limitations and has largely been replaced by MLST, which is based on the same principles, but assigns alleles at multiple housekeeping loci directly by nucleotide sequencing of an internal fragment of each locus *(10)*. MLEE identifies only that minority of the nucleotide sequence variation that leads to an alteration in the electrophoretic mobility of the gene product, whereas MLST identifies all genetic variation at the locus. Consequently, high levels of discrimination in MLST can be achieved using far fewer loci than are used in MLEE. Each different sequence at a locus is assigned as a distinct allele and, as in MLEE, the alleles at each locus provide an allelic profile (**Fig. 1**). Each allelic profile is described as a sequence type (ST), which

**A**

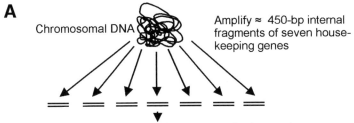

Chromosomal DNA

Amplify ≈ 450-bp internal
fragments of seven house-
keeping genes

Sequence the seven gene fragments on both strands

Compare the sequences of each gene fragment with the known
alleles at the locus, held in a central database on the web

Assign alleles at the seven loci, to give the allelic profile

Compare the allelic profile to those of isolates within a central
database on the web via the internet

**B**

|  | Allelic Profile | | | | | | | ST |
|---|---|---|---|---|---|---|---|---|
| Strain 1 | 1 | 5 | 3 | 7 | 2 | 8 | 4 | 7 |
| Strain 2 | 11 | 15 | 30 | 12 | 22 | 11 | 20 | 3 |
| Strain 3 | 6 | 5 | 9 | 1 | 2 | 13 | 17 | 5 |
| Strain 4 | 11 | 15 | 30 | 12 | 22 | 11 | 20 | 3 |
| Strain 5 | 9 | 10 | 1 | 19 | 12 | 18 | 14 | 2 |
| Strain 6 | 1 | 5 | 3 | 7 | 2 | 8 | 4 | 7 |
| Strain 7 | 20 | 11 | 29 | 9 | 21 | 13 | 11 | 4 |
| Strain 8 | 9 | 15 | 13 | 27 | 22 | 18 | 14 | 1 |
| Strain 9 | 20 | 11 | 29 | 9 | 21 | 13 | 11 | 4 |
| Strain 10 | 11 | 15 | 30 | 12 | 22 | 11 | 20 | 3 |
| Strain 11 | 1 | 5 | 3 | 5 | 2 | 8 | 4 | 6 |
| Strain 12 | 11 | 15 | 30 | 12 | 22 | 11 | 20 | 3 |

**C**

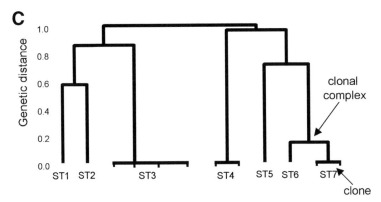

Fig. 1. Characterization of isolates of bacteria using MLST. **(A)** Internal fragments
of seven housekeeping loci are amplified by PCR and are sequenced on both strands.

provides a convenient descriptor for a strain and is equivalent to the electrophoretic type (ET) used in MLEE.

Using multiple housekeeping loci is crucial for a number of reasons. Firstly, neutral variation accumulates slowly, compared with the more poorly characterized variation that is indexed by methods such as pulsed-field gel electrophoresis (PFGE), or fluorescent amplified fragment length polymorphism (fAFLP). The number of alleles at a single housekeeping locus is therefore far too low for typing. The use of multiple loci allows very large numbers of allelic profiles to be distinguished. MLST schemes use seven housekeeping loci and, even if there are only ten alleles per locus, 10 million STs can be distinguished. Of more importance is the low probability that two unrelated isolates will, by chance, have the same or very similar STs; this depends on the number of alleles per locus, and the frequencies of the alleles at each locus *(10,23)*. The expected frequency of any ST can be calculated from the products of the frequencies of each allele and, in a discriminatory MLST scheme, even the ST with the most common allele at each locus should be unlikely to occur by chance within a sample of a few thousand isolates. The current MLST schemes have many alleles per locus (typically more than 30, allowing >20 billion different STs to be distinguished), and unrelated isolates with the same, or very similar, allelic profiles are extremely unlikely to occur by chance.

MLST and MLEE are unusual among typing procedures in that they provide high levels of discrimination, but use neutral genetic variation that accumulates slowly. Most other typing methods obtain high levels of discrimination by searching out variation in the genome that accumulates very rapidly. Those typing systems based on a single locus (typically, and inappropriately, these are based on a highly polymorphic locus subject to diversifying selection from the host immune system) are also unsatisfactory because they are sensitive to recombinational events. Two identical isolates will look completely unrelated if a recombinational event in one isolate changes the allele at the single locus that is used for typing. MLST provides buffering against this effect, as a recom-

---

Fig. 1. *(continued from previous page)* For each locus, the sequence from the new strain is compared with all of the known alleles (sequences) at that locus. Sequences that are not in the database are assigned new allele numbers. Each strain is defined by a string of seven integers, the allelic profile, which corresponds to the allele numbers at the seven loci. Each unique allelic profile is defined as a sequence type (ST), which corresponds to a clone. **(B)** The allelic profiles of 12 strains of a bacterial species are shown, along with their assigned STs. For each strain, the seven integers correspond to the alleles at the seven loci. **(C)** The similarities between the 12 strains are represented as a dendrogram, constructed by the unweighted pair-group method with arithmetic averages (UPGMA), using the matrix of pair-wise differences between the allelic profiles of the strains.

binational replacement at one locus will not obscure the similarity of the two strains, since they will still have identical sequences at six of the seven loci.

## 3.1. The Advantages of MLST for Strain Characterization

The great advantage of MLST is that sequence data are unambiguous, and any laboratory that sequences the seven gene fragments used in an MLST scheme can compare the sequences from their isolates with those in a central MLST database *(10)*. Isolates can therefore be characterized with precision and compared via the Internet. MLST schemes and databases are available for several major bacterial pathogens, including *Neisseria meningitidis (10)*, *Streptococcus pneumoniae (24)*, *Staphylococcus aureus (25)*, *Streptococcus pyogenes (26)*, *Campylobacter jejuni (27)*, *Enterococcus faecium (28)*, and *Haemophilus influenzae (29)*, and software for assigning alleles and allelic profiles online, for comparing the allelic profiles of strains with those in the database, and for displaying the relatedness of isolates, are available through the MLST website (*http://www.mlst.net*). Unlike MLEE and most other typing procedures, MLST can be applied directly to clinical samples and can therefore be used to type bacteria even when viable organisms cannot be recovered *(30,31)*. MLST can also be semiautomated using robotic platforms to provide higher throughput *(32)*.

## 4. What Is a Clone?

Molecular typing procedures are often employed to establish the relatedness of isolates from some epidemiological setting. Those that are indistinguishable are typically assigned as members of the same clone. The concept of a clone is derived from studies of asexual eukaryotes, where the absence of sex results in progeny that are genetically identical. In bacteria the situation is more dynamic, as their genomes are more plastic, and they are subject to genome re-arrangements (deletions, movement of insertion sequences, and so on) and, to a variable extent, to localized recombinational replacements. Thus, bacterial isolates assigned to the same clone may not be identical, as the recent descendents of the same common ancestor may differ somewhat in genotype. This can be seen clearly using PFGE, where, at least in some species, even isolates from an outbreak of disease may differ slightly in their fragment patterns. In microbiology, the strict definition of a clone therefore tends to be loosened slightly, and clones are defined pragmatically as isolates that are indistinguishable, or highly similar, using a particular molecular typing procedure.

This loose definition of a clone is unsatisfactory, but unfortunately several terms that are difficult to define precisely are used in microbiology to describe groups of identical, similar, or related isolates. Thus, besides clones, we have species, genomovars, lineages, clonal complexes, and strains, to name but some of them. Part of the problem in defining these terms is the very different level of genetic diversity within named species of bacteria, and the differences in their population biology *(13)*. Molecular typing procedures that are highly resolving in most bacterial species (e.g., MLEE or MLST) would indicate that almost all isolates of some species, such as *Mycobacterium tuberculosis (33)*, *Bordetella pertussis (34)*, *Yersinia pestis (35)*, *Bacillus anthracis (36)*, and *Burkholderia mallei (37)*, have the same genotype (i.e., are members of a single clone), while in others, such as *H. pylori (22)*, all isolates from epidemiologically unlinked individuals typically would have different genotypes. If we define a clone as a group of isolates that are indistinguishable by the typing procedure that is employed, using MLEE or MLST, the former named species would each be a single clone, or a few closely related clones, whereas the latter species would consist of an immeasurable number of clones.

The utility of the clone concept is that it attempts to recognize groups of isolates that have all descended from a recent common ancestor and that therefore may have broadly similar biological properties (pathogenicity, host adaptation, and so on). The problem is that bacterial clones are not real genetic entities, since the genotype of any fitter variant that arises will continually diversify as it becomes more prevalent in the population, to produce an increasingly variable cloud of descendent genotypes (**Fig. 2A**). These variant genotypes may fall into many clones using a procedure (e.g., fAFLP) that sensitively detects minor variation across the whole genome, whereas they may all be seen as a single clone by MLEE or MLST, which look at neutral variation within a very small proportion of the genome. The challenge is to provide a sensible nomenclature that is appropriate for the biology of the organism.

Species that are essentially monoclonal using MLEE or MLST typically appear, on population genetic grounds, to be single clones within another named species, that were given species status because of the distinctive diseases that they caused *(13)*. If we are to understand the epidemiology of diseases caused by monoclonal species, we need to use methods that discriminate the variant genotypes within these slowly diversifying clones. For example, isolates from typhoid fever have, historically, been given the species name *Salmonella typhi*. However, population genetic analysis of the Salmonellae has shown that there is no justification for defining isolates from typhoid fever as a distinct species, since using MLEE we find that they are members of two closely related clones within the assemblage of clones that defines the species *S. enterica (38)*. Similarly, the nomenclature of some other "species" could be rationalized on population genetic grounds. *Y. pestis* could be re-assigned as the clone of *Yersinia pseudotuberculosis* that causes plague *(35)*, and *B. mallei* as the clone of *B. pseudomallei* that causes glanders *(37)*. Using the terms *clone* or *clonal complex* to define the clusters of very similar genotypes

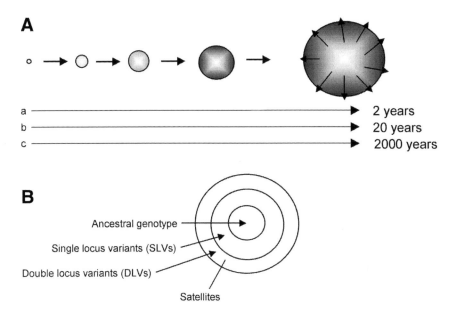

Fig. 2. Diversification of a bacterial clone to produce a clonal complex. (**A**) The rise in frequency of a fitter genotype is represented by the increasing size of the sphere. The ancestral genotype diversifies with time and this process is represented by the increasing density of shading and by the arrows in the expanding clone. The rate of diversification of a clone into a clonal complex varies, and may be so rapid that the clone or clonal complex cannot easily be recognized in a sample of the population (a), or may be more gradual so that diversification becomes apparent within decades (b), or may be very slow with little variation being introduced over many hundreds of years (c). The rate of diversification probably depends mainly on the extent of recombination in the population. (**B**) MLST indexes variation in only about 0.1% of the genome, and identifies those isolates where no variation has yet occurred within any of the sequenced fragments of the seven housekeeping loci (ancestral genotypes), and those where the variation has involved one of these loci (SLVs), or two of these loci (DLVs). These three classes of genotypes are represented by the BURST diagram (*see* examples in **Figs. 3** and **4**); isolates that are assigned to a clonal complex by BURST, but which are not the ancestral genotype, or SLVs or DLVs, are called satellites and are linked to the ST that is most similar in allelic profile.

that cause typhoid fever or plague or glanders is attractive, and has biological plausibility.

For typhoid fever there is enough genetic variation between isolates to discriminate adequately using established typing procedures (e.g., PFGE or fAFLP) that are more sensitive detectors of minor genetic variation than MLEE or MLST. This use of the genetic magnifying glass can go even further, as in the recent study of deliberate anthrax releases in the United States, where genome sequencing of *B. anthracis* isolates was used to identify some of the small number of single-nucleotide polymorphisms (SNPs) within this extremely uniform species *(36)*. Similarly, comparison of the genome sequences of multiple isolates of *M. tuberculosis* have revealed a number of SNPs *(39)* that are being used to produce an unambiguous digital typing system that can resolve the minor variants of this uniform species.

What we should call these very fine subdivisions within a monoclonal species is unclear, but if they are unlikely to have much biological meaning, it does not seem sensible to define them as clones. Thus, *clone* would be an inappropriate term for the very fine (and presumably biologically unimportant) subdivisions within *B. anthracis* made for forensic purposes following a deliberate release, but it would be appropriate for defining the subdivisions within *M. tuberculosis* if they corresponded to epidemiologically important entities that varied in their virulence, transmissibility, or drug resistance profile.

### 4.1. Identification of Clones and Clonal Complexes Using MLST

Using MLST, we define isolates that are identical in allelic profile as a clone, and clusters of isolates with closely related allelic profiles as clonal complexes. The identification of clones and clonal complexes within samples of a bacterial population is typically achieved by using the genetic variation indexed by a molecular typing procedure to construct a dendrogram. This procedure is simple and useful but has a number of problems. Firstly, dendrograms are cumbersome when very large numbers of isolates are analyzed, as occurs in MLST since results from many different laboratories may be submitted to a single central database. They also do not accurately represent the processes by which bacterial clones diversify to produce clonal complexes, and they fail to extract information about clonal ancestry and patterns of evolutionary descent that may be present in MLST data.

An algorithm, BURST, has been developed by Ed Feil to identify the isolates within a MLST dataset that are likely to be descended from the same recent ancestor (http://eburst.mlst.net). This program starts by identifying groups of isolates within a MLST dataset that have some predefined level of similarity (the default value is that all isolates are related to at least one other member of the group at five of the seven loci), and then predicts the ancestral genotype of each of these groups (clonal complexes), and the most likely patterns of evolutionary descent of all isolates within each clonal complex from the predicted ancestral genotype. The program therefore gets around the problem of trying to display the relatedness of all of the isolates, and focuses on those that are similar in genotype, without being concerned about the relationships between these groups of similar isolates, which in many species will be impossible to discern, owing to the impact of recombination (*see* **Subheading 5.**).

BURST predicts ancestral genotypes by assuming a simple but realistic model of bacterial evolution in which a genotype rises in frequency within the population (typically because it has some fitness advantage, such as increased transmission to new hosts, or antibiotic resistance) and, while doing so, accumulates point mutations and recombinational replacements (**Fig. 2**). The ancestral genotype therefore diversifies, firstly to produce variants that differ by MLST at just one of the seven loci (single-locus variants [SLVs]) and subsequently those that differ at two loci (double-locus variants [DLVs]). As ancestral genotypes initially diversify to produce a set of variants that differ at only one of the loci, the ancestral genotype is assumed to be the ST within the clonal complex defined by BURST that has the largest number of SLVs. The ST of the predicted ancestral genotype, and the SLVs and DLVs, and those isolates that are within the complex, but are neither SLVs nor DLVs (satellites), are represented as a BURST diagram (**Fig. 2B**). The representation of the genotypes of a group of similar isolates of pneumococci and meningococci are displayed in **Figs. 3** and **4** using both a dendrogram and a BURST diagram.

Some caution is required in using MLST for purposes other than for identifying clones and clonal complexes and predicting ancestral genotypes of clonal complexes. MLST identifies isolates that are identical at all seven loci and those that have accumulated variation at one or more of these loci, but this represents only approx 0.1% of a bacterial genome. For example, if we consider 1000 meningococcal isolates with an absolutely identical genotype and, to represent the genetic variation that has accumulated in contemporary descendents of this genotype, we were to pepper the genome with 200 small random recombinational replacements, some isolates would have no changes at any of the seven loci used in MLST (and would appear to have retained the ancestral genotype), whereas others would have alterations at one of the loci and would be recognized as SLVs. However, if we chose a different set of loci for the MLST scheme, the same distribution of replacements around the genome would result in some of the SLVs in the first MLST scheme appearing to be ancestral genotypes using the second MLST scheme, and some that were ancestral genotypes will be SLVs. In a real population, contemporary isolates with the ancestral genotype, and those that are SLVs, are derived from the same ancestor and, on average, will have received similar numbers of mutations or replacements, few of which will have altered a locus used in the MLST scheme. Isolates with the ancestral genotypes are, therefore, not necessarily a more uniform, natural, or fit group of isolates than the apparently diverse collection of SLVs.

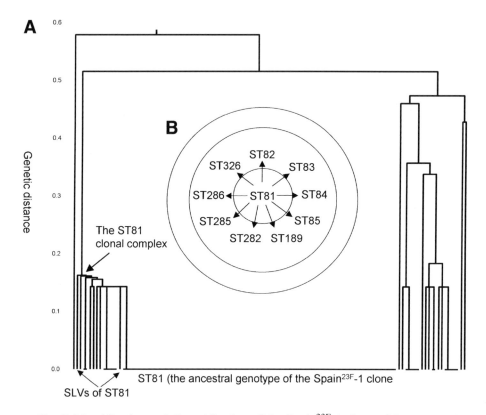

Fig. 3. Identification and diversification of the Spain[23F]-1 clone of *S. pneumoniae*. All isolates with allelic profiles that are similar (identity at $\geq$ 3 loci by MLST) to that of the reference isolate of the Spain[23F]-1 clone were extracted from the pneumococcal MLST database (http://spneumoniae.mlst.net). The relatedness of the resulting 119 strains was displayed as a dendrogram, constructed from the matrix of pairwise differences in their allelic profiles, by the UPGMA method (**A**). Isolates of the Spain[23F]-1 (ST81) clonal complex are easy to recognize, since the cluster of closely related genotypes is well resolved from all other isolates in the MLST database, presumably because resistance did not emerge in a genotype that was prevalent within the pneumococcal population. BURST also identifies the ST81 clonal complex and predicts that ST81 is the ancestral genotype (**B**). ST81 is therefore placed in the central circle of the BURST diagram. The other nine STs in the clonal complex are all shown to differ from the ancestral genotype at only one of the seven MLST loci (i.e., they are all SLVs of ST81), and are displayed in the middle circle. The arrows indicate that they have arisen from the diversification of the ancestral genotype. No isolates differ from ST81 at two loci, presumably because there has not yet been sufficient time for DLVs to accumulate. The relationships of the isolates within the clonal complex are far better displayed using BURST than a dendrogram, as the latter gives no indication that all the variant STs in the complex are the diversification products of the ancestral genotype, ST81.

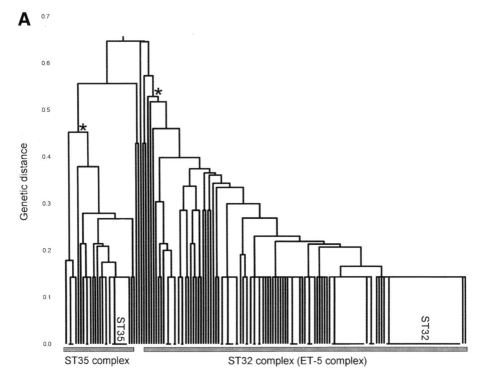

Fig. 4. Identification and diversification of the ST32 (ET-5) clonal complex of *N. meningitidis*. All isolates with allelic profiles that were similar to that of *N. meningitidis* strain 44/76 (identity at ≥ 3 loci by MLST), the reference isolate of ET-5, were extracted from the meningococcal MLST database (http://neisseria.org/nm/typing/mlst/). The relatedness of the resulting 163 isolates was assessed using a dendrogram **(A)**. The dendrogram is complex, and provides no objective way of distinguishing those isolates that are predicted to be the diversified descendents of the ancestral genotype of the ET-5 complex, from those that are similar in genotype, but are not derived from this ancestral genotype. It also provides an inadequate representation of the way in which an ancestral genotype diversifies to form a clonal complex.

BURST defines two clonal complexes among the 163 isolates, and there are only four isolates that do not belong to one of these clonal complexes. These isolates differ from each other, and from all isolates in the two clonal complexes, at ≥ 3 of the 7 MLST loci. The ancestral genotypes of the two complexes were predicted by BURST to be ST32 and ST35. The node that includes all isolates assigned to the ST32 and ST35 clonal complex by BURST is shown by an asterisk on the dendrogram. The two clonal complexes defined by BURST are separated on the dendrogram by the isolates that are not assigned by BURST to either clonal complex. The inferred relationships between the isolates within the ST32 (ET-5) clonal complex are shown as a BURST diagram **(B)**. ST32 is strongly predicted to be the ancestral genotype, and is surrounded by a large number of SLVs (middle circle; there are so many STs that the ST numbers cannot be read), and by three DLVs (outer circle). This is a highly diverse clonal complex and the BURST diagram is complex compared to that in **Fig. 3.** BURST has separated some SLVs into distinct subgroups, as these STs are themselves associated with a number of SLVs, and in the case of the ST33

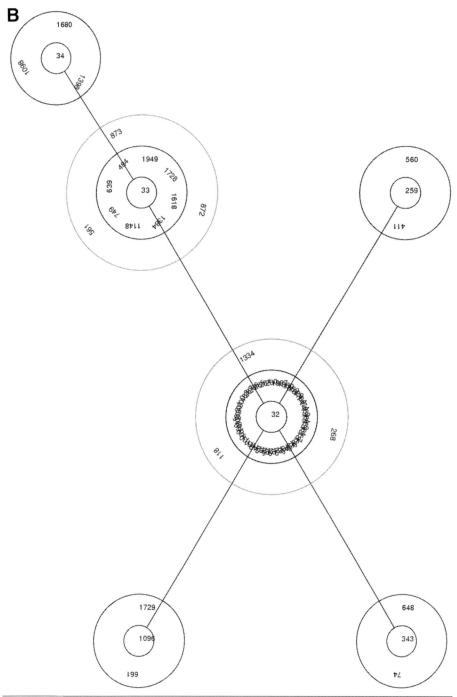

Fig. 4. *(continued from previous page)* subgroup, by three DLVs. BURST assigns isolates objectively into the ST32 (ET-5) clonal complex and shows the diversification of its predicted ancestral genotype far more clearly than a dendrogram. A greatly improved version of BURST (eBURST) is available at http://eburst.mlst.net.

BURST performs very well with species where rates of recombination are moderate. For example, it has proved excellent at resolving the clonal complexes and likely patterns of evolutionary descent within methicillin-resistant and methicillin-susceptible strains of *S. aureus (40)* and, more challengingly, is capable of objectively assigning isolates of meningococci to a clonal complex (**Fig. 4**). However, in the latter species, the rapid diversification of clones results in very complex dendrograms, and a complex BURST diagram. BURST would probably not be effective at analyzing the even more diffuse clonal complexes within populations of pathogens that have higher rates of recombination than the meningococcus. However, other approaches are also likely to be unsuccessful.

## 5. Recombination, Clonal Diversification, and Molecular Epidemiology

The relative impact of recombination and point mutation is a major determinant of the stability of bacterial clones. If, as a first approximation, we assume the mutation rate is constant, the stability of clones will be determined by the extent of recombination and the recombinational replacement size. There appears to be a large variation in the stability of clones in different pathogens. In some species, clones are so unstable that they cannot be observed, whereas in others species, clones appear to have persisted without much change over long periods of time. In reality, the mutation rate may not be constant across species, or even within a species (some pathogens may lack some DNA repair pathways, or some isolates may have mutator phenotypes), and diversification may be more easily observed in some species than others, depending on the relative fitness of the variant genotypes that arise from clonal diversification. However, very stable clones suggest low rates of recombination between genetically distinct strains, and very unstable clones imply frequent recombination between strains.

*Y. pestis* is a good example of clonal stability, and a sample of 36 isolates from diverse sources, which included isolates assigned as remnants of the last three major historical pandemics (antiqua, medievalis, and orientalis), showed no sequence variation at six loci *(35)*. There is evidence that *Y. pestis* was the cause of the Black Death *(41)*, and, from the descriptions of the disease, it is very likely that it caused some of the earlier historical plagues. The pestis clone is therefore certainly over 650 yr old and is probably substantially older. Similarly, tuberculosis is an ancient disease and yet there is almost no genetic variation in housekeeping genes of *M. tuberculosis (33)*. In contrast, many clones of multiply antibiotic-resistant isolates of *S. pneumoniae* are probably only about 20 yr old, and yet variants of these clones that differ at one of the seven loci used in MLST are already commonly encountered *(42)*.

Clones in some species appear to have remained uniform by MLST for over a thousand years, whereas those in other species diversify so rapidly that variants of clones that must be less than 50 yr old can be detected, and in a few species clones may be so unstable that they cannot even be discerned—all isolates from epidemiologically unlinked sources are different in genotype (**Fig. 2A**). It is likely that a large contributor to this variation in clonal stability is the rate of recombination between distinct strains of the species, although other factors are undoubtedly important.

Some care is required in assuming that clones are highly stable when there is almost no variation at neutral loci among all contemporary isolates of a species that causes an undoubtedly old disease. Worldwide pandemics of some diseases can eliminate the existing variation that has built up since the previous pandemic. For example, influenza is likely to have been around for many hundreds of years, but finding that all influenza viruses isolated during a worldwide flu pandemic are closely related in genotype does not imply that the virus has remained unchanged over hundreds of years. Influenza has unusual features and, for bacterial pathogens, the availability of isolates from diverse geographic sources, and of old isolates stored in culture collections, or of sequences obtained from archaeological material, can provide supporting evidence for assigning clonal stability over substantial periods of time.

The rate of diversification of clones has implications for the way molecular epidemiology is carried out *(15)*. It is convenient to distinguish two types of epidemiological investigation. Short-term epidemiology is concerned with recent events, such as whether a cluster of cases of disease in a community or hospital are unrelated and caused by different strains, or are an outbreak caused by the transmission of a single strain. This type of epidemiology is usually straightforward as, for an outbreak, all of the isolates are derived from a very recent common ancestor (the isolate from the index case), and no variation is likely to have accumulated among the isolates within a period of a few weeks. Provided the typing method is sufficiently discriminatory, so that isolates from epidemiologically unlinked individuals within the community are usually distinguishable, the identity of the genotypes from the cases of disease is sufficient to establish the epidemiological link. In many cases, MLST or MLEE may be unsuitable for short-term epidemiology, as a significant percentage of isolates circulating in the community may have the same allelic profile as those from the outbreak, and an additional more discriminatory method is needed to establish it is an outbreak rather than independent acquisitions of a prevalent clone.

Epidemiological investigations may also be concerned with longer-term events, such as the worldwide spread of important clones (global epidemiology), where the relatedness of isolates recovered from different countries in different decades may need to be assessed. In this situation, the common ancestor of all of these isolates may have arisen several decades ago, and considerable genetic variation may have accumulated in this time, such that it is not always easy to establish whether or not contemporary isolates from different countries are closely related and share the same common ancestor. This problem becomes greater as the rate of diversification of clones increases, and at sufficiently high recombination rates it will be impossible to establish that isolates descended from the same common ancestor that existed several decades ago are related.

For global epidemiology of pathogens where rates of recombination are high and clones diversify relatively rapidly, it is essential to use genetic variation that accumulates slowly. MLST is ideal for this purpose as it provides the required combination of an ability to discriminate between billions of genotypes and the use of genetic variation that accumulates slowly *(10)*. Methods that are more sensitive at detecting minor variation are much less appropriate for global epidemiology, as the accumulation of far too much variation may obscure the underlying relatedness of strains derived from a recent common ancestor.

## 6. Can I Use MLST for Everything?

There is a large spectrum of genetic variation at neutral loci among different species of pathogens, and this can impact upon the ability of MLST to produce a highly discriminatory typing scheme. For example, *M. tuberculosis* is so genetically uniform *(33)* that a typing scheme would resolve the whole species into far too few STs to make the scheme useful for epidemiological purposes, and other typing methods need to be employed *(39)*. Similar uniformity at housekeeping loci occurs in *Y. pestis (35)*, *B. anthracis (36)*, *B. pertussis (43)*, and, to a slightly lesser extent, in *Neisseria gonorrhoeae (44)*.

Even where there is adequate variation at housekeeping loci, MLST may produce a scheme that is of limited utility. For example, in a highly clonal species, where clones are stable over hundreds of years, MLST may readily identify these clones, but may provide little advantage over other established methods that achieve this aim. For example, MLST would undoubtedly provide an excellent procedure for identifying the clone within *S. enterica* that causes the great majority of typhoid fever *(38)*, but as clonal diversification at neutral loci appears to be very slow in this species, it would not provide the further resolution within the typhi clone that is needed to investigate the spread of typhoid fever.

There are also limitations at the other end of the spectrum of population structures. Very frequent recombination results in an absence of clones. Thus, if a fitter strain arises within the population, and begins to increase in frequency in the population, frequent recombinational replacements will lead to a rapid diversification of its genotype, to produce an increasingly diverse array of genotypes that soon cannot be recognized as having been derived from the same recent ancestor. These populations therefore consist of a vast number of unique genotypes; all isolates from epidemiologically unlinked individuals will be different by MLST, and only very recent transmission events may be discerned. The absence of clones, or of clusters of related clones, therefore makes it impossible to address longer-term epidemiological questions, and we cannot expect to be able to observe the spread of particular clones within countries or worldwide *(15,45)*. *H. pylori* provides an example of such a nonclonal population; there is ample genetic variation for a highly discriminatory MLST scheme, but extremely frequent recombination precludes long-term epidemiology, and only very recent epidemiological events can be explored *(22,46)*. Thus, recent transmission of a strain within a family may be detected by the presence in family members of very similar genotypes, but the relationships between these strains and those in the wider community cannot be established. Of course, this is an inherent problem with nonclonal species, rather than with MLST as a typing procedure, and other methods would also fail to provide useful epidemiological information over longer timescales.

However, the majority of bacterial pathogens have levels of genetic diversity at housekeeping loci that are adequate for a discriminatory MLST scheme, and have rates of recombination that allow the identification of clones and of clusters of related clones (clonal complexes). For example, the MLST procedure was developed using a challenging species for global epidemiology, *N. meningitidis,* as this species (or at least serogroup B and C isolates) was known from previous studies to have a high rate of recombination, leading to clones which diversify very rapidly *(10)*. The fact that MLST works extremely well for meningococci, and that clonal complexes can be discerned using BURST, suggests that it will be useful for most bacterial species where there is adequate genetic diversity.

MLST may also have applications for typing eukaryotic microbes. These organisms can raise additional problems, as some may be diploid, or they may be difficult to culture routinely. An MLST scheme using six loci has been published for the diploid species *Candida albicans (47)*. In this scheme, heterozygous nucleotide sites can be identified as there are two peaks of approximately equal amplitude on the sequencer trace, and these can be assigned the standard ambiguity codes. A more difficult problem arises if the organism cannot be cultured. For example, in malaria, an MLST scheme might be possible, providing there was sufficient diversity at neutral loci, but it would probably have to be applied directly to blood samples, as purification to single clones is not realistic for large numbers of samples. Allelic profiles could then be assigned only where there was a single or a predominant parasite strain present in the blood. This would preclude the use of the scheme for the most interesting epidemiological situations, where there are high rates of disease transmission and blood samples often contain multiple strains of the parasite.

Multilocus typing schemes have recently been reported which mix antigen-encoding and housekeeping loci, or include rRNA loci *(43,48)*. The mixing of genes that are subject to very different evolutionary constraints is unwise, as it complicates the ways in which the data can be analyzed, and is likely to provide less reliable indications of the true relatedness between isolates than schemes based on seven housekeeping loci. These schemes should not be referred to as MLST.

## 7. Analyzing Population Structures

Bacterial population structures are often described as clonal or nonclonal, to indicate that clones can, or cannot, be observed within the population. In reality there is a continuous gradation in population structures, from those in which recombination is very rare and clones are stable; through populations where recombination is a major contributor to clonal diversification, but unstable clones can be observed; to populations where recombination is so frequent that clones are not observed *(9)*. How do we decide whether bacterial populations are highly clonal, weakly clonal, or nonclonal, or perhaps more meaningfully, how do we measure the extent and impact of recombination in different bacterial species?

### 7.1. Linkage Disequilibrium

The extent of recombination within bacterial populations has usually been addressed by examining the level of linkage disequilibrium between the alleles at different loci within the population *(3,4,9,14)*. If rates of recombination are very high, the alleles at different loci are randomly associated (linkage equilibrium), and the population consists of a vast array of unique genotypes. Conversely, if recombination is very rare, those fitter genotypes that arise periodically within the population will increase in frequency and lead to the loss of other genotypes from the population (periodic selection). These fitter genotypes may rise to high frequency, and will diversify only slowly in the absence of recombination, resulting in a number of clusters of isolates with identical or very closely related genotypes *(4)*. The alleles at different loci in the population are therefore not randomly associated; i.e., there is significant linkage disequilibrium between alleles.

Bacteria divide by binary fission, and a high frequency of recombinational replacements is required to break down the inevitable associations between alleles at different loci that occur as a consequence of this mode of reproduction. A statistically significant departure from linkage equilibrium occurs in most populations, since complete randomization of alleles is rarely achieved. Recombination can therefore be considered to be very frequent if linkage equilibrium is observed, but the presence of significant disequilibrium establishes only that rates of recombination are not very high *(14)*.

Significant departures from linkage equilibrium can occur because recombination cannot completely break down linkage associations, but it may also occur as a consequence of poor sampling *(9)*. Artifacts introduced by sampling are likely to be particularly evident in those species where carriage is very common but rarely proceeds to disease (e.g., meningococci, pneumococci, *S. aureus*). If in these species only 3 in 100,000 children get invasive disease, although 1 in 10 children carry the organism in the nasopharynx or nose, the vast majority of isolates in a truly representative sample of the population would be from carriers. In practice, most isolates of these species that have been studied are from disease, and are therefore a highly biased sample of the real population (**Fig. 5**).

Sampling problems may also occur in situations where some isolates of a species gain particular medical importance and thus are greatly oversampled. For example, methicillin-resistant *S. aureus* and multiply antibiotic-resistant pneumococci have been studied extensively, with much less attention being paid to antibiotic-susceptible isolates of these species, which in almost all countries still cause the majority of disease; and even less attention has been paid to the carried population.

These types of sampling bias will tend to make populations appear to be more highly clonal than they really are, and will inflate the level of linkage disequilibrium, as often a restricted set of genotypes is associated with most cases of invasive disease, and with resistance to antibiotics *(9,14)*.

### 7.2. The Presence of Clones

One of the original criteria of a clonal population was the presence of indistinguishable isolates in temporally and geographically separated samples of the population *(8)*. The presence of clones is still often used to justify the description of a population or species as clonal. However, clones may be observed in most populations, including those in which rates of recombination are high, and the term *clonal* lumps many rather different population structures together. For example, rates of recombination are considered to be high in serogroup B and C meningococci (the impact of recombination appears to be much smaller in serogroup A isolates), but clones are very obvious *(6,10)*. In some studies this may be due to the analysis of invasive isolates, which are highly enriched for the few genotypes that have a substantially increased ability to cause meningococcal disease, but clones are also observed among the

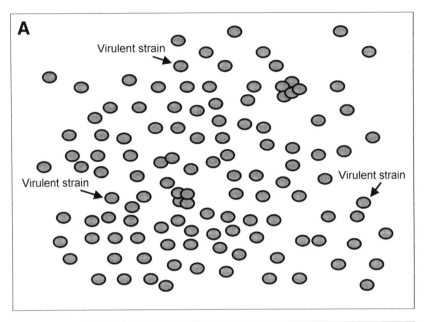

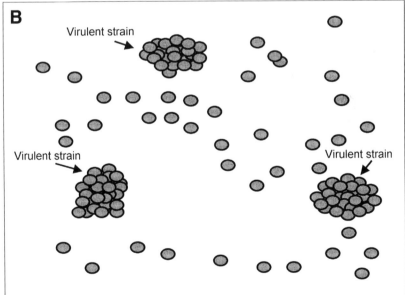

Fig. 5. Representative sampling of a bacterial population. (**A**) shows the isolates present in a representative sample of the population of a species in which disease is rare compared to carriage. Since disease is very rare, these isolates are all from carriage. The similarities in the genotypes of the isolates are reflected by their distance apart; the great majority of isolates are spatially separated and therefore have unique

diverse collection of genotypes obtained from asymptomatic nasopharyngeal carriage *(49,50)*. In a recombining species of this type, recombination over evolutionary timescales may randomize the alleles at different loci in the population, but, within human timescales, diversification may still be slow enough for some genotypes to increase in frequency in the carried population as a consequence of a transmission advantage (increased fitness) or stochastics (drift). Multiple isolates with the same allelic profile by MLST may therefore be observed within a small representative sample of the total meningococcal population.

These transient clones will diversify over human timescales to produce increasingly variable clonal complexes, and the presence of the decaying remnants from the emergence and rapid breakdown of fitter genotypes will introduce significant linkage disequilibrium into all but the most highly recombinogenic populations. A population in which recombination is the dominant mode of evolution may therefore still show the presence of clones, and may exhibit a significant departure from linkage equilibrium. Only at very high rates will these two features be absent from a bacterial population.

### 7.3. Rates of Clonal Diversification

Measures of linkage disequilibrium are essentially qualitative—there is linkage equilibrium or there is not; and significant linkage disequilibrium will often be present in populations where recombination is a major mode of clonal diversification. Similarly, the presence of clones is an insensitive indicator of the extent of recombination within a population. A more quantitative measure would be to know the average rate of diversification of bacterial clones in different species. Unfortunately, obtaining a "clonal half-life" from analyzing a contemporary sample of the population is problematic, and historical samples of most pathogens provide little help as they go back only a few decades, which is far too short for measuring this type of parameter. However, we do have a few estimates of the stability of clones as, in some cases, we can relate clonal

---

Fig. 5. *(continued from previous page)* genotypes. Two small clusters of isolates are present, which represent clones that are overrepresented in the carried population, as they have a slight fitness advantage. The great majority of genotypes in this diverse carried population are unlikely to cause disease. However, three genotypes, which are relatively rare in the carried population, have a greatly increased ability to cause disease (virulent strains). **(B)** shows a sample of the isolates of this species obtained exclusively from cases of disease. Most disease is caused by the three virulent strains circulating in this community, and these genotypes are therefore greatly overrepresented in the sample of isolates taken from individuals with disease. Inappropriate sampling of the real population can make a diverse weakly clonal population appear to be much more highly clonal than it really is, and can introduce highly significant linkage disequilibrium.

diversity to some measure of clonal age. For example, Falush et al. *(51)* analyzed isolates of *H. pylori* recovered at intervals from the stomachs of the same individuals. This species is one of the best examples of a nonclonal species, and rates of recombination, and thus rates of clonal diversification, are expected to be high. Examination of the sequences of ten loci in sequential samples showed that diversification occurs very quickly, predominantly by small (approx 400 bp) recombinational replacements (presumably from coinfecting *H. pylori* isolates).

The minimum time since a pair of isolates from the same stomach shared a common ancestor can be taken as the interval between their isolation dates, and the maximum time (unless both variants were already present in the source of the *H. pylori* infection) is based on the more conservative assumption that colonization occurred when the patients were children *(51)*. Using the minimum estimate of the time between isolates, the authors calculate that approx 60 recombinational replacements occur in the genome per year, and that half of the *H. pylori* genome may be replaced within about 40 yr. On the maximum divergence time, half of the genome would be replaced in about 2200 yr.

In a few other cases we can estimate, with various degrees of certainty, the age of clones and relate this to the diversity observed within the clone. The most robust measure of age is obtained for multiply antibiotic-resistant clones, which cannot predate the introduction of antibiotics. The multiresistant clones of *S. pneumoniae* that have been identified since the late 1970s provide a good example *(42)*. **Figure 3** shows the relatedness of the allelic profiles of the 97 isolates in the pneumococcal MLST database assigned to the Spain[23F]-1 clonal complex. Although 83 of these isolates have an identical allelic profile (the ancestral genotype, ST81), there are nine different variant allelic profiles (which arise predominantly by recombination) that differ at one of the seven loci used in MLST (SLVs), but none that differ at two loci (DLVs). Thus, within about 25 yr, variation is readily observed at one of the seven neutral loci indexed by MLST.

In other cases we have a date before which isolates of a particular clone have not been identified, as in the case of the ST32 clonal complex (ET-5 complex) of *N. meningitidis,* the first isolate of which was recovered in the 1970s, although of course it may have arisen earlier but remained undetected *(6)*. In the meningococcal MLST database there are 31 isolates with the ancestral genotype of this clonal complex (ST32), and a large number of SLVs and DLVs (**Fig. 4**). Diversification of this clone into a highly variable clonal complex appears to have occurred very rapidly, which agrees with other data indicating a high rate of recombination in this species (*see* **Subheading 8.**). In other species, clones are far more stable and appear to remain virtually unchanged in allelic profile over centuries.

## 8. Measuring Recombination Rates in Bacterial Species

Another approach to understanding the impact of recombination is to measure recombination rates in different species. This is not a trivial task, but MLST data provide a way of measuring the relative contribution of recombination, compared with point mutation, in clonal diversification *(52,53)*. Clonal complexes within the population are first identified using BURST, and the ancestral allelic profiles of each are predicted. The sequence changes that have occurred as the ancestral allelic profiles diversify to produce SLVs can then be examined, and can be assigned as the result of a point mutation or a recombinational replacement. Summing the number of mutational and recombinational events during these initial stages of clonal diversification provides a recombination/mutation ratio for the species. These ratios can be expressed per allele (how often an allele defined by MLST changes to a variant allele by recombination or mutation), or per individual nucleotide site. The per allele ratios vary from about 10:1 in favor of recombination in meningococci *(52)* and pneumococci *(53)* to about 15:1 in favor of point mutation in *S. aureus* (E. J. Feil and B. G. Spratt, unpublished data). Presumably these values would be substantially lower than the *S. aureus* value in those species where clones appear to be stable over many centuries.

## 9. The Impact of Recombination

The relative contribution of recombination to clonal diversification varies greatly among bacterial species. What is the impact of recombination on the long-term evolution of different bacterial populations and our ability to extract information about the phylogenetic history of a species? If recombination is absent, the genetic variation at neutral loci can be used to reconstruct the phylogeny of the species, and variation at other loci (e.g., the presence or absence of virulence genes) can be mapped onto the resulting phylogenetic tree, to understand the origins of pathogenic lineages, host adaptation, and so on *(54)*. However, frequent recombination will preclude any attempt to construct a meaningful phylogeny of the species, and only relatively recent evolutionary events can be examined. In order to understand how much phylogenetic information can be extracted from MLST data, some idea of the impact of recombination within the species is required.

Statistical tests of congruence between the trees obtained for each of the seven loci used in MLST have been used to assess the impact of recombination *(55)*. If recombination is rare, the tree obtained using sequence data from one locus should be similar (congruent) to that derived from each of the other loci. However, if recombination is very frequent, the trees obtained from each locus will be different (noncongruent), and none will reflect the true phylogenetic relationships between the isolates. For this type of analysis, isolates that are very different in allelic profile should be selected to minimize any congruence that would be introduced by including multiple isolates of the same clonal complex. The approach determines whether the maximum likelihood tree obtained for each of the MLST loci is a better fit to the sequence data from the other six loci than it is to 200 trees of random topologies. In some species (e.g., *N. meningitidis* and *S. pneumoniae*) the similarities between the trees are, in all cases, no better than random, indicating a total lack of congruence between loci; the elimination of all phylogenetic information, such that patterns of evolutionary descent cannot be discerned, is a long-term consequence of frequent recombination *(55)*. In other cases there are a few tree comparisons which show a similarity that is significantly greater than random (*H. influenzae*), and in other species most tree comparisons show significant congruence (*E. coli*).

The amount of phylogenetic information present within a dendrogram therefore differs among bacterial species; in some cases only the clusters of isolates at the tips of a dendrogram may have any validity, whereas in others even the relatedness between the deepest branches of the tree may approximate the evolutionary history of the species. For example, in the meningococcus, the relatedness of very similar STs is likely to be meaningful, since these are the result of the recent diversification of a clone into a clonal complex, but the phylogenetic relationships between the different clonal complexes that are implied by a dendrogram are likely to be meaningless *(55,56)*.

## 10. Concluding Remarks

Bacterial species vary considerably in the amount of genetic variation they contain and in the extent and impact of recombination. In many cases highly uniform bacterial species are misnomers, where a distinctive stable clone within a more appropriately constituted species has been raised to species status because the disease it causes is so distinctive. Recombination in some species is the major contributor to the diversification of clones and, in a few species, recombination may be so frequent, compared with point mutation, that it prevents the emergence of observable clones. In other species, recombination may play a minor role, either because colonization with multiple genotypes rarely occurs, or because sexual exchanges are rare. Recombination appears rarely to be so frequent that it prevents the appearance of clones; in this sense, most bacterial species can be called clonal. However, rather than classifying species as clonal or nonclonal, it is more appropriate to consider the relative impact of recombination in each species, or in different subpopulations of the species. Low rates of recombination produce more stable clones, and allow intra-species phylogeny to be reconstructed which, in an era of genome sequencing and microarrays, holds out the hope of understanding the changes in the genome along the different branches of an evolutionary tree—changes that have influenced important biological properties, such as virulence or host- or niche-adaptation. High rates of recombination result in unstable clones, a fact that needs to be considered when molecular epidemiological studies are undertaken. Unstable clones are likely to invalidate any attempts to produce an intra-species phylogeny, and may make it far more difficult to understand the pathways and processes by which virulent strains emerge within populations.

## Acknowledgments

I thank Dr. Bill Hanage for comments on the manuscript, and the Wellcome Trust for a Principal Research Fellowship.

## References

1. Milkman, R. (1973) Electrophoretic variation in *Escherichia coli* from natural sources. *Science* **182,** 1024–1026.
2. Selander, R. K. and Levin, B. R. (1980) Genetic diversity and structure in *Escherichia coli* populations. *Science* **210,** 545–547.
3. Selander, R. K., Caugant, D. A., Ochman, H., Musser, J. M., Gilmour, M. N., and Whittam, T. S. (1986) Methods of multilocus enzyme electrophoresis for bacterial population genetics and systematics. *Appl. Environ. Microbiol.* **51,** 873–884.
4. Whittam, T. S., Ochman, H., and Selander, R. K. (1983) Multilocus genetic structure in natural populations of *Escherichia coli. Proc. Natl. Acad. Sci. USA* **80,** 1751–1755.
5. Selander, R. K., McKinney, R. M., Whittam, T. S., Bibb, W. F., Brenner, D. J., Nolte, F. S., et al. (1985) Genetic structure of populations of *Legionella pneumophila. J. Bacteriol.* **163,** 1021–1037.
6. Caugant, D. A., Froholm, L. O., Bovre, K., Holten, E., Frasch, C. E., Mocca, L. I., et al. (1986) Intercontinental spread of a genetically distinctive complex of clones of *Neisseria* meningitidis causing epidemic disease. *Proc. Natl. Acad. Sci. USA* **83,** 4927–4931.
7. Musser, J. M., Kroll, J. S., Moxon, E. R., and Selander, R. K. (1988) Evolutionary genetics of the encapsulated strains of *Haemophilus influenzae. Proc. Natl. Acad. Sci. USA* **85,** 7758–7762.
8. Selander, R. K., Musser, J. M., Caugant, D. A., Gilmour, M. N., and Whittam, T. S. (1987) Population genetics of pathogenic bacteria. *Microb. Pathogenesis* **3,** 1–7.
9. Maynard Smith, J., Smith, N.H., O'Rourke, M., and Spratt, B.G. (1993) How clonal are bacteria? *Proc. Natl. Acad. Sci. USA* **90,** 4384–4388.
10. Maiden, M. C. J., Bygraves, J. A., Feil, E., Morelli, G., Russell, J. E., Urwin, R., et al. (1998) Multilocus sequence typing: a portable approach to the identification of clones within populations of pathogenic microorganisms. *Proc. Natl. Acad. Sci. USA* **95,** 3140–3145.
11. Levin, B. R., Lipsitch, M., and Bonhoeffer, S. (1999) Population biology, evolution, and infectious disease: convergence and synthesis. *Science* **283,** 806–809.

12. Levin, B. R. and Bergstrom, C. T. (2000) Bacteria are different: observations, interpretations, speculations, and opinions about the mechanisms of adaptive evolution in prokaryotes. *Proc. Natl. Acad. Sci. USA* **97,** 6981–6985.

13. Lan, R. and Reeves, P. R. (2001) When does a clone deserve a name? A perspective on bacterial species based on population genetics. *Trends Microbiol.* **9,** 419–424.

14. Feil, E. J. and Spratt, B. G. (2001) Recombination and the population biology of bacterial pathogens. *Ann. Rev. Microbiol.* **55,** 561–590.

15. Spratt, B. G., Feil, E. J., and Smith, N. H. (2001) The population biology of bacterial pathogens, in *Molecular Medical Microbiology* (Sussman, M., ed.), Academic Press, NY, pp. 445–484.

16. Gupta, S. and Maiden, M. C. J. (2001) Exploring the evolution of diversity in pathogen populations. *Trends Microbiol.* **9,** 181–185.

17. Ochman H. (2001) Lateral and oblique gene transfer. *Curr. Opin. Genet. Dev.* **11,** 61–69.

18. Ochman, H., Lawrence, J. G., and Groisman, E. A. (2000) Lateral gene transfer and the nature of bacterial innovation. *Nature* **405,** 299–304.

19. Dobrindt, U. and Hacker, J. (2001) Whole genome plasticity in pathogenic bacteria. *Curr. Opin. Microbiol.* **4,** 550–557.

20. Maynard Smith, J., Dowson, C. G., and Spratt, B. G. (1991) Localized sex in bacteria. *Nature* **349,** 29–31.

21. Levin, B. R. (1981) Periodic selection, infectious gene exchange and the genetic structure of *E. coli* populations. *Genetics* **99,** 1–23.

22. Go, M. F., Kapur, V., Graham, D. Y., and Musser, J. M. (1996) Population genetic analysis of *Helicobacter pylori* by multilocus enzyme electrophoresis: extensive allelic diversity and recombinational population structure. *J. Bacteriol.* **178,** 3934–3938.

23. Spratt, B. G. (1999) Multilocus sequence typing: molecular typing of bacterial pathogens in an era of rapid DNA sequencing and the Internet. *Curr. Opin. Microbiol.* **2,** 312–316.

24. Enright, M. C. and Spratt, B. G. (1998) A multilocus sequence typing scheme for *Streptococcus pneumoniae:* identification of clones associated with serious invasive disease. *Microbiology* **144,** 3049–3060.

25. Enright, M. C., Day, N. P. J., Davies, C. E., Peacock, S. J., and Spratt, B. G. (2000) Multilocus sequence typing for the characterization of methicillin-resistant (MRSA) and methicillin-susceptible (MSSA) clones of *Staphylococcus aureus. J. Clin. Microbiol.* **38,** 1008–1015.

26. Enright, M. C., Spratt, B. G., Kalia, A., Cross, J. H., and Bessen, D. E. (2001) Multilocus sequence typing of *Streptococcus pyogenes* and the relationships between *emm*-type and clone. *Infect. Immun.* **69,** 2416–2427.

27. Dingle, K. E., Colles, F. M., Wareing, D. R., Ure, R., Fox, A. J., Bolton, F. E., et al. (2001) Multilocus sequence typing system for *Campylobacter jejuni. J. Clin. Microbiol.* **39,** 14–23.

28. Homan, W. L., Tribe, D., Poznanski, S., Li, M., Hogg, G., Spalburg, E., et al. (2002) Multilocus sequence typing scheme for *Enterococcus faecium. J. Clin. Microbiol.* **40,** 1963–1971.

29. Meats, E., Feil, E. J., Stringer, S., Cody, A. J., Goldstein, R., Kroll, J. S., et al. (2003) Characterization of encapsulated and non-capsulated *Haemophilus influenzae,* and determination of phylogenetic relationships, using multilocus sequence typing. *J. Clin. Microbiol.,* **41,** 1623–1636.

30. Enright, M. C., Knox, K., Griffiths, D., Crook, D. W. M., and Spratt, B. G. (2000) Multilocus sequence typing of *Streptococcus pneumoniae* directly from cerebrospinal fluid. *Eur. J. Clin. Microbiol. Infect. Dis.* **19,** 627–630.

31. Kriz, P., Kalmusova, J., and Felsberg, J. (2002) Multilocus sequence typing of *Neisseria meningitidis* directly from cerebrospinal fluid. *Epidemiol. Infect.* **128,** 157–160.

32. Clarke, S. C., Diggle, M. A., and Edwards, G. F. (2001) Semiautomation of multilocus sequence typing for the characterization of clinical isolates of *Neisseria meningitidis. J. Clin Microbiol.* **39,** 3066–3071.

33. Sreevatsan, S., Pan X., Stockbauer, K. E., Connel, N. D., Kreiswirth, B. N., Whittam, T. S., et al. (1997) Restricted structural gene polymorphism in the *Mycobacterium tuberculosis* complex indicates evolutionarily recent global dissemination. *Proc. Natl. Acad. Sci. USA* **94,** 9869–9874.

34. van der Zee, A., Mooi, F., Van Embden, J., and Musser, J. M. (1997) Molecular evolution and host adaptation of *Bordetella* spp.: phylogenetic analysis using multilocus enzyme electrophoresis and typing with three insertion sequences. *J. Bacteriol.* **179,** 6609–6617.

35. Achtman, M., Zurth, K., Morelli, G., Torrea, G., Guiyoule, A., and Carniel, E. (1999) *Yersinia pestis,* the cause of plague, is a recently emerged clone of *Yersinia pseudotuberculosis. Proc. Natl. Acad. Sci. USA* **96,** 14,043–14,048.

36. Read, T. D., Salzberg, S. L., Pop, M., Shumway, M., Umayam, L., Jiang, L., et al. (2002) Comparative genome sequencing for discovery of novel polymorphisms in *Bacillus anthracis. Science* **296,** 2028–2033.

37. Godoy, D., Randle, G., Simpson, A. J., Aanensen, D. M., Pitt, T. L., Kinoshita, R., and Spratt, B. G. (2003) Multilocus sequence typing and evolutionary relationships among the causative agents of melioidosis and glanders, *Burkholderia pseudomallei* and *B. mallei. J. Clin. Microbiol.,* **41,** 2068–2079.

38. Selander, R. K., Beltran, P., Smith, N. H., Helmuth, R., Rubin, F. A., Kopecko, D. J., et al. (1990) Evolutionary genetic relationships of clones in Salmonella serovars that cause human typhoid and other enteric fevers. *Infect. Immun.* **58,** 2262–2275.

39. Gutacker, M. M., Smoot, J. C., Migliaccio, C. A., Ricklefs, S. M., Hua, S., Cousins, et al. (2003) Genome-wide analysis of synonymous single nucleotide polymorphisms in *Mycobacterium tuberculosis* complex organisms. Resolution of genetic relationships among closely related microbial strains. *Genetics* **162,** 1533–1544.

40. Enright, M. C., Robinson, D. A., Randle, G., Feil, E. J., Grundmann, H., and Spratt, B. G. (2002) Evolutionary history of methicillin-resistant *Staphylococcus aureus* (MRSA). *Proc. Natl. Acad. Sci. USA* **99,** 7687–7692.

41. Raoult, D., Aboudharam, G., Crubezy, E., Larrouy, G., Ludes, B., and Drancourt, M. (2000) Suicide amplification of the medieval Black Death bacillus. *Proc. Natl. Acad. Sci. USA* **97,** 12,800–12,803.

42. Zhou, J., Enright, M. C., and Spratt, B. G. (2000) Identification of the major Spanish clones of penicillin-resistant pneumococci via the Internet using multilocus sequence typing. *J. Clin. Microbiol.* **38,** 977–986.

43. Van Loo, I. H., Heuvelman, K. J., King, A. J., and Mooi, F. R. (2002) Multilocus sequence typing of *Bordetella pertussis* based on surface protein genes. *J. Clin. Microbiol.* **40,** 1994–2001.

44. Vázquez, J. A., de la Fuente, L., Berrón, S., O'Rourke, M., Smith, N. H., Zhou, J., et al. (1993) Ecological separation and genetic isolation of *Neisseria gonorrhoeae* and *Neisseria meningitidis. Curr. Biol.* **3,** 567–572.

45. O'Rourke, M., Ison, C. A., Renton, A. M., and Spratt, B. G. (1995) Opa-typing— a high resolution tool for studying the epidemiology of gonorrhoea. *Mol. Microbiol.* **17,** 865–875.

46. Achtman, M., Azuma, T., Berg, D. E., Ito, Y., Morelli, G., Pan, Z. J., et al. (1999) Recombination and clonal groupings within *Helicobacter pylori* from different geographical regions. *Mol. Microbiol.* **32,** 459–470.

47. Bougnoux, M. E., Morand, S., and d'Enfert, C. (2002) Usefulness of multilocus sequence typing for characterization of clinical isolates of *Candida albicans. J. Clin. Microbiol.* **40,** 1290–1297.

48. Kotetishvili, M., Stine, O. C., Kreger, A., Morris, J. G., and Sulakvelidze, A. (2002) Multilocus sequence typing for characterization of clinical and environmental salmonella strains. *J. Clin. Micro.* **40,** 1626–1635.

49. Caugant, D. A., Kristiansen, B. E., Froholm, L. O., Bovre, K., and Selander, R. K. (1988) Clonal diversity of *Neisseria meningitidis* from a population of asymptomatic carriers. *Infect. Immun.* **56,** 2060–2068.

50. Jolley, K. A., Kalmusova, J., Feil, E. J., Gupta, S., Muselik, M., Kriz, P., et al. (2000) Carried meningococci in the Czech Republic: a diverse recombining population. *J. Clin. Microbiol.* **38,** 4492–4498.

51. Falush, D., Kraft, C., Taylor, N. S., Correa, P., Fox, J. G., Achtman, M., et al. (2001) Recombination and mutation during long-term gastric colonization by *Helicobacter pylori:* estimates of clock rates, recombination size, and minimal age. *Proc. Natl. Acad. Sci. USA* **98,** 15,056–15,061.

52. Feil, E. J., Maiden, M. C. J., Achtman, M., and Spratt, B. G. (1999) The relative contributions of recombination and mutation to the divergence of clones of *Neisseria meningitidis. Mol. Biol. Evol.* **16,** 1496–1502.

53. Feil, E. J., Maynard Smith, J., Enright, M. C., and Spratt, B. G. (2000) Estimating recombinational parameters in *Streptococcus pneumoniae* from multilocus sequence typing data. *Genetics* **154,** 1439–1450.

54. Reid, S. D., Herbelin, C. J., Bumbaugh, A. C., Selander, R. K., and Whittam, T. S. (2000) Parallel evolution of virulence in pathogenic *Escherichia coli. Nature* **406,** 64–67.

55. Feil, E. J., Holmes E. C., Bessen, D. E., Chan, M.-S., Day, N. P. J., Enright, M. C., et al. (2001) Recombination within natural populations of pathogenic bacteria:

short-term empirical estimates and long-term phylogenetic consequences. *Proc. Natl. Acad. Sci. USA* **98,** 182–187.

56. Holmes, E. C., Urwin, R., and Maiden, M. C. J. (1999) The influence of recombination on the population structure and evolution of the human pathogen *Neisseria meningitidis. Mol. Biol. Evol.* **16,** 741–749.

# 16

## Bacterial Taxonomics

*Finding the Wood Through the Phylogenetic Trees*

### Robert J. Owen

### Summary

Bacterial taxonomy comprises systematics (theory of classification), nomenclature (formal process of naming), and identification. There are two basic approaches to classification. Similarities may be derived between microorganisms by numerical taxonomic methods based on a range of present-day observable characteristics (phenetics), drawing in particular on conventional morphological and physiological test characters as well as chemotaxonomic markers such as whole-cell protein profiles, mol% G+C content, and DNA–DNA homologies. By contrast, phylogenetics, the process of reconstructing possible evolutionary relationships, uses nucleotide sequences from conserved genes that act as molecular chronometers. A combination of both phenetics and phylogenetics is referred to as polyphasic taxonomy, and is the recommended strategy in description of new species and genera. Numerical analysis of small-subunit ribosomal RNA genes (rDNA) leading to the construction of branching trees representing the distance of divergence from a common ancestor has provided the mainstay of microbial phylogenetics. The approach has some limitations, particularly in the discrimination of closely related taxa, and there is a growing interest in the use of alternative loci as molecular chronometers, such as *gyrA* and RNAase P sequences. Comparison of the degree of congruence between phylogenetic trees derived from different genes provides a valuable test of the extent they represent gene trees or species trees. Rapid expansion in genome sequences will provide a rich source of data for future taxonomic analysis that should take into account population structure of taxa and novel methods for analysis of nonclonal bacterial populations.

**Key Words:** Bacteria; classification; taxonomy; phenetics; phylogeny; trees.

From: *Methods in Molecular Biology, vol. 266: Genomics, Proteomics, and Clinical Bacteriology: Methods and Reviews*
Edited by: N. Woodford and A. Johnson © Humana Press Inc., Totowa, NJ

## 1. Introduction

The discovery of new genera and species is at the core of understanding biological diversity within the framework of a stable classification system. The natural environment, particularly the marine biosphere, provides a rich source of many new microbial species. Sampling of such environments suggests that the diversity of prokaryotes is probably 100 times higher than previously expected *(1)*. By contrast, the discovery of completely novel bacterial species of human origin has become an increasingly difficult and demanding task, and the emergence of new species of proven clinical significance is now a comparatively rare event. In recognition of these difficulties, microbiologists are encouraged to proceed in taxonomic studies by use of the "Candidatus" concept, which entails publication of descriptions of novel but as yet uncultured organisms, provided they have been well characterized by molecular techniques and by electron microscopy *(2)* (*see* Chapter 14). Some recent examples using this approach include *Helicobacter suis (3)*, *Helicobacter bovis (4)*, and *Cam- pylobacter hominis (5)*. Interestingly, a culture method was subsequently developed for the latter species *(6)*.

For the nonspecialist, taxonomy can sometimes prove a confusing subject, because basic terms may not necessarily have been agreed upon, while the interpretation of some initially unambiguous and precise definitions may have changed with time, as a result of new developments and the availability of new information. To assist the reader to overcome these potential difficulties, key terms that are generally consistent with the sense and views expressed by Cowan *(7)* will be defined at the outset. *Systematics* is the scientific discipline that embraces the many approaches involved in the characterization of the diversity of organisms and their arrangement in an orderly scheme of relationships. Central to systematics is *classification,* the process by which organisms are ordered on the basis of their relationships into groups (taxa) within a hierarchical scheme, from the basic unit, the species, through the different ranks of genus, family, order and class. Phylum and kingdom, and likewise subspecies and variety, have no formal standing in bacterial nomenclature, so are not recognized by the International Code of Nomenclature of Bacteria *(8)*, which provides a formal set of Rules and Statutes, agreed upon by the International Committee of Systematic Bacteriology, to standardize the naming of bacteria. The Code is used in conjunction with the Approved Lists of Bacterial Names, published in 1980 *(9)*, and subsequent notifications of new approved and validated names are published on a regular basis in the *International Journal of Systematic and Evolutionary Microbiology.*

The key requirements of any classification are that it should be stable, reproducible, scientifically based, and have high information content. However, the purpose of classification varies among different groups of microbiologists. For example, in practical medical bacteriology, the main objective is to facilitate a better understanding of relationships at the species and genus level, whereas relationships at higher levels remain very much the domain of more academic/theoretical taxonomists. Apparently "natural" classifications may well be an ideal to aim for, rather than the artificial classifications developed for special purposes. However, the distinction between these extremes may not necessarily be clear-cut, and may lead to different interpretations of taxonomic analyses, and the development of possibly incompatible nomenclature.

*Taxonomy* is the scientific theory of classification and, in practice, is applied in conjunction with nomenclature (the process of allocating names), and with identification (the means by which unknown microorganisms are allocated to previously described taxa). Taxonomy has an important function in clinical bacteriology in providing a stable framework for the description and identification of bacterial species that cause disease, although it is recognized that names must change occasionally as new and improved information is gathered, despite the short-term confusion it may cause *(10)*. It is also recognized that the ecological role of bacteria can, in certain cases, take priority in deciding species status. Consequently some medical organisms associated with defined clinical symptoms may continue to bear names that are not in accord with their genomic relatedness *(11)*.

Despite the numerous detailed systematic investigations undertaken since the publication of the Bacteriological Code, there are no formal procedures for the classification of bacteria. The process of classification, although dynamic, continues to be subjective and dependent on the particular type of organism under study, as well as the current state of the art of relevant methods for strain characterization. Recommendations have been made from time to time to reconcile and reevaluate approaches to bacterial systematics *(11,12)*. Attempts have also been made at standardization by the publication of minimal standards for particular genera, but these are by no means comprehensive, and only provide guidelines and are not mandatory for the description of novel taxa.

The aim in this chapter is to highlight possible paths that provide strategies when considering the description and validation of a new bacterial species originating from human clinical specimens. Taxonomy, like many aspects of clinical microbiology, has moved into the postgenomic era, so it is essential to encompass data on individual genes as well as on complete genome organization, and to assess how such developments might impact on the more traditional concepts of bacterial speciation. These aspects will be considered here by reference to the Campylobacteriaceae and related organisms that include helicobacters, as they provide a good model group to illustrate the application of different taxonomic approaches. Another new area that needs to be factored into contemporary taxonomy is the growing information on the population structure of individual species, with its importance in relation to the species definition.

Production of trees to illustrate relationships or similarities is one of the basic strategies involved in defining new species. As investigators use ever enlarging datasets, and a growing number of analytical tools, taxonomists have a key role to play in interpreting such analyses for the benefit of the scientific community. The purpose here is to consider possible paths through the veritable forest of potential microbial trees, and to identify those that might bear fruit.

## 2. Following the Phenotypic Path

### 2.1. Phenetic Classification

A classification derived from detailed study of the current observable or expressed properties of the organism (the phenotype) is described as a *phenetic* classification. It is important to recognize that the term *phenetic* does not have any evolutionary implications except in the sense of showing the present-day product of evolution *(7)*. The use of phenotypic traits, such as the biochemical tests employed for the characterization of aerobic metabolically active bacteria, such as *Salmonella, Staphylococcus,* or *Pseudomonas,* has been a traditional approach, but it has a number of inherent limitations, both conceptual and practical. These are particularly evident when applied to the characterization of more fastidious and intrinsically metabolically inactive microorganisms, such as slow-growing bacteria (e.g., mycobacteria), microaerobes (e.g., *Campylobacter* spp.), or strict anaerobes, such as *Clostridium* spp. The taxonomic validity of species and genera defined on traditional characters might be open to question, as such tests sample only a minute fraction of the properties encoded by the genomes of such microorganisms.

### 2.2. Trees From Conventional Test Characteristics

#### 2.2.1. Descriptive Analysis

Until the mid-1960s, classification of bacteria largely relied on the description of species based on readily observable phenotypes, represented by morphological (microscopic, cellular, and cultural), physiological, and biochemical characteristics. This tradition followed in the footsteps of the work of the nineteenth-century founders of medical microbiology, such as Robert Koch. The application of this systematic observational approach, familiar to generations of microbiologists, is illustrated extensively in classic bacteriology texts, such as Bergey's *Manual of Determinative Bacteriology (13)*. Some strain features, such as antibiotic susceptibility and host pathogenicity, are generally not regarded as valid taxonomic features at the species level, although until recently, animal tests were used for determination of strain pathogenicity for some species (e.g., toxigenic variants of *Corynebacterium diphtheriae* and *Escherichia coli*). Similar approaches were adopted in other fields, such as insect pathology and plant pathology, where particular variants (pathovars) were typically described on the basis of the disease caused by infection of a susceptible host.

Despite the limitations of traditional test characteristics, they nevertheless continue to provide the starting point for the formal description of new bacterial genera or species, which should include as many basic phenotypic properties as possible. Typical test categories include culture medium requirements, gaseous requirements, growth temperature range, Gram stain, cell shape and size, cell motility, and enzymic indicators of biochemical activity, such as cytochrome oxidase, catalase, nitrate reduction, and ability to ferment different carbohydrates (sugars). Such tests form a key part of the formal genus and species descriptions, and are used in the development of dichotomous keys and tables for practical identification purposes in medical microbiology *(14)*. Although it is difficult to apply standard criteria to all bacteria, there are now minimal standards, drawn up by relevant experts, recommended for defining new species for a growing number of bacterial genera, including, for example, *Campylobacter (15)* and *Helicobacter (16)*. Such standards emphasize the importance of basing descriptions on a number of individual isolates that represent species diversity in terms of ecological, host, and geographic origins.

### 2.2.2. Numerical Analysis of Conventional Tests

The rapid developments in computing technology in the 1960s facilitated the foundation of numerical taxonomy, which now provides a rigorous mathematical basis for the analysis of biological data *(17)*. The basic stages in performing numerical phenetic taxonomy for bacteria comprise:

- Strain selection
- Test selection
- Recording of results for qualitative and quantitative tests
- Data coding
- Numerical analysis to produce a similarity matrix and hierarchical clustering
- Interpretation—notably definition of clusters

The principal outcome from any such numerical analysis is one or more tree-like structures (dendrograms/phenograms), which provide a two-dimensional representation based on degrees of similarity or distances between different strains, defined as operational taxonomic units (OTUs). Trees form a part of mathematics called graph theory, and in simple terms are composed of lines (branches), that intersect at internal nodes and terminate at external nodes at the tips of branches that represent the OTUs. The tree topology is determined by the order in which the different branches bifurcate. A tree is described as rooted if there is a particular node (the root) from which a unique directional path leads to each OTU. By making comparisons of large numbers of characters (ideally in excess of 50), it is possible to define groups (phena) with the greatest overall similarity of characters.

The advantage of the numerical taxonomic approach is that it should avoid subjectivity if every character is given equal weight. The introduction of numerical (statistical) methods based on overall similarity or distance marked an important departure from the traditional approaches to taxonomy, which were more subjective and essentially concerned with the definition of monothetic taxa, in which certain characters were heavily weighted. Initially, the data used for such numerical analyses relied mainly on traditional bi- or multistate phenotypic features derived from basic attributes of the organism such as morphology, and physiological and biochemical reactions. However, a problem of performing numerical analysis based only on conventional tests, is to have an adequate number of discriminatory tests. This problem is exemplified by metabolically unreactive organisms such as *Campylobacter* spp., which typically give mostly negative results in metabolic tests. An attempt was made to circumvent this practical difficulty by using a range of alternative tests, such as determinations of carbon-source utilization, which in the case of *Campylobacter* have proved a specialized but nevertheless useful basis for numerical phenetic analysis *(18)*.

### 2.3. Trees From Molecular Phenetic Data: Protein and Other Expressed Molecules

#### 2.3.1. Selection of Chemotaxonomic Parameters

Because a phenetic classification is based on current observable characters, it can make use of macromolecular data, including genetic information, wherever available. From the mid-1960s, there were rapid developments in analytical chemical techniques and instrumentation, so that many new methodologies became available for the detailed chemical analysis of microorganisms. The application in systematics of these new types of data is broadly referred to as chemotaxonomy, and the topic has been reviewed and documented extensively *(19)*. The main categories of strain characteristics, particularly those based on the comparison of the structural properties of component macromolecules of cells are listed in **Table 1**.

Study of the chemical structure of individual cellular components has provided many valuable chemotaxonomic markers, at both the genus and species level. Such markers have included components of the cell wall (e.g., pepti-

**Table 1**
**Main Phenetic Characters Used in Bacterial Systematics
and Their Applicability for Use in Numerical Analysis**

Conventional microbiological tests[a]
One-dimensional (1D) electrophoretic protein profiling[a]
Whole-cell fingerprinting (PyMS and/or GLC)[b]
Fingerprinting of cell surface components (MALDI)[b]
Cell wall analysis
Cell membrane analysis
End products of metabolism (cellular fatty acid analysis)
DNA % G+C content and genome size
DNA–DNA hybridization[c]
RAPD/ribotyping/PCR-RFLP fingerprints[a]

[a]Used for cluster analysis by phenetic methods.
[b]Used for analysis by ordination methods.
[c]Complete data sets to construct matrices are rarely available.

doglycan, polysaccharides) and cell membranes (e.g., polar lipids, isoprenoid quinones), and metabolic end-products such as cellular fatty acids. Of these chemotaxonomic approaches, data derived from cellular fatty acid profiles have proved generally useful and applicable across many different bacterial genera, and databases have been developed that allow for rapid species identification. However, in most instances, such data, while important in descriptive studies of taxa, do not lend themselves readily for use in numerical taxonomy and the generation of taxonomic trees. Nevertheless, certain types of chemical data have been applied in numerical taxonomy with the successful generation of phenetic trees as indicated in **Tables 1** and **2**.

## 2.3.2. Electrophoretic Protein Profiles

One-dimensional SDS-polyacrylamide gel electrophoresis (PAGE) of whole-cell proteins has provided a widely used approach for phenetic analysis, particularly at the level of bacterial species *(20,21)*. The technique, if performed under carefully standardized conditions, provides excellent data for numerical analysis *(22)*. Protein profiles typically comprise 30 to 40 discrete bands of varying intensity, which, after staining and drying of gels, can be scanned with a densitometer. The absorbancies are digitized, then stored electronically, and numerical analysis is performed with similarities between profiles being calculated using a correlation coefficient (Pearson product-moment). Trees are constructed by a distance method, such as the unweighted pair group method with arithmetic mean (UPGMA). An example to illustrate the application of this

**Table 2**
**Numerical Taxonomic Analyses Commonly Used to Infer Taxonomic Structure of Bacteria**

| | Comment |
|---|---|
| **Phenetic analyses** | |
| • Dissimilarity/similarity matrix determined (Simple matching or Dice coefficients) + hierarchical cluster analysis by UPGMA + representation (dendrogram) | Distance method applicable to conventional phenotypic test data |
| • Dissimilarity/similarity matrix determined (Pearson Product Moment Correlation Coefficient) + hierarchical cluster analysis by UPGMA + representation (dendrogram) | Applied to electrophoretic band profiles, e.g., protein or AFLP patterns |
| • Dissimilarity/similarity matrix determined (either as above) + multivariate analysis (principle component analysis) + representation (ordination diagram) | Usually applied to complex data sets, e.g., PyMS but also applicable to further detailed analysis of taxonomic structure evident from dendrograms to investigate relationships between individual OTUs |
| *Phylogenetic inference analyses* | |
| • Pairwise distance methods from multiple sequence alignments such as neighbor joining (NJ) + representation (rooted or unrooted tree) | Most widely used approach in bacteriology. Trees are constructed from a series of matrices and distances are directly calculated to internal nodes. |
| • As above, but with clustering by UPGMA | Rarely used as the clustering assumes all taxa are equally distant from the root |
| • Maximum parsimony methods<br>• Maximum likelihood | Rarely used in bacterial phylogenetics |

*Note:* a wide range of software is available for molecular phylogenetics. For details of some 194 packages, *see* http://evolution.washington.edu/phylip/software.html.

approach used in the description of *H. pullorum,* a new species from human feces, is shown in the form of a rooted dendrogram in **Fig. 1** *(23)*. Dedicated analysis software packages, such as GelCompar (Applied Maths, Korttrijk, Belgium), are now available, that allow corrections for background and for distortions in the electrophoretic gel profiles.

## 2.3.3. Whole-Cell Fingerprinting

The direct analysis of components of whole bacterial cells is another approach that can be used to generate an organism fingerprint. Favored analytic methods for this purpose have included pyrolysis mass spectrometry (PyMS) *(24)* and, more recently, Fourier-transform infrared (FT-IR) spectroscopy *(25)* and matrix-assisted laser desorption time of flight (MALDI TOF) spectrometry *(26,27)*. Basically, the techniques involve whole-cell pyrolysis to generate volatile low molecular weight products, or ionization of cell-surface components, after the organisms are grown under carefully controlled conditions. Each method provides a complex fingerprint for each strain. The data are then subjected to numerical analysis, usually by ordination methods because of their complexity, to determine the key discriminatory features of OTUs that can be used in identification of taxa at the species and subspecies levels. Ordination methods produce multidimensional diagrams and are non-hierarchic; they use multivariate analyses such as principle component analysis, and canonical variate analysis *(17)*. While these approaches may have some merit as rapid means of strain identification, particularly in situations where high throughput of samples is required (e.g., in the analysis of food contaminants), they have not been developed and established as primary methods for taxonomic research or in clinical microbiology.

## 2.4. Trees From Molecular Phenetic Data: Genomic DNA Analysis

The second source of information, and the most fundamental for use in systematics, is provided by the investigation of cellular DNA and RNA—the so-called genomic approach. Development of this approach was greatly facilitated by the availability of rapid techniques for the isolation of bacterial DNA *(28)* and its analysis by a variety of methods, including buoyant density centrifugation, thermal denaturation, DNA–DNA and DNA–ribosomal RNA (rRNA) hybridization, agarose gel electrophoresis, blot hybridization, PCR, and DNA sequencing *(29)* (**Table 1**).

From the perspective of microbial classification, the two key genomic parameters used in phenetic classifications are DNA base composition and DNA–DNA hybridization (binding) levels. However, neither of these produce information that can be used readily in the construction of phenetic trees,

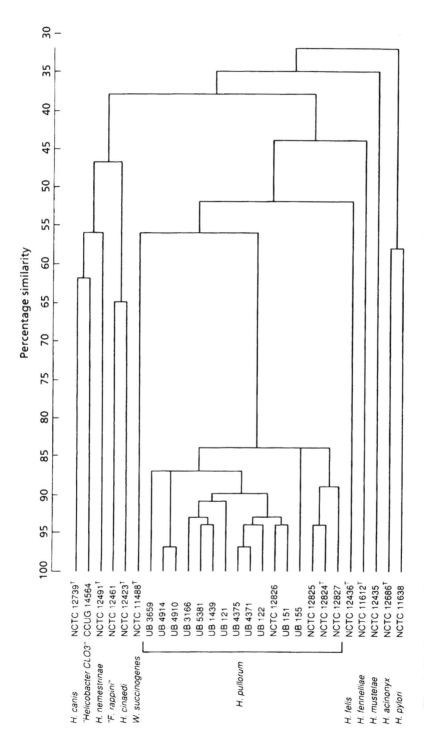

Fig. 1. Dendrogram of numerical analysis of the total protein pattern of *H. pullorum* isolates and reference strains of other helicobacters and closely related organisms. From Stanley et al. (**23**) with permission.

although they have proved of crucial importance as a basis for testing the validity of taxa defined from numerical analysis based on other data.

## 2.4.1. DNA Base Composition

The overall base composition of the complete genome, usually expressed as a mol% guanine plus cytosine (mol% G+C), was one of the first extensively documented, highly reproducible, and stable genomic features recorded for bacteria, as it provided a valuable constant irrespective of growth conditions *(30)*. Microbial values were observed to range from 25% to 75% G+C, with members of the same species generally having values within 2–3% of each other. The same G+C does not necessarily indicate relatedness, as many completely different organisms can have similar G+C values. Its value in taxonomy, therefore, is as a criterion on which to *exclude* a strain from a taxon, if its G+C content is significantly different. As individual strains have a single specific G+C value, it is not of relevance for tree construction, although it is of value in tree interpretation. Minimal standards recommend inclusion of G+C content as a primary defining feature of the type strain of the type species of a new genus, but it is optional to provide values for new species of an established genus *(11)*. Lack of access to equipment for thermal denaturation or centrifugation may deter investigators from performing G+C determinations. However, an alternative is provided by the availability of a novel real-time PCR assay for the estimation of G+C content *(31)*. Most G+C values are empirical estimates, so it is interesting, with the availability of complete genome sequencing, that the accuracy of original estimates can be validated.

## 2.4.2. DNA–DNA Reassociation

DNA–DNA similarity and, wherever determinable, the stability of resultant DNA hybrids, currently provide the acknowledged standards for species delineation *(11)*. Differences in DNA relatedness of between 5 and 10% can be detected by current hybridization assays, but the methods are not sensitive enough to distinguish point mutations and small additions and deletions. As complete matrices of DNA–DNA similarities between all strains in a study are rarely available, their use in tree construction, although theoretically feasible, is uncommon in practice. DNA–DNA hybridization data are nevertheless, extremely valuable for interpreting phenons constructed by numerical methods. There are no definite rules to correlate DNA–DNA relatedness and levels in the taxonomic hierarchy, but the guidelines indicated below *(32)* are now generally accepted:

- Hybridization levels of 70% or more, and low levels of mismatching (high thermal stability) indicate almost identical sequences and that strains should be classified in the same species

- Hybridization levels of less than 20% indicate strains should belong in different species, or even in different genera
- DNA–DNA hybridization data between 20% and 70% are more difficult to interpret, and may depend on the method used and the particular group of organisms

### 2.4.3. DNA–rRNA Hybridization

Saturation hybridization between rRNA and rRNA cistrons, which form a minor part of the total DNA (generally up to 0.4%), was an important tool for the elucidation of bacterial relationships at the intergeneric and suprageneric levels, prior to the widespread availability of DNA sequencing *(33)*. In this technique, the thermal stabilities of the DNA–rRNA hybrids formed on membrane filters were used to establish relationships based on the parameter $T_{m(e)}$, the thermal melting point of the elution profile of the hybrids. The values were used mainly for clustering different bacterial groups, with groupings presented typically in the form of two-dimensional similarity maps with $T_{m(e)}$ plotted against % rRNA binding. The data were not suitable for use in numerical analyses, but were extremely valuable for interpretation of relationships inferred by such methods. The methods were complex and have now largely been replaced by direct sequencing of rRNA cistrons (*see* **Subheading 3.2.2.**).

### 2.4.4. Relationships Derived From Other DNA Parameters

A wide range of DNA-based criteria have been used in identification of predefined taxa, or in the typing of strains within species (**Table 1**), but most of these do not provide information of direct value in systematics and, with a few exceptions, are not suitable for numerical analysis. The two most widely used approaches from which fragment patterns (fingerprints) lend themselves for analysis by numerical methods are ribotyping *(34)* and amplified fragment length polymorphism analysis (AFLP) *(35,36)*. Another fingerprinting approach that has proven taxonomic application is restriction fragment length polymorphism (RFLP) analysis of the 16S rRNA *(37)* and 23S rRNA genes *(38)*, where profiles may provide criteria for species-specific identification, although they have limited application in numerical analysis. The strength of RFLP-based approaches, however, is generally in sub-specific discrimination (strain typing), rather than in establishing inter-specific relationships. The application of numerical methods to such data is considered elsewhere *(39)*.

## 3. Following the Phylogenetic Path

### 3.1. Concepts of Phylogenetic Classification

Phylogenetics is the study of the ancestry of organisms over long time scales, involving many thousands or even millions of years. In 1859 in *The Origin of Species,* Darwin described a tree-like structure to illustrate diversification of species *(40)*. Then, in 1866, Haeckel proposed possibly the first phylogeny and classification of life illustrated in the form of a tree *(41)*. Traditionally such studies in biology have relied on the availability of fossil records, but in the case of bacteria no such evidence was available. Nevertheless, that did not deter occasional early speculative attempts based on purely descriptive data. For instance in 1921, Orla-Jensen *(42)* proposed a phylogenetic scheme based on physiology, while in 1936, Kluyver and van Neil *(43)* developed a scheme based on morphology.

However, it was not until the 1960s that a conceptual leap was made, when Zuckerkandl and Pauling *(44)* introduced the idea that macromolecules such as proteins can act as documents of evolutionary history, containing a record of changes since divergence from a common ancestor. This concept marked the birth of molecular systematics in biology, which may now be defined as the study of phylogenies and processes of evolution by the analysis of DNA or amino-acid sequence data *(45)*. This original idea was taken forward in 1980 by Kimura *(46)*, who introduced the concept of the evolutionary clock, and it was some 7 yr later that Woese *(47)* proposed the idea of an independent evolutionary clock embedded in the genotype for prokaryotic organisms, for the first time giving microbiologists the capacity to infer evolutionary histories and relationships. Central to this approach was the selection of a suitable molecular chronometer that was not restricted to particular taxonomic levels or bacterial groups, and had not undergone subtle functional changes. As there is no fossil record for microbes, phylogenetic inferences depend almost entirely on the *indirect* evidence provided by the analysis of sequence data that are considered to express rates of genetic change with time *(48)*. It was proposed by Woese *(47)* that the most useful molecular chronometers for microbial taxonomy were rRNAs, as they were universal, constant, often present in multiple copies, and had highly constrained functions that were not changed by the organism's environment. Furthermore, there were three types of rRNA present in all bacteria, namely 16S rRNA, comprising approx 1500 nt and found in the small (30S) ribosomal subunit, and 5S and 23S rRNA, comprising about 115 and 2900 nt, respectively, and both found in the large (50S) ribosomal subunit. These macromolecules were easy to isolate, could be sequenced directly, and in the case of 16S and 23S rRNA, contained considerable taxonomic information that was typically present in multiple copies in the genome. As each nucleotide in the RNA represents a multiple-character test with four alternative states (namely the bases, A, G, C, and U), the sequences could be subjected to rigorous mathematical analysis according to the methods developed for numerical taxonomy.

Another advantage of rRNA is that, because of functional constraints, it is a nonlinear chronometer. The primary structures are alternating sequences, invariant, more or less conserved to highly variable regions. In addition, rRNAs are capable of forming higher-order structures (helices) by short- and long-distance intramolecular interactions involving inverse complementary sequence stretches. The frequencies of compositional changes at different positions in the molecules vary greatly, so it is feasible to investigate phylogenetic relationships from the domain to the species level. Slowly evolving positions provide information on the deepest groups, whereas positions of higher variability may indicate more recent branchings.

### 3.2. Gathering the Data and Analysis

#### 3.2.1. Sequence Databases

Since the 1980s, phylogenetic analysis based on oligonucleotide catalogs, then on sequences of 16S rRNA, and, more recently, on high-quality sequencing of 16S rRNA genes (rDNA) has become widely accepted as the reference method for defining phylogenetic lineages (*see* Chapter 14). In 1987, Woese *(47)* presented the first comprehensive 16S rRNA phlyogenetic tree, and provided evidence for division of living organisms into three primary domains—the Eucarya, the Bacteria, and the Archaea. The phylogenetic trees based on 16S rRNA have led to the division of *Bacteria* into 18 major lines of descent (traditionally called phyla, but more recently, divisions), which fall into two primary and structurally distinct groups—the Gram-positives, which appear to be ancestral and structurally allied to the *Archaea,* and the Gram-negatives.

With the availability of PCR-based amplification/sequencing methods, the number of sequences has increased dramatically, with publicly accessible databases containing over 16,000 sequences of small-subunit sequences *(49)* (*http://rrndb.cme.msu.edu/rrndb/*). Furthermore, species-specific probes and PCR assays targeting 16S rRNA genes are also widely used in detecting and identifying species in clinical samples. Another major use of 16S rRNA gene sequences has been in exploring and surveying the diversity of environmental sources, by isolation of target sequences from bulk samples of environmental DNA. The importance of such analyses is that they provide the only means of determining the phylogenetic position of nonculturable organisms *(2)* (*see* Chapter 14).

#### 3.2.2. 16S rRNA Tree Construction

Why construct trees? The simple answer is that a tree is the most convenient way currently available of representing cluster analysis–derived relationships in two-dimensional space, particularly for the analysis of large amounts of complex data. No single tree is ideal, so it is important to appreciate, therefore, that trees are drawn to illustrate and to improve understanding of the biological relationships between the OTUs, although some information is lost in the reductive process. This applies equally to both phenetic and phylogenetic trees. Trees are essentially attempts at the reconstruction of phylogenies, and should be viewed only as models of possible events in evolutionary history, as they are based on indirect evidence from sequence data with a number of assumptions. As there are numerous strategies for constructing phylogenies, many different trees may be obtained, depending on the algorithm used and the sequences included. Furthermore, the number of possible trees increases dramatically as the number of taxa increases; it can be calculated that while four taxa could give three unrooted trees, fifty taxa could give $3 \times 10^{74}$ possible trees *(50)*.

The first step in extracting phylogenetic information from 16S rRNA genes is to make multiple alignments, preferably of high-quality nucleotide sequences. These may be new sequences as well as sequences downloaded from a public database. An ideal alignment procedure would ensure that homologous positions were placed in correspondence, in order to maximize their overall similarity. Homology is assumed when two sequences are similar due to common ancestry, but it should not be overlooked that two sequences could also be similar because they have evolved independently by chance (homoplasy). In general, two main types of analysis are used to infer phylogenetic trees from sequence alignments, namely evolutionary distance methods such as UPGMA and Neighbour Joining, which are algorithmic in that they use a specific series of calculations to estimate a tree, and character-based (tree searching) methods, that include maximum likelihood, maximum parsimony, and Bayesian methods (**Table 2**). More detailed overviews of the methods and programs for analyzing DNA sequence data can be obtained elsewhere *(50–52)* (*see* Chapter 2). In the case of 16S rDNA phylogenetic analysis, the most commonly adopted approach is to construct a distance matrix from all aligned sequences, where distances are expressed as the fraction of sites that differ in the alignment. Corrections for multiple base changes are applied, as the number of observed differences between DNA sequences may be underestimated, due to a number of superimposed substitutions that may have occurred during the evolution of the lineages from a common ancestor. In a simple model with only one parameter, such as that of Jukes and Cantor *(53)*, it is assumed that all four bases occur at equal frequencies, and that all types of substitutions are equally likely. The phylogenetic trees are then constructed from the distance matrix; in the case of 16S rRNA gene sequences, the technique usually applied is neighbour-joining *(54)*, in which the tree is reconstructed from a series of matrices. An example of this approach for type species of the genus *Helicobacter* is shown in **Fig. 2**.

### 3.3. Interpretation of rRNA Gene Trees

#### 3.3.1. Reliability of Tree Topology

The reconstruction of evolutionary trees using sequence divergence from a common ancestor is based on a number of assumptions about substitution rates, irrespective of the analytical methods. A key question for any particular method and dataset is how reliable a representation is the tree in terms of its topology? It is important that the tree should allow us to see the order in which existing taxa diverged from a hypothetical common ancestor. Branch lengths are then proportional to evolutionary distance, and distances are typically expressed as the number or fractions of substitutions per site. Tree reliability is usually assessed by means of nonparametric bootstrapping, in which data are resampled to give a series of pseudoalignments *(50)*. Typically 100 to 1000 bootstrap repetitions are required. The value at each node then indicates the number of times the clade occurred in the bootstrap replicates, and provides a measure of confidence in the clade.

#### 3.3.2. Biological Relevance at Genus and Species Level

The interpretation and validation of 16S rRNA phylogenies in terms of their biological relevance can be facilitated by direct comparison with other categories of data, such as those derived from phenetic analysis that could include conventional bacteriological tests, protein profiles, or DNA–DNA hybridizations. Application of a comparative approach is feasible, however, only where relevant information is available, which is usually at species and genus levels. Comparative studies using alternative taxonomic criteria may yield results that are ambiguous or that apparently contradict the 16S rRNA phylogenies and cast doubt on the reliability of that approach in certain applications. A major limitation of 16S rRNA phylogeny is the lack of discrimination between closely related species within genera. This problem has been noted from studies of a number of genera, and is illustrated by the analysis of the various species classified in the genus *Helicobacter,* within which there is a continuum without any evidence of a distinct phylogenetic substructure, despite clear distinctions in phenotypic characteristics and host specificities (**Fig. 2**).

Another limitation in the application of 16S rRNA phylogeny is that sequences are generally available only for single isolates of each species, often the type strain of the species, which may not actually be representative of that species. For instance, 16S rRNA phylogenetic analysis of multiple strains delineated *Campylobacter hyointestinalis* into two principal groups corresponding to subspecies *lawsonii* and *hyointestinalis,* which would not have

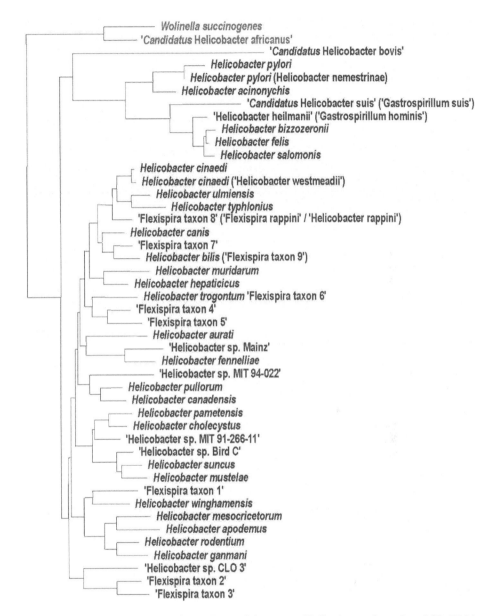

Fig. 2. Phylogenetic tree of members of the genus *Helicobacter* based on 16S rDNA sequences. (Provided by Dr. A. Lawson.)

been otherwise discernable using only type strains *(55)*. The fact that there may well be some intra-operon sequence variation between strains within a species, provides another reason why multiple isolates are needed to obtain more meaningful phylogenies *(56)*. Other factors that may contribute to anomalies between 16S rRNA gene phylogenies and alternative taxonomic data are outlined below:

- Inter-operon variation in 16S rRNA copy number between strains within species, and between strains within different species. Direct sequencing of a PCR product will produce a mean sequence that may mask subtle inter-operon variation *(56)*. This effect may increase with the number of copies present. For instance, *H. pylori* has 2 operons, whereas *E. coli* has 7 operons, and *Bacillus cereus* has 12 (http://rrndb.cmc.msu.edu/rrndb).
- Presence of inserts such as intervening sequences (IVS) that result in enlarged 16S rRNA genes. For example, there are reports of a 148-bp IVS present in some strains of *C. helveticus* and a 235-bp IVS in *H. canis*. IVSs have also been detected in strains of *C. hyointestinalis, C. sputorum* and *C. rectus (57)*.
- Recombination. There is strong evidence that horizontal gene transfer is common in some bacterial species—for example, frequent recombination is a feature of *H. pylori (58,59)*. Furthermore, transfer of parts of 16S rRNA sequences between species of *Aeromonas* has been suggested *(60)*. If this phenomenon is widespread among unrelated bacterial species, it might lead to convergence of evolutionary lineages that could confound orthodox taxonomic interpretations *(61)*.
- Adaptation of certain species to a particular ecological niche or host may have an effect on clock rate (*see* **Subheading 3.1.**), although it is generally assumed that 16S rRNA sequences are highly constrained and are independent of the environment.
- Poor quality of sequence data in public databases, which could include chimeric sequences due to PCR-generated artifacts, and may lead to errors in alignments and the resultant trees *(62)*.

### 3.3.3. Biological Relevance at Higher Taxonomic Levels

A critical issue in bacterial phylogeny is the relative branching orders and rooting of the three primary domains (Archaea, Bacteria, and Eucarya), as it is now recognized they cannot be deduced from 16S rRNA sequence data *(63,64)*. Uncertainties also exist about the criteria for division of the Bacteria into specific phyla, which have been described as arbitrary and ill defined *(65)*. Furthermore, the 16S rRNA phylogenies do not show any correlation with the main structural groups within the Bacteria, namely the Gram-positive and the Gram-negative forms.

## 4. Alternative Phylogenetic Markers

Although 16S rRNA gene phylogenies have become the standard in microbiology, their identified limitations mean there is growing interest in the use of alternative gene targets as evolutionary chronometers. The expansion in datasets from the growing number of genomic sequencing projects has raised the possibility of identifying alternative markers for inferring phylogenetic relationships. New markers might avoid the dependence on 16S rRNA and the assumption that its content reflects the evolutionary history of an organism. For example, protein-encoding genes may evolve faster than rDNA, and phylogenies based on such genes may have a higher resolution than those based only on 16S rRNA sequences. Such analyses can then be used to test whether the gene tree corresponds to the species tree. In an attempt to overcome some of the limitations of 16S rRNA phylogeny, Gupta *(65)* proposed a novel approach that involved identification of conserved inserts and deletions (indel signature sequences) in genes encoding various proteins. Results from analysis of some 60 completed bacterial genomes provide strong evidence that the phylogenetic placements and branching orders of taxa deduced by indel analysis are highly reliable and internally consistent. The indels are in genes that encode key proteins involved in highly conserved housekeeping functions, such as glutamate semialdehyde aminomutase, the Hsp 60 and Hsp 70 heat shock proteins, 5-aminoimidazole–4-carboxamide formyltransferase, and phosphoribosyl pyrophosphate synthetase. The rapidly expanding datasets of multilocus sequence types (*see* Chapter 15) will provide an important source for identifying polymorphisms in housekeeping gene sequences, which may be linked to nonsynonymous mutations that might provide valuable phylogenetic as well as typing markers (http://www.mlst.net/new).

### 4.1. Large Ribosomal Subunit 23S rRNA Sequences

In theory, the 23S rRNA, with a size of about 3 kb, should be the most informative marker, as it is estimated that under optimal circumstances, it should provide a resolution of one evolutionary event per 0.2–0.3 million years *(63,64)*. However, there are only a limited number of sequences of this gene available for analysis, compared with the many now available for the smaller 16S rRNA gene. Nevertheless, despite the differences in the degree of information content, detailed and comprehensive phylogenies inferred from equivalent datasets of both subunit sequences are in good agreement for the three primary domains, and for phyla within the domain Bacteria *(63,64)*. Interestingly, the branches within the 23S rRNA tree were generally longer than those of the 16S rRNA tree, indicating a higher degree of variability in the large subunit. The taxonomic information in the large subunit sequences is demonstrated by its use in a rapid identification scheme for *Helicobacter* species based on RFLP analysis *(38)*. However, 23S rRNA genes of several bacterial genera, including *Bacteroides, Salmonella, Leptospira, Yersinia, Coxiella, Helicobacter,* and *Campylobacter,* have been shown to contain IVSs *(57)*, and the effect of these enlarged genes on phylogenic analysis has not yet been established.

## 4.2. The Gyrase B Subunit Gene

PCR amplification and sequencing of the DNA gyrase B subunit gene (*gyrB*) of *Acinetobacter* strains provided a greater degree of resolution than that based on 16S rDNA *(66)*. Phylogenetic trees showing the relationships of various genera and species in the Enterobacteriaceae were also constructed from analysis of *gyrB*, and compared with 16S rRNA genes *(67,68)*. Differences between the gene trees were noted in areas of low resolution or low bootstrap values for one or both of the molecules. Overall, the 16S rRNA sequences were useful for describing relationships between distantly related taxa, whereas *gyrB* sequence comparisons were more appropriate for inferring intra- and some intergeneric relationships. It was suggested that a combination of both genes might be useful for future evolutionary systematic studies. Another study demonstrated *gyrB* sequence analysis was a useful alternative to 16S rRNA analysis for investigating phylogenetic relationships of *Oceanospirillum* and other *Pseudomonas*-like organisms *(69)*.

## 4.3. The urel Gene

The *H. pylori urel* gene encodes a urea accessory (transporter) protein thought to be an integral cytoplasmic membrane protein, forming a proton-gated urea channel regulating cytoplasmic urease *(70)*. A *urel* PCR assay developed for *H. pylori* has been applied to detect the locus in other *Helicobacter* species, which comprised the non-urealytic enterohepatic species *H. pullorum, H. fennelliae, H. cinaedi,* and *H. pametensis,* as well as in representatives of the urealytic gastric species *H. mustellae, H. felis, H. acinonychis,* and *H. heilmannii (71)*. The intestinal urealytic species *H. muridarum* was an exception, in appearing not to contain a *urel* homolog. Phylogenetic analysis showed that the nucleotide sequence homology levels for the *urel* homologs present in these other species of *Helicobacter* were within the range of variation observed for multiple isolates of *H. pylori*. *H. pullorum* was closely linked to *H. pylori,* even though the species differed in urealytic ability and gastric habitat. The *urel* tree topologies from analysis of nucleotide and amino acid sequences (**Fig. 3**), however, were not congruent with that based on 16S rRNA analysis. To investigate in further detail and validate the relationships among

**A**

0.1 substitutions/site

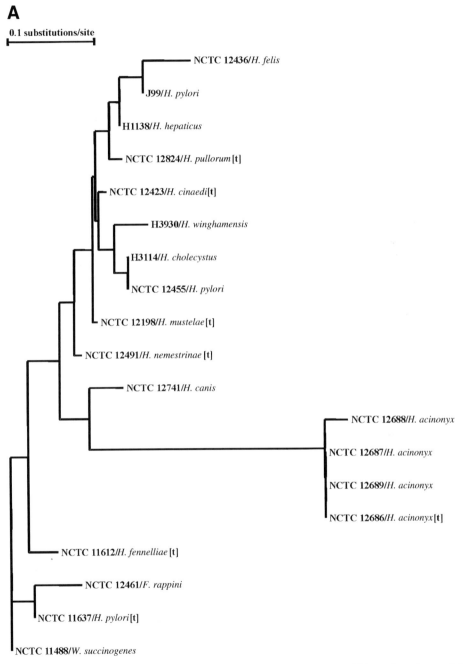

Fig. 3. Phylogenetic trees based on *ureI* DNA sequence (**A**) and amino acid sequence alignments (**B,** *next page*) of species within *Helicobacter.* (Provided by J. Xerry.)

**B**

0.1 substitutions/site

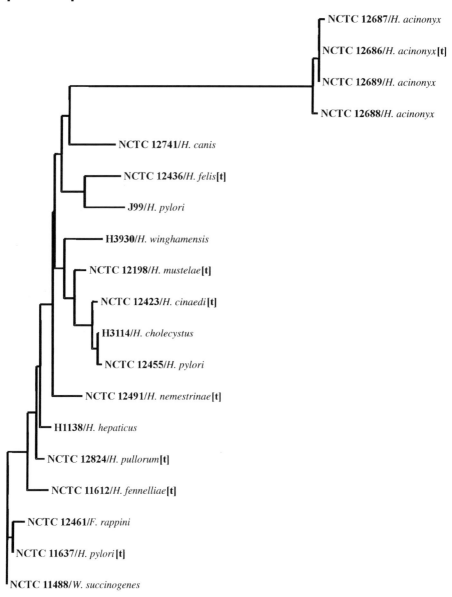

NCTC 12687/*H. acinonyx*

NCTC 12686/*H. acinonyx*[t]

NCTC 12689/*H. acinonyx*

NCTC 12688/*H. acinonyx*

NCTC 12741/*H. canis*

NCTC 12436/*H. felis*[t]

J99/*H. pylori*

H3930/*H. winghamensis*

NCTC 12198/*H. mustelae*[t]

NCTC 12423/*H. cinaedi*[t]

H3114/*H. cholecystus*

NCTC 12455/*H. pylori*

NCTC 12491/*H. nemestrinae*[t]

H1138/*H. hepaticus*

NCTC 12824/*H. pullorum*[t]

NCTC 11612/*H. fennelliae*[t]

NCTC 12461/*F. rappini*

NCTC 11637/*H. pylori*[t]

NCTC 11488/*W. succinogenes*

key *H. pylori* reference strains and type strains of several other species of *Helicobacter*, a multivariate analysis was performed, and strains grouped by multidimensional scaling are shown in **Fig. 4.**

### 4.4. Ribonuclease P RNA

Ribonuclease P (RNase P) is a key enzyme (ribozyme) in the biosynthesis of tRNA; it is present in all cells that carry out RNA synthesis, including representatives of all three phylogenetic domains. In the Bacteria, the RNase P holoenzyme is a ribonucleoprotein composed of RNA (usually 350–400 nt) and a single molecule of a small protein, comprising approx 120 amino acids *(72)*. The RNA and protein components of RNase P are encoded by *rnpB* and *rnpA*, respectively. The evidence indicates that RNase P RNA is an ancient molecule, and the fact that the holoenzymes from *Archaea* and *Eucarya* differ significantly from their bacterial counterparts is suggestive of its potential as a basis for inferring phylogenies. Sequences have been obtained from representatives of all of the 11 major bacterial phylogenetic lineages; these are available from the Ribonuclease P Database (*http://www.mbio.ncsu.edu/RnaseP/*). The evolutionary relationships of bacterial species based on RNase P RNA analysis were consistent with those inferred on the basis of 16S rRNA, with the exception of the placement of the photosynthetic Gram-positive bacteria. Phylogenetic analysis indicated that most bacterial RNase P RNAs have a common secondary structure (type A), whereas it has undergone a dramatic restructuring in the ancestry of Gram-positive bacteria (type B), especially in genera (*Clostridium, Eubacterium, Micrococcus* and *Streptomyces*) with a low G+C content *(73)*. Interestingly, each genus was shown to have an RNase P RNA with a distinctive secondary structure. Conserved and variable domains were detected within the *rnpB* sequences of members of the *Prochlorococcus* group. The RNase P RNA and 16S rRNA phylogenetic trees were not fully congruent, but the application of *rnpB* sequences allowed a better resolution between clades of very closely related genotypes *(74)*.

### 4.5. Other Genes as Potential Molecular Chronometers

Other protein-encoding genes have been investigated with regard to their potential for construction of bacterial phylogenetic trees; they include the genes encoding *recA* elongation factors, ATPase subunits, RNA polymerases, translation initiation factor (*infB*), and *groEL* and other heat shock proteins. As an example, *groEL* has been used in the classification of *Bartonella* spp. *(75)*.

### 5. The Third Way—The Polyphasic Path

The term *polyphasic* was introduced by Colwell *(76)* to refer to classifications based on a consensus of all available data (both phenotypic and genomic),

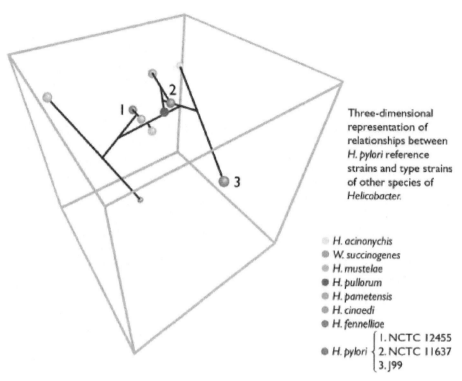

Three-dimensional
representation of
relationships between
*H. pylori* reference
strains and type strains
of other species of
*Helicobacter.*

⊙ *H. acinonychis*
◐ *W. succinogenes*
◑ *H. mustelae*
● *H. pullorum*
◑ *H. pametensis*
◐ *H. cinaedi*
● *H. fennelliae*

● *H. pylori* { 1. NCTC 12455 2. NCTC 11637 3. J99

Fig. 4. Relationships in the form of three-dimensional plots (grouped by multidimensional scaling) derived from *ureI* sequences for three reference strains of *H. pylori* and type strains of other species of *Helicobacter*. (Provided by J. Xerry and A. Lawson.)

and would now be considered analogous to phenetic classification. With the growth of molecular phylogenetics and the recognition of its key role in classification, it was subsequently suggested by Vandamme and colleagues *(77)* that polyphasic classification should be reinterpreted to include phylogenetic data, although it was left open how it should be best applied to derive relationships at the different taxonomic levels *(62)*.

The polyphasic and pragmatic species definition was recently re-evaluated *(11)*, in view of many recent innovations, and it was concluded that the present system was generally sound and that the community was served well by the bacterial species defined by Rossello-Mora and Amann *(78)* as "a category that circumscribes a (preferably) genomically coherent group of individual isolates/strains sharing a high degree of similarity in (many) independent features comparatively tested under highly standardized conditions."

## 6. Future Prospects and Goals

Over the past 20 yr, application of numerical taxonomy to many different categories of bacterial strain data has resulted in the generation of a multiplicity of taxonomic trees, with topologies that vary depending on the computational methods used and the input data. Significantly, 16S rRNA gene trees have provided the mainstay of efforts in reconstructing bacterial phylogeny and, as a result, have provided many novel insights and significant advances in our understanding of the classification of bacteria that were not apparent by the application of phenetics alone. Nevertheless, there is evidence that 16S rRNA gene phylogenies have limitations, which indicate that future strategies should not be reliant on equating single-gene evolution with species evolution. A single gene may be misleading if constraints or specific selections differ between organisms. A more rigorous test should be for congruence between trees constructed with data for several genes encoding different cellular functions, which should be combined to build a "species phylogenetic tree." This would be consistent with a polyphasic taxonomic approach.

The selection of alternative phylogenetic markers independent of rDNA genes will be aided by the present rapid developments in bacterial population genetics (*see* Chapter 15). These have resulted from the availability of increasing numbers of genome sequences, and from developments in high-throughput sequencing, as well as from new algorithms for analysis of population structure from multiple-sequence alignments, such as Markov chain, Monte Carlo, and Bayesian inference *(79)*. The availability of sequences on more strains and longer sequences, particularly of genes that have optimal rates of evolution for making inferences, will increase the robustness of phylogenetic estimates. Bacterial taxa at all levels appear to be currently defined largely by activities dependent on the presence or absence of genes relative to other taxa. The availability of increased numbers of complete genomes, and of tools for comparative genome analysis, such as DNA microarrays, should facilitate the development of the species genome concept based on the presence of a core set of known genes *(80)*. These could be used in combination with measures of the extent of polymorphism within genes and also within amino acid sequences, so that rates of synonymous and nonsynonymous substitutions are taken into account. This approach could provide a new way of defining bacterial taxa that would avoid reliance on a single 16S rDNA sequence, usually from one strain. Robust methods of statistical analysis are an essential requirement for obtaining the best phylogenetic estimates, and although there are many software packages available, there is a need for continued development and application of improved phylogenetic methodologies.

## Dedication

This chapter is dedicated to the memory of Dr. Jan Ursing, whose practical ideas and intellectual insights on taxonomy were a great source of stimulation.

## References

1. Blaxter, M. (2003) Molecular systematics: counting angels with DNA. *Nature* **421**, 122–124.
2. Murray, R. G. E. and Stackebrandt, E. (1994) Taxonomic notes: implementation of the provisional status Candidatus for incompletely described procaryotes. *Int. J. Syst. Bacteriol.* **45**, 186–187.
3. De Groote, D., van Doorn, L. J., Ducatelle, R., Verachuuren, A., Haesebrouch, F., Quint, W. G., et al. (1999) "Candidatus Helicobacter suis," a gastric helicobacter from pigs, and its phylogenetic relatedness to other gastrospirilla. *Int. J. Syst. Bacteriol.* **49**, 1769–1777.
4. De Groote, D., van Doorn, L. J., Ducatelle, R., Verschuuren, A., Tilmant, K., Quint, W. G., et al. (1999) Phylogenetic characterization of "Candidatus Helicobacter bovis," a new gastric Helicobacter in cattle. *Int. J. Syst. Bacteriol.* **49**, 1707–1715.
5. Lawson, A. J., Linton, D., and Stanley, J. (1998) 16S rRNA gene sequences of "Candidatus Campylobacter hominis," a novel uncultivated species, are found in the gastrointestinal tract of healthy humans. *Microbiol.* **49**, 2063–2071.
6. Lawson, A. J., On, S. L., Logan, J. M., and Stanley, J. (2001) *Campylobacter hominis* sp. nov., from the human gastrointestinal tract. *Int. J. Syst. Evol. Microbiol.* **51**, 651–660.
7. Cowan, S. T. (1978) *A Dictionary of Microbial Taxonomy* (Hill, L. R., ed.), Cambridge University Press, Cambridge.
8. Lapage, S. P., Sneath, P. H. A., Lessel, E. F., Skerman, V. B. D., Seeliger, H. P. R., and Clark W. A., eds. (1992) *International Code of Nomenclature of Bacteria* (1990 Revision). Bacteriological Code. American Society for Microbiology, Washington, DC.
9. Skerman, V. B. D., McGowan, V., and Sneath, P. H. A. (1980) Approved lists of bacterial names. *Int. J. Syst. Bacteriol.* **30**, 225–420.
10. Holmes, B. (1995) Why do bacterial names change? *PHLS Microbiol. Digest* **12**, 195–197.
11. Stackebrandt, E., Frederiksen, W., Garrity, G. M., Grimont, P. A. D., Kämpfer, P., Maiden, C. J., et al. (2002) Report of the ad hoc committee for the re-evaluation of the species definition in bacteriology. *Int. J. Syst. Evol. Microbiol.* **52**, 1043–1047.
12. Wayne, L. G., Brenner, D. J., Colwell, R. R., Grimont, P. A. D., Kandler, O., Krichevsky, M. I., et al. (1987) International Committee on Systematic Bacteriology: report of the ad hoc committee on reconciliation of approaches to bacterial systematics. *Int. J. Syst. Bacteriol.* **37**, 463–464.

13. Garrity, G. M. and Holt, J. G. (2001) The road map to the manual, in *Bergey's Manual of Systematic Bacteriology, 2nd ed.* (Boone, D. R., Castenholz, R. W., and Garrity, G. M., eds.), Springer, New York, pp. 119–166.

14. Cowan, S. J., Steel, K. J., Barrow, G. I., and Feltham, R. K. A. (1993) *Cowan and Steel's Manual for the Identification of Medical Bacteria, 3rd ed.* Cambridge University Press, Cambridge, UK.

15. Ursing, J. B., Lior, H., and Owen, R. J. (1994) Proposal of minimal standards for describing new species of the family Campylobacteraceae. *Int. J. Syst. Bacteriol.* **44,** 842–845.

16. Dewhurst, F. E., Fox, J. G., and On, S. L. W. (2000) Recommended minimal standards for describing new species of the genus *Helicobacter. IJSEM* **50,** 2231–2237.

17. Sokal, R. R. and Sneath, P. H. A. (1963) *Principles of Numerical Taxonomy.* W. H. Freeman, San Francisco, CA.

18. On, S. L. W. and Holmes, B. (1995) Classification and identification of campylobacters, helicobacters and allied taxa by numerical analysis of phenotypic characters. *Syst. Appl. Microbiol.* **18,** 374–390.

19. Goodfellow, M. and Minnikin, D. E., eds. (1985) *Chemical Methods in Bacterial Systematics.* The Society for Applied Bacteriology, Technical Series No. 20. Academic Press, New York.

20. Kersters, K. (1985) Numerical methods in the classification of bacteria by protein electrophoresis, in *Computer-Assisted Bacterial Systematics* (Goodfellow, M., Jones, D., and Priest, F. G., eds.), American Press, London, pp. 337–365.

21. Pot, B., Vandamme, P., and Kersters, K. (1994) Analysis of electrophoretic whole organism protein fingerprints, in *Chemical Methods in Prokaryotic Systematics* (Goodfellow, M. and O'Donnell, A. G., eds.), John Wiley & Sons, Chichester, UK, pp. 493–521.

22. Costas, M., Pot, B., Vandamme, P., Kersters, K., Owen, R. J, and Hill, L. R. (1990) Interlaboratory comparative study of the numerical analysis of one-dimensional sodium dodecyl sulphate–polyacrylamide gel electrophoretic protein patterns of *Campylobacter* strains. *Electrophoresis* **11,** 467–474.

23. Stanley, J., Linton, D., Burnens, A. P., Dewhirst, E., On, S. L., Porter, A., et al. (1994) *Helicobacter pullorum* sp. nov.—genotype and phenotype of a new species isolated from poultry and from human patients with gastroenteritis. *Microbiol.* **140,** 3441–3449.

24. Goodacre, R. (1994) Characterization and quantification of microbial systems using pyrolysis mass spectrometry: introducing neural networks to analytical pyrolysis. *Microbiol. Eur.* **2,** 16–22.

25. Naumann, D., Helm, D., and Schultz, C. (1994) Characterization and identification of microorganisms by FT-IR spectroscopy and FT-IR microscopy, in *Bacterial Diversity and Systematics* (Priest, F. G., Ramos Cormenzana, A., and Tindall, B. J., eds.), Plenum, New York, pp. 67–85.

26. Claydon, M., Davey, S. N., Edwards-Jones, V., and Gordon, D. B. (1996) The rapid identification of intact microorganisms using mass spectrometry. *Nat. Biotechnol.* **14,** 1584–1586.

27. Conway, G. C., Smole, S. C., Sarracino, D. A., Arbeit, R. D., and Leopold, P. E. (2001) Phyloproteomics: species identification of Enterobacteriaceae using matrix-assisted laser desorption/ionization time-of-flight mass spectrometry. *J. Mol. Microbiol. Biotechnol.* **3,** 103–112.

28. Marmur, J. (1961) A procedure for the isolation of deoxyribonucleic acid from microorganisms. *J. Molec. Biol.* **3,** 208–218.

29. Stackebrandt, E. and Goodfellow M., eds. (1991) *Nucleic Acid Techniques in Bacterial Systematics.* John Wiley & Sons, Chichester, UK.

30. Owen, R. J. and Pitcher, D. G. (1985) Current methods for estimating DNA base composition and levels of DNA–DNA hybridization. *Chemical Methods in Bacterial Systematics.* Goodfellow, M. and Minnikin, D. E. Soc. Appl. Bacteriol.—Technical Series 20, pp.67–93.

31. Xu, H. X., Kawamura, Y., Li, N., Zhao, L., Li, T. M., Li, Z. Y., et al. (2000) A rapid method for determining the G+C content of bacterial chromosomes by monitoring fluorescence intensity during DNA denaturation in a capillary tube. *Int. J. Syst. Evol. Microbiol.* **50,** 1463–1469.

32. Stackebrandt, E. and Goebel, B. M. (1994) Taxonomic note: a place for DNA–DNA reassociation and 16S rRNA sequence analysis in the present species definition in bacteriology. *Int. J. Syst. Bacteriol.* **44,** 846–849.

33. DeSmedt, J. and DeLey, J. (1977) Intra- and intergeneric similarities of *Agrobacterium* ribosomal ribonucleic acid cistrons. *Int. J. Syst. Bacteriol.* **27,** 222–240.

34. Owen, R. J. and Hernandez, J. (1993) Ribotyping and arbitrary-primer PCR fingerprinting of Campylobacters, in *New Techniques in Food and Beverage Microbiology* (Kroll, R. G., Gilmour, A., and Sussman, M.), Soc. Appl. Bacteriol. Technical Series 31, pp. 265–285.

35. Owen, R. J., Ferrus, M., and Gibson, J. (2001) Amplified fragment length polymorphism genotyping of metronidazole-resistant *Helicobacter pylori* infecting dyspeptics in England. *Eur. J. Clin. Microbiol. Infect. Dis.* **7,** 244–253.

36. Desai, M., Logan, J. M., Frost, J. A., and Stanley, J. (2001) Genome sequence-based fluorescent amplified fragment length polymorphism of *Campylobacter jejuni,* its relationship to serotyping, and its implications for epidemiological analysis. *J. Clin. Microbiol.* **39,** 3823–3829.

37. Marshall, S. M., Melito, P. L., Woodward, D. L., Johnson, W. M., Rodgers, F. G., and Mulvey, M. R. (1999) Rapid identification of *Campylobacter, Arcobacter* and *Helicobacter* isolates by PCR-restriction fragment length polymorphism analysis of the 16S rRNA gene. *J. Clin. Microbiol.* **37,** 4158–4160.

38. Hurtado, A. and Owen, R. J. (1997) A rapid identification scheme for *Helicobacter pylori* and other species of *Helicobacter* based on 23S rRNA gene polymorphisms. *Syst. Appl. Microbiol.* **20,** 222–231.

39. Clewley, J. P. (1998) A user's guide to producing and interpreting tree diagrams in taxonomy and phylogenetics. Part 3: Using restriction fragment length polymorphism patterns of bacterial genomes to draw trees. *Comm. Dis. Publ. Hlth.* **1,** 208–210.

40. Darwin, C. (1859) *On the Origin of Species,* John Murray, London.

41. Haeckel, E. (1866) *Generelle Morphologie der Organismen: Allgemeine Grundzuge der organischen Formen-Wissenschaft, mechanisch begrundet durch die von Charles Darwin reformirte Descendenz-Theorie.* Georg Riemer, Berlin.

42. Orla-Jensen, S. (1921) The main lines of the natural bacterial system. *J. Bacteriol.* **6,** 263–273.

43. Kluyver, A. J. and van Neil C. B. (1936) Prospects for a natural system of classification of bacteria. *Zentrabl. Bakteriol. Parasitenkd. Infektionskr. Hyg. Abt.* II **94,** 369–403.

44. Zuckerkandl, E. and Pauling, L. (1965) Molecules as documents of evolutionary history. *J. Theoret. Biol.* **8,** 357–366.

45. Whelan, S., Liò, P., and Goldman, N. (2001) Molecular phylogenetics: state-of-the-art methods for looking into the past. *Trends Genet.* **17,** 262–272.

46. Kimura, M. (1980) A simple method for estimating evolutionary rates of base substitutions through comparative studies of nucleotide sequences. *J. Mol. Evol.* **16,** 111–120.

47. Woese, C. R. (1987) Bacterial evolution. *Microbiol. Rev.* **51,** 221–271.

48. Young, J. M. (2001) Implications of alternative classifications and horizontal gene transfer for bacterial taxonomy. *Int. J. System. Evol. Microbiol.* **51,** 945–953.

49. Maidak, B. L., Cole, J. R., Lilburn, T. G., Parker, C. T., Jr., Saxman, P. R., Farris, R. J., et al. (2001) The RDP-II (Ribosomal Database Project). *Nucleic Acids Res.* **29,** 173–174.

50. Hall, B. G. (2001) *Phylogenetic Trees Made Easy. A How-to Manual for Molecular Biologists.* Sinauer Associates, Sunderland, Massachusetts.

51. Clewley, J. P. (1998) A user's guide to producing and interpreting tree diagrams in taxonomy and phylogenetics. Part 2. The multiple alignment of DNA and protein sequences to determine their relationships. *Comm. Dis. Publ. Hlth.* **1,** 132–134.

52. McCormack, G. P. and Clewley, J. P. (2002) The application of molecular phylogenetics to the analysis of viral genome diversity and evolution. *Rev. Mod. Virol.* **12,** 221–238.

53. Jukes, T. H. and Cantor, C. R. (1969) Evolution of protein molecules, in *Mammalian Protein Metabolism* (Munro, H. N., ed.), Academic Press, New York, pp. 21–132.

54. Saitou, N. and Nei, M. (1987) The neighbor-joining method: a new method for reconstructing phylogenetic trees. *Mol. Biol. Evol.* **4,** 406–425.

55. Harrington, C. S. and On, S. L. W. (1999) Extensive 16S rRNA gene sequence diversity in *Campylobacter hyointestinalis* strains: taxonomic and applied implications. *Int. J. System. Bacteriol.* **49,** 1171–1175.

56. Clayton, R. A., Sutton, G., Hinkle, P. S., Jr., Bult, C., and Fields, C. (1995) Intraspecific variation in small-subunit rRNA sequences in GenBank: why single sequences may not adequately represent Prokaryotic taxa. *Int. System. Bacteriol.* **45,** 595–599.

57. Hurtado, A., Clewley, J. P., Linton, D., Owen, R. J., and Stanley, J. (1997) Sequence similarities between large subunit ribosomal RNA gene intervening sequences from different *Helicobacter* species. *Gene* **194,** 69–75.

58. Suerbaum, S. and Achtman, M. (1999) Evolution of *Helicobacter pylori:* the role of recombination. *Trends Microbiol.* **7,** 182–184.

59. Falush, D., Kraft, C., Taylor, N. S., Correa, P., Fox, J. G., Achtman, M., et al. (2001) Recombination and mutation during long-term gastric colonization by *Helicobacter pylori:* estimates of clock rates, recombination size and minimal age. *PNAS* **98,** 15,056–15,061.

60. Sneath, P. H. (1993) Evidence from *Aeromonas* for genetic crossing-over in ribosomal sequences. *Int. J. Syst. Bacteriol.* **43,** 626–629.

61. Young, J. M. (2001) Implications of alternative classifications and horizontal gene transfer for bacterial taxonomy. *Int. J. Syst. Evol. Microbiol.* **51,** 945–953.

62. Hugenholtz, P. and Huber, T. (2003) Chemeric 16S rDNA sequences of diverse origin are accumulating in the public databases. *Int. J. Syst. Evol. Microbiol.* **53,** 289–293.

63. Ludwig, W. and Schleifer, K. H. (1994) Bacterial phylogeny based on 16S and 23S rRNA sequence analysis. *FEMS Microbiol. Rev.* **15,** 155–173.

64. Ludwig, W. and Schleifer, K. H. (1999) Phylogeny of Bacteria beyond the 16S rRNA Standard. *ASM News* **65,** 752–757.

65. Gupta, R. S. (2002) Phylogeny of bacteria: are we now close to understanding it? *ASM News* **68,** 284–291.

66. Yamamoto, S. and Harayama, S. (1996) Phylogenetic analysis of *Acinetobacter* strains based on the nucleotide sequences of *gyrB* genes and on the amino acid sequences of their products. *Int. J. Syst. Bacteriol.* **46,** 506–511.

67. Dauga, C. (2002) Evolution of the *gyrB* gene and the molecular phylogeny of Enterobacteriaceae: a model molecule for molecular systematic studies. *Int. J. Syst. Evol. Microbiol.* **52,** 531–547.

68. Fukushima, M., Kakinuma, K., and Kawaguchi, R. (2002) Phylogenetic analysis of *Salmonella, Shigella* and *Escherichia coli* strains on the basis of *gyrB* gene sequence. *J. Clin. Microbiol.* **40,** 2779–2885.

69. Satomi, M., Kimura, B., Hamada, T., Harayama, S., and Fujii, T. (2002) Phylogenetic study of the genus *Oceanospirillum* based on 16S rRNA and *gyrB* genes: emended description of the genus *Ocenaospirillum,* description of *Pseudospirillum* gen. nov., *Oceanobacter* gen. nov. and *Terasakiella* gen. nov. and transfer of *Oceanospirillum jannaschii* and *Pseudomonas stanieri* to *Marinobacterium* as *Marinobacterium jannaschii* comb. nov. and *Marinobacterium stanieri* comb. nov. *Int. J. Syst. Evol. Microbiol.* **52,** 739–747.

70. Weeks, D. L., Eskandari, S., Scott, D. R., and Sachs, G. (2000) A H+-gated urea channel: the link between *Helicobacter pylori* urease and gastric colonization. *Science* **287,** 482–485.

71. Owen, R. J., Xerry, J., and Chisholm, S. A. (2001) Sequence diversity within the *Helicobacter pylori urel* locus and identification of homologues in other ureolytic and nonureolytic species of *Helicobacter* of human and animal origin. *Gut* **49 (Suppl 11),** A9.
72. Pace, N. R. and Brown, J. W. (1995) Evolutionary perspective on the structure and function of ribonuclease P, a ribozyme. *J. Bacteriol.* **177,** 1919–1926.
73. Haas, E. S., Banta, A. B., Harris, J. K., Pace, N. R., and Brown, J. W. (1996) Structure and evolution of ribonuclease P RNA in Gram-positive bacteria. *Nucleic Acids Research* **24,** 4775–4782.
74. Schön, A., Fingerhut, C., and Hess, W. R (2002) Conserved and variable domains within divergent Rnase P RNA gene sequences of *Prochlorococcus* strains. *J. Syst. Evol. Microbiol.* **52,** 1383–1389.
75. Zeaiter, Z., Fournier, P. E., Ogata, H., and Raoult, D. (2002) Phylogenetic classification of *Bartonella* species by comparing *groEL* sequences. *Int. J. Syst. Evol. Microbiol.* **52,** 165–171.
76. Colwell, R. R. (1970) Polyphasic taxonomy of the genus *Vibrio:* numerical taxonomy of *Vibrio cholerae, Vibrio parahaemolyticus,* and related *Vibrio* species. *J. Bacteriol.* **104,** 410–433.
77. Vandamme, P., Pot, B, Gillis, P, De Vos, P, Kersters, K., and Swings, J. (1996) Polyphasic taxonomy, a consensus approach to bacterial systematics. *Microbiol. Rev.* **60,** 407–438.
78. Rossello-Mora, R. and Amann, R. (2001) The species concept for prokaryotes. *FEMS Microbiol. Rev.* **25,** 39–67.
79. Falush, D., Wirth, T., Linz, B., Pritchard J. K., Stephens, M., Kidd, M., et al. (2003) Traces of human migrations in *Helicobacter pylori* populations. *Science* **299,** 1582–1585.
80. Lan, R. and Reeves, P. R. (2000) Intraspecies variation in bacterial genomes: the need for a species genome concept. *Trends Microbiol.* **8,** 396–401.

# Index

Printed in Great Britain by
Amazon.co.uk, Ltd.,
Marston Gate.